Biotechnologies for Gene Therapy

Yang H. Yun • Kristine E. Yoder
Editors

Biotechnologies for Gene Therapy

RNA, CRISPR, Nanobots, and Preclinical Applications

Springer

Editors
Yang H. Yun
Department of Biomedical Engineering,
College of Engineering and
Polymer Science
The University of Akron
Akron, OH, USA

Kristine E. Yoder
Department of Cancer Biology and Genetics
College of Medicine
The Ohio State University
Columbus, OH, USA

ISBN 978-3-030-93335-7 ISBN 978-3-030-93333-3 (eBook)
https://doi.org/10.1007/978-3-030-93333-3

This Springer imprint is published by the registered company Springer Nature Switzerland AG
The registered company address is: Gewerbestrasse 11, 6330 Cham, Switzerland

Preface

Gene therapy for many investigators is an amazing technological platform for developing novel therapies against life-threatening diseases that are previously untreatable using traditional pharmaceuticals. It is highly unfortunate that controversies and ethical concerns also have been associated with gene therapy. As an example, the discovery of CRISPR-Cas9 and its application as a genome-editing tool has been transformative. It allows investigators quickly and more efficiently to modify genes which is invaluable for studying cellular functions and for developing therapeutics that addresses the underlying problems associated with diseases instead of treating their symptoms. In 2018, a scientist shocked and horrified the world by applying CRISPR to embryos in vitro and creating the world's first gene-edited human babies. Germ-line gene editing of humans has had a clear ethical line of caution, transparency, and rigorous oversight before it can be crossed because unforeseen consequences to the human genome could cascade throughout the generations. Since this and other controversies have been documented in scientific journals and disseminated to the public by news outlets, the ethics of gene therapy continues to be debated, and the social trust in gene therapy seems to be further eroded.

We sincerely hope 2021 marks a turning point in gene therapy. The rapid development of novel vaccines in response to the COVID-19 pandemic has been unprecedented. As of September 27, 2021, over 390 million doses of both mRNA and adenoviral vaccines have been administered in the USA alone. In addition to showing efficacy of greater than 94% for mRNA vaccines and protection against the delta variant, monitoring the safety of these vaccines has resulted in only rare cases of anaphylaxis. Once our global society recovers from the deaths due to COVID-19, we hope these exceptional achievements will be celebrated, mRNA vaccines will be recognized as the first therapeutics based on a non-viral gene therapy to be widely administrated in a clinical setting, and gene therapy will once again inspire other therapeutic advancements.

While the vaccines against COVID-19 appear to be a resounding success, other diseases could benefit from advancements in gene therapy. The purpose of this book is to highlight some of the latest developments and applications of RNA, CRISPR, and DNA to treat diseases ranging from cancers to degenerative disorders. This book also features innovations of the delivery methods for nucleic acids ranging from nanodevices made from DNA and pseudo amino acids to viral vectors. We hope some of these proposed therapies can successfully transition from basic science research to clinical trials resulting in new medical breakthroughs.

Akron, OH, USA Yang H. Yun
Columbus, OH, USA Kristine E. Yoder

Contents

Contents

About the Editors

Yang H. Yun is an Associate Professor in the Biomedical Engineering Department at the University of Akron. His research interests include biomaterials, cancer therapy, drug and gene delivery, electrospinning, nanotechnology, tissue engineering, and the treatment of infectious diseases.

Kristine E. Yoder is an Associate Professor at the department of Cancer Biology and Genetics in Ohio State University.

Therapeutic Potential of MicroRNAs and Their Nanoparticle-based Delivery in the Treatment of Liver Fibrosis

Jieun Kim, Yang H. Yun, and Youngmi Jung

Abstract Chronic liver disease is a global health problem owing to its high morbidity and the limited available treatment options. Liver fibrosis, the most common feature of chronic liver disease, is characterized by excessive accumulation of extracellular matrix (ECM) in the liver, eventually leading to cirrhosis. Hepatic stellate cells (HSCs), the major contributors to hepatic fibrosis, undergo transdifferentiation from a quiescent to an activated/myofibroblastic state, resulting in the accumulation of ECM. MicroRNAs (miRNAs) are small noncoding RNAs that are involved in the regulation of gene expression at the post-transcriptional level. Because miRNAs mediate a broad range of biological functions, dysregulation of miRNAs is strongly associated with various diseases, including liver fibrosis. Therefore, modulation of miRNAs by supplementing or inhibiting them represents a novel therapeutic strategy for liver fibrosis. With recent advances in our understanding of nanomedicines, nanoparticles are regarded as promising candidates for efficient delivery methods for miRNAs because of their biological and technical advantages. In this chapter, we review the pathogenesis of liver fibrosis, the roles of miRNAs in liver fibrosis, the therapeutic potential of miRNAs and their nanoparticle-based delivery for liver fibrosis, and the development of novel miRNA-based therapeutics for liver diseases.

J. Kim
Institute of Systems Biology, Pusan National University, Pusan, Republic of Korea

Department of Integrated Biological Science, College of Natural Science, Pusan National University, Pusan, Republic of Korea

Y. H. Yun
Department of Biomedical Engineering, College of Engineering and Polymer Science, The University of Akron, Akron, OH, USA

Y. Jung (✉)
Department of Integrated Biological Science, College of Natural Science, Pusan National University, Pusan, Republic of Korea

Department of Biological Sciences, College of Natural Science, Pusan National University, Pusan, Republic of Korea
e-mail: y.jung@pusan.ac.kr

Y. H. Yun, K. E. Yoder (eds.), *Biotechnologies for Gene Therapy*,
https://doi.org/10.1007/978-3-030-93333-3_1

1 Introduction

Liver is the largest internal human organ in adults and has vital roles in maintaining homeostasis by regulating metabolism, bile production, and detoxification [1]. The liver has the greatest regenerative capacity of any organ in the body. Even with 70% surgical removal (partial hepatectomy) of the liver mass, the remnant tissue has the ability to grow into the original mass and to recover its functions [2, 3]. Because of this outstanding capacity, the liver can self-repair and restore itself after mild injury. However, when the liver damage is repetitive and/or severe, the regenerative capacity is impaired owing to massive death of hepatocytes, which triggers the proliferation of nonparenchymal cells, including hepatic stellate cells (HSCs), and replaces the damaged liver tissue [4, 5].

In the normal liver, HSCs are quiescent and function as the major storage facility for vitamin A metabolites known as retinoids [6]. However, when the liver is injured, HSCs undergo transdifferentiation from quiescent HSCs to activated/myofibroblastic HSCs, which are the principal cell types that produce extracellular matrix (ECM) proteins in the liver [7]. Excessive deposition of fibrous ECM components replaces the parenchyma with fibrotic tissue, which causes severe structural and functional alterations, leads to liver dysfunction, and eventually develops into liver fibrosis and cirrhosis [4, 5]. Cirrhosis, an end-stage disease, results in liver failure in many patients, leading to high mortality worldwide [8, 9]. The global incidences of cirrhosis and other chronic liver disease has been estimated at 1.5 billion and accounts for two million deaths per year [9]. Hence, many researchers have focused on the clearance, deactivation, or inactivation of HSCs as a therapy because of the essential role that HSCs play in pathogenesis [10]. Nevertheless, the therapies available for liver fibrosis are still limited [11]. Therefore, further investigation and development into new therapeutic strategies for liver fibrosis/cirrhosis are urgently needed.

MicroRNAs (miRNAs) are a class of short (approximately 18–24 nucleotides) endogenous noncoding RNA molecules that regulate gene expression during post-translation [12]. Since the discovery of miRNAs by Victor Ambros and his group in 1993 [13], their biological importance has rapidly emerged in recent decades. Usually, miRNAs bind to the 3′ untranslated region (UTR) of their target mRNAs and result in the translational inhibition, degradation, or cleavage of miRNAs depending on the degree of complementarity [14]. As a single miRNA can target hundreds of messenger RNAs and a single messenger RNA can also be targeted by numerous miRNAs, miRNAs influence complex networks of signaling pathways associated with almost all biological/cellular processes, including cell growth, differentiation, immune response, tissue remodeling, and cancer development [15–18]. Accumulating evidence has demonstrated that alterations in miRNA expression are intimately associated with the initiation and progression of human diseases, including chronic liver disease [19–21]. Consequently, modulating miRNA expression could be a key strategy for developing novel therapies for liver fibrosis/cirrhosis [22]. The systemic dosage of therapeutic miRNAs, unfortunately, has short half-lives in circulation and can induce toxicity; therefore, designing a delivery platform that protects the therapeutic miRNAs, efficiently transporting therapeutic miRNAs to the liver and modulating the levels of specific miRNAs in vivo within activated HSCs, remains a challenge [23].

To address these obstacles, researchers are increasingly applying techniques developed in nanomedicines to deliver miRNA as a therapy for chronic liver disease [24, 25]. Our research group has developed a ‘pseudo’ poly[amino acid] polymer and has engineering L-tyrosine polyurethane (LTU) into biodegradable nanoparticles (NPs) as a delivery system for miRNA. The biological and technical advantages of these NPs include protection of miRNAs from enzymatic activity, absence of cellular toxicity by the NPs’ degradation products, surface decoration of the NP with polyethylene glycol (PEG) to minimize the immune response, optimization of size distribution for endocytosis, and the ability to induce the proton sponge effect so that NPs can escape from endosomes and release nucleic acids into the cell’s cytoplasm [26–28]. These features (Fig. 1) make LTU NPs an ideal delivery system for miRNA therapy and should be explored further as a therapeutic option for liver diseases [27, 29]. In this chapter, we summarize the general pathogenesis of liver

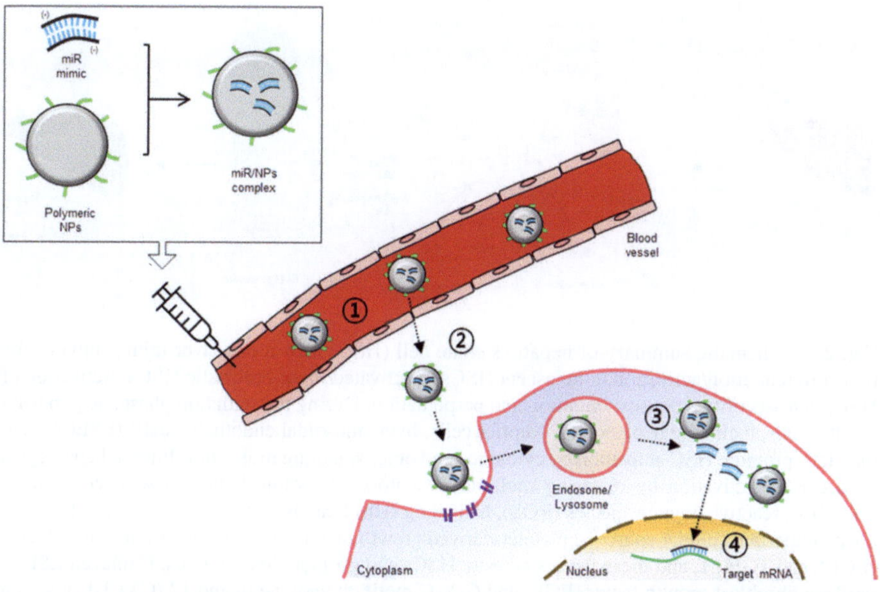

Fig. 1 A simplified model of nanoparticle (NP)-based delivery of microRNAs to target cells. Polymeric NPs are used as carriers of (miRNA) mimics and have substantial advantages over the administration of naked miRNA mimics. When miRNA is encapsulated into polymeric NPs and systemically administered to the body, the encapsulation prevents miRNA degradation by serum nucleases and results in a long period of circulation in the blood (①). Injury to tissues facilitates blood vessel dysfunction and gives NPs an opportunity to leak into diseased tissue and become endocytosed by hepatic stellate cells (②). Incorporating linear polyethylenimine into NPs can induce the proton sponge effect and allow NPs to escape from the endosomal/lysosomal pathway, and the NPs can be degraded and release miRNA in the cell’s cytoplasm (③) so that it can enter the nucleus to induce gene knockdown (④)

fibrosis/cirrhosis, review the current roles of miRNAs in HSC activation and liver fibrosis, and evaluate the therapeutic potential of miRNAs encapsulated in LTU NPs as a treatment for liver fibrosis.

2 The Role of Hepatic Stellate Cells in Liver Fibrosis

Liver fibrosis is a wound-healing process in response to liver injury [4]. To date, substantial progress has been made in understanding the process of hepatic fibrosis, such as characterization of ECM components in fibrotic liver, identification of HSCs as the major source of ECM in liver fibrosis, and characterization of key signaling pathways in liver fibrosis (Fig. 2) [4, 10]. Of them, the identification and establishment of HSCs as the key cellular source of ECM in the liver have been a major advancement in elucidating liver fibrosis. The main fibrogenic cell type in the liver is activated or myofibroblastic HSCs [7, 10]. In the liver, HSCs reside in the

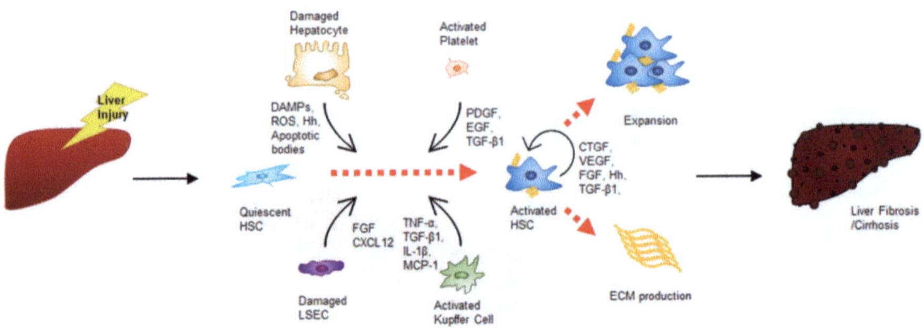

Fig. 2 A schematic summary of hepatic stellate cell (HSC) activation. Liver injury initiates the transdifferentiation/activation of quiescent HSCs to activated/myofibroblastic HSCs. Activation of HSCs consists of two phases: initiation and perpetuation. During the initiation phase, neighboring hepatic cells, including hepatocytes, Kupffer cells, liver sinusoidal endothelial cells (LSECs), and platelets, promote HSC activation by cytokines and other signaling molecules. Injured hepatocytes induce HSC activation by releasing multiple mediators, including damage-associated proteins (DAMPs), reactive oxygen species (ROS), hedgehog (Hh) ligands, and apoptotic bodies. Platelets are an important cellular source of platelet-derived growth factor (PDGF), epidermal growth factor (EGF), and TGF-β1, and these factors activate HSCs and promote liver fibrosis. Damaged LSECs produce fibroblast growth factor (FGF) and C-X-C motif chemokine ligand 12 (CXCL12), which are paracrine stimuli of HSCs. Activated Kupffer cells produce cytokines and chemokines, such as tumor necrosis factor alpha (TNF-α), transforming growth factor beta 1 (TGF-β1), interleukin-1 beta (IL-1β), and monocyte chemoattractant protein-1 (MCP-1), which directly influence HSC activation. During the perpetuation phase, autocrine and paracrine stimulations maintain the activated HSC phenotype and promote the production of fibrotic extracellular matrix (ECM). In addition to paracrine stimulation, activated HSCs produce and secrete connective tissue growth factor (CTGF), vascular endothelial growth factor (VEGF), FGF, Hh ligands, and TGF-β1 in an autocrine manner, which are known to promote the activation, maintenance, and expansion of HSCs and ECM production. These diverse paracrine/autocrine signals that converge upon HSCs promote HSC activation, leading to liver fibrosis and cirrhosis

subendothelial space of Disse between hepatocytes and sinusoidal endothelial cells and represent approximately 5–8% of the total number of resident cells [6]. In a normal healthy liver, HSCs exist in a quiescent state and serve as the principal storage site for retinoids by storing retinyl esters within lipid droplets present in the cytoplasm of HSCs. Following liver injury of any etiology, HSCs undergo an activation process, which involves cell transdifferentiation from quiescent cells into fibrogenic myofibroblasts [10]. This change is characterized by the loss of lipid droplets, increased proliferative and migratory activities, and accumulation of contractile filaments, including α-smooth muscle actin (α-SMA) [10]. Quiescent HSCs are also known to lose epithelial markers, such as E-cadherin, and gain mesenchymal markers, such as Snail1, thus undergoing an epithelial-to-mesenchymal transition (EMT)-like process to acquire myofibroblastic features during HSC activation [30]. Although other cell types, such as portal fibroblasts, also contribute to hepatic fibrogenesis, fate-tracing studies have confirmed that activated HSCs are the major source of ECM in chronically injured livers [31, 32].

Activation of HSCs consists of two phases: initiation and perpetuation [10]. During the initiation phase, paracrine stimulation from neighboring cells, including platelets, endothelial cells, and Kupffer cells, causes alterations in the gene expression and phenotype of HSCs that render them more responsive to other profibrogenic cytokines and stimuli [7, 10]. In addition, injured/dying hepatocytes release paracrine factors, such as damage-associated molecular patterns (DAMPs) and hedgehog (Hh) ligands, which promote the activation of HSCs [33]. Autocrine and paracrine stimulations during the perpetuation phase maintain the activated/myofibroblastic HSC phenotype and promote the production of fibrotic ECM components such as collagen, glycoprotein, and proteoglycans [7, 10]. Initiation is largely due to paracrine stimulation, whereas perpetuation involves autocrine and paracrine loops. Various growth factors, including platelet-derived growth factor (PDGF), epidermal growth factor, fibroblast growth factor, connective tissue growth factor (CTGF), and vascular endothelial growth factor), are known to promote the expansion and activation of HSCs [10]. A simplified illustration of the complex inter- and intracellular events in HSC activation and liver fibrosis is depicted in Fig. 1.

The intracellular processes by which HSCs regulate the initiation and perpetuation of fibrosis are complex and multifactorial involving various signal transduction pathways, such as the transforming growth factor-β (TGF-β), phosphatidylinositol 3-kinase (PI3K)/protein kinase B (AKT), Wnt/β-catenin and Hh pathways [10]. TGF-β, a well-known profibrotic cytokine, is produced by several cell types, including activated HSCs, platelets, and Kupffer cells, and promotes HSC activation through the mitogen-activated protein kinase and c-Jun N-terminal kinase pathways [34–36]. The PI3K/AKT pathway is activated and is required for the survival and proliferation of HSCs [37]. The Hh pathway, a well-characterized signal transduction pathway, is implicated in HSC activation and liver fibrosis. Hh signaling is critically involved in the proliferation and activation of HSCs, leading to liver fibrosis [33]. Inhibition of this pathway leads to decreased HSC activation and reduced hepatic fibrosis [38]. Fibrosis, if not treated, eventually progresses to advanced fibrosis and cirrhosis, which are the major causes of liver-related morbidity and mortality; therefore, the development of antifibrotic treatments that prevent and/or reverse liver fibrosis is urgently needed [11]. An effective method of preventing or halting liver fibrosis is to attenuate the activation of HSCs in response to chronic hepatic injuries [10]. There are several predominant strategies that contribute to the clearance of activated HSCs and resolution of fibrosis, such as induction of HSC apoptosis or senescence and reversion of HSCs to an inactivated state [39–41]. Although the antifibrotic activities of many drugs or compounds have been demonstrated in vitro and in vivo, none has been clinically validated or commercialized as a therapy for liver fibrosis [11]. Therefore, further research is required to develop novel antifibrotic therapies in the treatment of chronic liver disease.

3 The Roles of MicroRNAs in Liver Fibrosis

As the liver has essential functions in the human body that require highly orchestrated and regulated processes, hepatic physiology is tightly controlled by a complex maze of regulatory networks. Hence, disruption of these regulatory networks is associated with the progression of liver diseases [11]. MiRNAs are known to be involved in the regulation of liver homeostasis, development, regeneration, and metabolic functions by modulating the gene expression of their targets [19]. Increasing evidence suggests that alterations of intrahepatic miRNA levels have been associated with almost every aspect of liver disease, including liver fibrosis/cirrhosis [19, 20]. Given that miRNAs are involved in regulating cell homeostasis and functions and that specific expression of miRNAs reflects the current state of cells, expression changes in miRNAs in liver tissues are closely associated with the progression of liver fibrosis. To date, more than 2500 miRNAs have been identified in humans [42]. MiRNA expression signatures are known to be highly tissue specific in human tissues, and approximately 300 miRNAs have been reported to be present in the human liver [43]. MiR-122 is the best-studied miRNA in hepatic miRNA pools. Most importantly, miR-122 is liver specific, and it is one of the most abundant miRNAs in the normal liver, making up 70% and 52% of the whole miRNA pools in adult mice and humans respectively [44–46]. MiR-122 is specifically expressed by healthy hepatocytes, which are the major parenchymal cells in the liver [44, 47] and primarily involved in normal hepatocyte functions to maintain liver homeostasis [48]. The level of miR-122 decreases in the injured liver and is related to the development of liver diseases [49–51]. MiR-122 was strongly decreased in the fibrotic livers of human patients with non-alcoholic steatohepatitis and liver fibrosis and in the hepatotoxin fibrosis model, where mice are injected with carbon tetrachloride (CCl_4) [52, 53]. In experimental animal models, knockout (KO) of miR-122 in mice promotes liver fibrosis and these responses are alleviated and reversed by the restoration of miR-122 levels in these mice [48, 54]. The molecular targets of miR-122 include profibrogenic factors, such as Krüppel-like factor (Klf6), TGF-β receptor, and Wnt-1 [48, 55–58]. These data suggest that miR-122 might have antifibrotic properties; however, Schueller et al. demonstrated that miR-122 expression is neither regulated nor relevant to HSC activation [59]. As detectable levels of miRNAs in liver tissue vary significantly depending on the conditions of the liver, the relative abundance or scarcity of miRNAs is directly influenced by variations in the cell populations. Given that activated HSCs are highly proliferative after liver injury and produce large amounts of ECM proteins, relating expression changes of miRNAs to the activation status of HSCs could be critical to understanding and treating pathological conditions of the liver.

Many reports have profiled the differences in miRNA expression between quiescent and activated HSCs or between healthy and fibrotic liver tissues to identify miRNAs closely associated with HSC activation and liver fibrosis [60–64]. Examples of miRNAs that have been implicated in the development of liver fibrosis, activation of HSC, and deposition ECM are given Table 1. These miRNAs can be broadly categorized into either profibrotic or antifibrotic, which are either upregulated or downregulated, respectively, during fibrogenesis [114]. However, inconsistencies in dysregulated miRNAs have been reported because of the different methods that have used to analyze microarray results, to activate HSCs, and to induce hepatic injuries using different animal models. Nevertheless, several key miRNAs have been clearly demonstrated to play either profibrotic or antifibrotic roles in the regulation of HSCs and liver fibrosis, and the targets of miRNAs have been identified and confirmed by multiple studies [19, 20, 114]. In this section, we focus on miRNAs that have been well documented to have a close association with HSC activation and present their therapeutic potential in liver fibrosis.

3.1 Potential Antifibrotic miRNAs

MiR-29 family members, including miR-29a, miR-29b, and miR-29c, are one of the best-studied miRNAs for HSCs and have been shown to be antifibrogenic [114, 115]. The downregulation of miR-29s have been linked to human cirrhotic livers and rodent models of liver fibrosis induced by carbon tetrachloride (CCl_4) (Fig. 3b) and by bile duct ligation [64]. MiR-29s are highly expressed in primary quiescent HSCs isolated from rodents but downregulated in activated HSCs by in vitro culture-induced activation and in vivo activation by CCl_4 injection [64, 116]. Furthermore, restoration of miR-29b by the administration of miR-29b mimic or miR-29b-expressing adeno-associated virus suppresses HSC activation and liver fibrosis in CCl_4-treated mice [76, 117]. Wang et al. showed that miR-29b inhibits HSC proliferation by arresting the cell cycle in the G1 phase and induces HSC apoptosis by inhibiting the PI3K/AKT pathway [118]. Zhang et al. reported that miR-29b suppresses heat shock protein 47 (HSP47) and lysyl oxidase (LOX), which are necessary for ECM maturation, and inhibits the maturation and production of collagens by HSCs [119]. In addition, various target genes of miR-29s have been identified, and the majority of them are involved in HSC activation, such as Col1α1, Col4α5, Col5α3, elastin, TGF-β, PI3K receptor 1, AKT3, PDGF-C, insulin-like growth factor I (IGF-I), and histone deacetylase 4 (HDAC4) [76, 77, 115, 118, 120, 121].

Table 1 List of microRNAs associated with liver fibrosis

Name	Pro- or anti-fibrotic	Target(s) in liver fibrosis	References
let-7/Lin28	Anti	HMGA2	[65]
miR-15b	Anti	LOXL1	[66]
miR-16	Anti	LOXL1, Gα 12	[66, 67]
miR-19b	Anti	CCR2, CTGF, TGFβRII	[68, 69]
miR-21	Pro	HNF4α, PDCD4, SMAD7, SPRY2	[70–73]
miR-25	Anti	ADAM17, FKBP14	[74]
miR-27	Pro	LXRα, SREBP1c	[75]
miR-29	Anti	CD36, COL1α1, HDAC4, PDGFC, SMAD3	[64, 76–79]
miR-30	Anti	BECLIN1, KLF11, SNAI1	[80–82]
miR-34	Pro	ACSL1, PPARγ	[83, 84]
miR-101	Anti	KLF6, TGFβRI	[85, 86]
miR-122	Anti	CTGF, PACT, P4HA1	[51, 87, 88]
miR-125b	Anti	SMO	[89]
miR-130a	Anti	TGFβRI, TGFβRII	[90]
miR-133a	Anti	COL1α1	[91]
miR-142	Anti	TGFβRI	[92]
miR-145	Anti	ZEB2	[93]
miR-146a	Anti	IRAK1, TRAF6, WNT1, WNT5A	[94, 95]
miR-185	Anti	RHEB, RICTOR	[96]
miR-193	Anti	CAPRIN1, TGFβ2	[97]
miR-195	Pro	SMAD7	[98]
miR-199	Pro	KGF	[99]
miR-200a	Anti	GLI2, GLI3, SIRT1	[100–102]
miR-200c	Pro	FOG2	[103]
miR-214	Pro	MIG6, SUFU	[104, 105]
miR-222	Pro	CDKN1B, PPP2R2A, TIMP3	[106–108]
miR-378a	Anti	GLI3, PRKAG2	[62, 109]
miR-486	Anti	SMO	[110]
miR-542	Pro	BMP7	[111]
miR-942	Pro	BAMBI, PPARγ	[112, 113]

ACSL1 Acyl-CoA Synthetase Long Chain Family Member 1, *ADAM17* ADAM Metallopeptidase Domain 17, *BAMBI* BMP and Activin Membrane Bound Inhibitor, *BMP7* Bone Morphogenetic Protein 7, *CAPRIN1* Cell Cycle-Associated Protein 1, *CCR2* C-C Motif Chemokine Receptor 2, *CDKN1B* Cyclin-Dependent Kinase Inhibitor 1B, *COL1α1* Collagen Type I Alpha 1 Chain, *CTGF* Connective Tissue Growth Factor, *FKBP14* FKBP Prolyl Isomerase 14, *FOG2* Friend Of GATA 2, *Gα 12* Guanine nucleotide-binding a-subunit 12, *GLI* GLI-Krüppel Family Member, *HDAC4* Histone Deacetylase 4, *HMGA2* High Mobility Group AT-Hook 2, *HNF4α* Hepatocyte Nuclear Factor 4 Alpha, *IRAK1* Interleukin 1 Receptor-Associated Kinase 1, *KLF* Krüppel-Like Factor, *LOXL1* Lysyl Oxidase-Like 1, *LXRα* Liver X Receptor-Alpha, *miR* microRNA, *MIG-6* Mitogen-Inducible Gene 6, *PACT* PKR-Activating Protein, *PDCD4* Programmed Cell Death 4, *PDGFC* Platelet-Derived Growth Factor C, *PPARγ* Peroxisome Proliferator-Activated Receptor Gamma, *PPP2R2A* Protein Phosphatase 2 Regulatory Subunit Alpha, *PRKAG2* Protein Kinase AMP-Activated Noncatalytic Subunit Gamma 2, *P4HA1* Prolyl 4-Hydroxylase Subunit Alpha 1, *RHEB* Ras Homolog Enriched In Brain, *RICTOR* RPTOR Independent Companion Of MTOR Complex 2, *SMAD* SMAD Family Member, *SIRT1* Sirtuin 1, *SMO* Smoothened, *SNAI1* Snail Family Transcriptional Repressor 1, *SPRY2* Sprouty RTK Signaling Antagonist 2, *SREBP1c* Sterol Regulatory Element Binding Transcription Factor 1, *SUFU* Suppressor of Fused Homolog, *TGFβ* Transforming Growth Factor Beta, *TGFβR* TGFβ Receptor, *TIMP3* Tissue Inhibitor of Metalloproteinases 3, *TRAF6* TNF Receptor-Associated Factor 6, *WNT* Wnt Family Member, *ZEB2* Zinc Finger E-Box Binding Homeobox 2

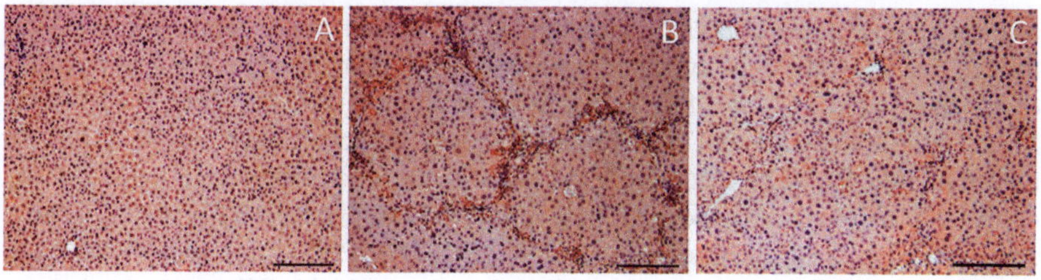

Fig. 3 Immunohistochemistry of liver sections for (**a**) normal, (**b**) carbon tetrachloride treatment, (**c**) L-tyrosine polyurethane (LTU) nanoparticle (NP) treatment for 3 weeks. Positive staining for α-smooth muscle actin can be observed for mice treated with carbon tetrachloride. LTU NP treatment prevented the matrix protein deposition. These images originally appeared in *Nature Communications* [62]

Other antifibrogenic miRNAs in the liver are miR-30 family members including miR-30a, miR-30b, miR-30c-1, miR-30c-2, miR-30d, and miR-30e. Abundant miR-30 levels in healthy liver are also reduced in fibrotic livers in human patients and in experimental mice [80, 122]. In primary HSCs isolated from mice, the expression of miR-30 decreases during in vivo and in vitro activation of HSCs [80]. Re-establishment of miR-30s by administration of miR-30-expressing lentivirus inhibits HSC activation and prevents CCl_4-induced liver fibrosis in rodents [80]. Direct targets of miR-30 include the following: CTGF, which is a profibrogenic cytokine; KLF11, which is a mediator of TGF-β signaling; and Snail1, which is a well-known EMT-stimulating transcription factor [80, 81, 123].

Our research group also reported that three members of the miR-378 family, mi-378a-3p, miR-378b, and miR-378d, are downregulated in both activated HSCs and cirrhotic livers of CCl_4-treated mice compared with quiescent HSCs and livers from corn oil-treated mice [62]. Among the three miR-378 family members, miR-378a-3p directly inhibits GLI-Krüppel family member 3 (Gli3), which is a transcriptional activator of the Hh pathway, and suppresses the activity of this pathway, leading to inactivation of HSCs [62]. In addition, restoration of miR-378a-3p in vivo by administration of biodegradable NPs releasing miR-378a-3p mimic attenuates CCl_4-induced liver fibrosis by downregulating Gli3 expression and suppressing HSC activation [62]. Therefore, these findings clearly demonstrate the inhibitory role of miRNA in HSC activation and liver fibrosis, suggesting them as potential therapeutics for treating liver fibrosis.

3.2 *Potential Profibrotic miRNAs*

Upregulation of miRNA during HSC activation usually plays a profibrotic role in liver fibrosis. MiR-222 and its paralog miR-221 are known to have oncogenic functions in the liver [124] and are also associated with liver fibrosis [106–108]. Ogawa et al. first showed that miR-221/222 expression is upregulated in patients with HCV infection and non-alcoholic steatohepatitis (NASH) with liver fibrosis, and miR-221/222 expressions correlate positively with the messenger RNA expression of col1α1 and α-SMA [108]. Increased expression of miR-221/222 is also confirmed in a thioacetamide (TAA)-induced mouse model of liver fibrosis [108]. In addition, miR-221/222 is upregulated during HSC activation and regulates the expression of cyclin-dependent kinase inhibitor 1B (CDKN1B) [108]. Recently, Jiang et al. demonstrated that liver-specific miR-221/222 KO mice treated with CCl_4 exhibit a significant reduction in liver fibrosis compared with CCl_4-treated wild type mice [107]. In contrast, the reinduction of miR-221/222 by adenovirus infection worsened liver fibrosis in the CCl_4-treated miR-221/222 KO mice, suggesting the profibrotic potential of miR-221/222 [107]. However, these researchers employed the albumin-cre/LoxP system to produce liver-specific miR-221/222 KO mice, and miR-221/222 was abolished in only albumin-expressing cells, such as hepatocytes, but not in activated HSCs. Given that activated HSCs express miR-221/222 in mice with fibrotic liver, these cell-specific miR-221/222 KO mice should be further investigated. The direct targets of miR-221/222 have been identified as protein phosphatase 2A subunit B (PPP2R2A) and tissue inhibitor of metalloproteinase-3 (TIMP-3) [106, 107].

Another profibrogenic messenger RNA is miR-214. Fibrotic liver shows increased expression of miR-214 in human patients and CCl_4-injected mice, demonstrating a positive correlation with the degree of liver fibrosis [104, 125]. The level of miR-214 is elevated during both in vitro and in vivo activation of primary murine HSCs [104, 105, 125]. Furthermore, inhibition of miR-214 by antagomir-214 suppresses the proliferation and activation of HSCs in vitro and has ameliorated CCl_4-induced liver fibrosis in mice [104]. Ma et al. demonstrated that miR-214 directly inhibits the expression of suppressor-of-fused homolog (Sufu), a negative regulator of the Hh signaling pathway, and the knockdown of miR-214 expression in vivo enhances the expression of Sufu, which alleviates hepatic fibrogenesis [104].

4 Nanoparticle-based Delivery of MiRNA for Liver Fibrosis Therapy

Given the importance of miRNAs in modulating HSC activation and their implications in liver fibrosis, major efforts have been made to develop miRNA-based therapeutics to prevent and/or cure liver fibrosis [19–21]. As abnormal expression of miRNA is intimately associated with pathogenesis of liver fibrosis/cirrhosis, many researchers have reported that modulation of miRNA levels through restoration of antifibrotic miRNAs or suppression of profibrotic miRNAs leads to the recovery of liver fibrosis in various experimental animal models [19, 20]. Despite the therapeutic potential of miRNAs, the progress of miRNA-based therapeutics is hampered by an inability to effectively deliver miRNAs in vivo [29, 126, 127]. The major limitation of miRNA delivery includes its lack of stability in a circulatory system, difficulty in reaching the target tissues, and immunotoxicity [128]. Naked miRNAs are degraded within seconds by an abundance of serum nucleases in the blood, or they are cleared rapidly via renal excretion, resulting in a short half-life in systemic circulation [127, 128]. The presence of naked miRNAs in the circulation also triggers secretion of inflammatory cytokines and type I interferons through Toll-like receptors, which provoke an inflammatory response and may cause systemic immune toxicity [29, 127, 128]. Even if miRNAs reach their target tissues, the negative charge of miRNA limits their ability to cross the cell membranes, [127] and any miRNAs that are endocytosed become trapped in endosomes and can be degraded by lysosomes [128, 129].

Viral vectors are frequently used as carriers to deliver miRNAs because high infection rates can be achieved [130]. Viruses have an innate ability to protect gene material within their capsids, recognize specific cells, transport their genetic material across cellular nuclear membranes, and escape from endosomes. These characteristics are attractive for miRNA delivery [130, 131]. However, these vectors are associated with diseases and raise significant medical concerns, and the process of generating recombinant viruses does not reduce their potential for immunogenicity or inducing cancer for retroviruses [27, 130].

In contrast, nonviral approaches are becoming attractive alternatives because many of the beneficial viral functions can be artificially replicated in the design of NPs without having the possibilities of inducing the diseases associated with viruses. Depending upon the materials and design, NPs could have low toxicity, low immune responses, surface decoration for targeting cellular receptors, biodegradability, and cost-efficient production [29, 131]. Although many types of polymer are available, the degradation rate and toxicity of the degradation products are critical to the success of NPs for gene therapy. Our research group has developed a polymer by modifying L-tyrosine [26, 28, 132], as amino acids are the building blocks of proteins. Although amino acids can be polymerized using peptide bonds and folded within cells into secondary and tertiary structures, folding limits the amount of nucleic acids that can be encapsulated, which also limits manufacturing on a large scale. These limitations have been overcome by chemically modifying the structure of L-tyrosine with two linkages (Fig. 4). Desaminotyrosine, an L-Dopa analog, is linked to the amide functional group using a peptide bond and connects to aromatic functional group on the L-tyrosine through polyurethane polymerization. The carboxyl terminal of L-tyrosine has been protected to prevent unwanted branching and undesired byproducts [132]. The polymerization results in LTU with a molecular weight of approximately 115 kDa [132] and is classified as a 'pseudo' poly [amino acid]. LTU is soluble in chloroform and can be easily processed into NPs using standard techniques. Previous studies show that LTU films degrade [133] through hydrolytic and enzymatic linkages in LTU's backbone. The degradation rate makes LTU an ideal candidate for the treatment of liver as it provides continuous release of nucleic acids for approximately 1 month. LTU and its degradation products also have been shown to be noncytotoxic. Human dermal fibroblasts incubated with 800 mg/ml of degradation products of LTU for 24 h show no significant reduction in cell viability compared with cells incubated with cell culture media [28].

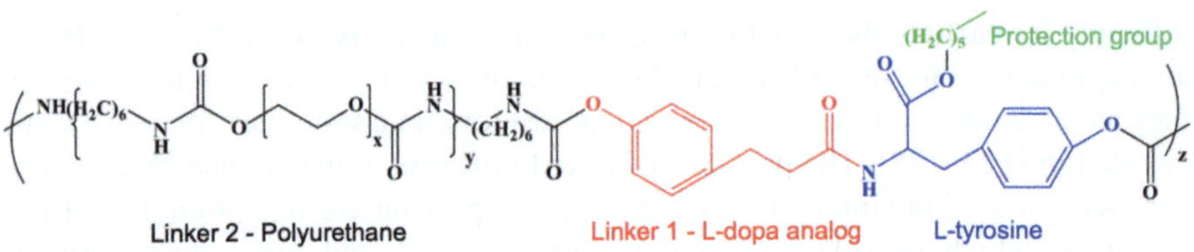

Fig. 4 Chemical structure of L-tyrosine polyurethane. L-tyrosine is modified with L-dopa analog and polyurethane linkages

Nanoparticles made with LTU are encapsulated with either FITC (LTU-FITC NPs, Fig. 5a) or miRNA-378a-3p (LTU-miRNA NP) against Gli3 messenger RNA using water-in-oil-in-water emulsion technique. Prior to the formation of these emulsions, nucleic acids are complexed with linear polyethylenimine (LPEI, MW 25 KDa) at a ratio of 1:1 and 5:1 respectively, which minimizes shear degradation when exposed to high mixing conditions. PEG-PLA, an amphiphilic copolymer that accumulates at the oil–water interfaces, is added to decorate the surface of NPs with PEG. After solvent evaporation, the NPs are washed and lyophilized. The resulting NPs are spherical with heterogeneous size distributions with a mean diameter of 1647 and 340 nm, respectively. As the size of the NPs is appropriate, the mechanism of uptake could be through endocytosis, and fluorescence microscopy shows LTU-FITC NPs were taken up by human hepatic stellate (LX2) cells (Fig. 5b and c).

Prior to the formation of an emulsion, nucleic acids are complexed to LEPI to prevent their degradation from exposure to high levels of shear stress during NP formation [26]. The analysis of the release studies (Fig. 5d) shows an initial burst release of mRNA; the release rate is initially slow (0.13 μg of mRNA per mg of NPs) between days 4 and 21. Afterward, the release rate rises sharply (1.2 μg of mRNA per mg of NPs) from week 3 to week 5 and reaches a steady state after week 5. Overall, a biphasic release of mRNA is observed. Thus, a sustained release of bioactive mRNA has been observed, and these NPs should be able to provide continuous release of miRNA for 5 weeks.

Liver fibrosis has been induced in a mouse model (male C57Bl/6) by intraperitoneal (IP) injections of carbon tetrachloride (CCl_4) at a concentration of 0.4 ml/kg. Figure 6a shows the injection schedule of both CCl_4 and LTU-miRNA NPs. Alanine transaminase (ALT), which is a clinical marker for liver disease, increased by 270% compared with control (no CCl_4 injections) after the first week of CCl_4 injections and remained high for the third week (230%, Fig. 6b). Animals injected with LTU-miRNA NPs show attenuation of ALT levels (23% higher than normal at week 3, Fig. 6b). The results for aspartate transaminase (AST), also a clinical marker for liver disease, mirrored the ALT results (increased by 26 and 46% for weeks 2 and 3, respectively, for CCl_4-injected mice and levels returning to 4% for LTU-miRNA NP-treated mice at week 3, Fig. 6c). Mice treated with CCl_4 show downregulation of miRNA-378a-3p (−82%, −72%, and −62% for weeks 1, 2, and 3, respectively,

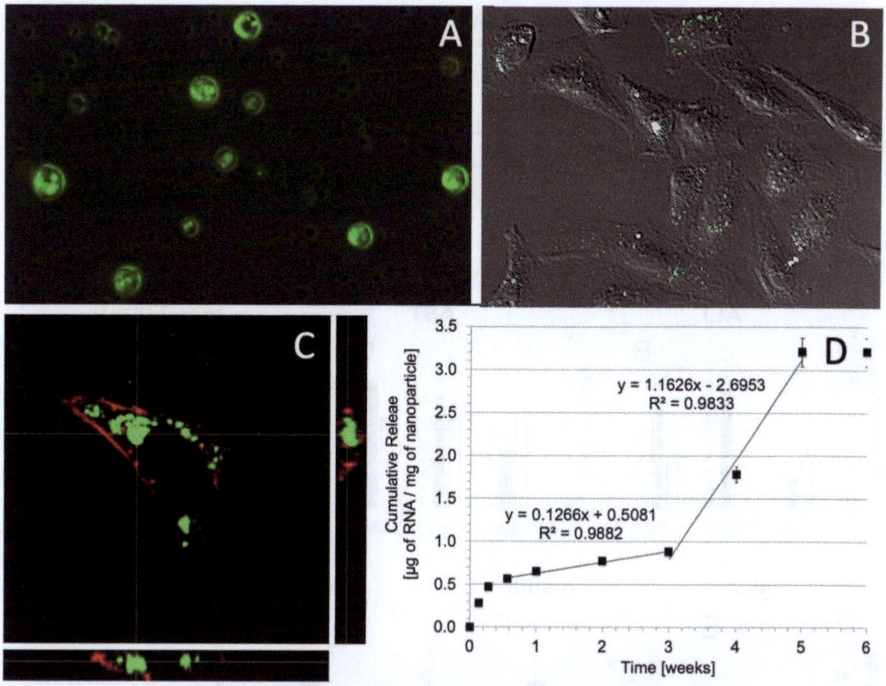

Fig. 5 Fluorescence microscopy and release profile of (**a**) L-tyrosine polyurethane fluorescein isothiocyanate LTU-FITC nanoparticles (NPs), (**b**) LX2 cells incubated with LTU-FITC NPs. (**c**) Confocal microscopy of LX2 cells with actin staining (red) incubated with LTU-FITC NPs, and (**d**) release profiles of LTU NPs loaded with miRNA. Images **a–c** originally appeared in *Nature Communications* [62]

compared with control tissues, Fig. 6d). In contrast, mice treated with LTU-miRNA NPs show −45% at week 1 (Fig. 6d). However, miRNA-378a-3p levels increase to 42% and 68% for weeks 2 and 3, respectively (Fig. 6d). These enzyme markers and miRNA-378a-3p levels correlated with increased matrix deposition (Fig. 3b) for CCl_4-treated mice. These results are confirmed by QRT-PCR results showing 35-, 19-, 12-, and 67-fold increases for gli3, collagen, α-smooth muscle actin, vimentin, and collagen, respectively. In contrast, mice treated with LTU-miRNA NPs show only 11-, 5-, 5-, and 18-fold increases, respectively. Together, exposure to CCl_4 initiates inflammation and fibrosis, but a single treatment of LTU NPs continuously releasing miRNA-378a-3p over a period of 3 weeks offsets the negative effects of CCl_4, prevents the deposition of matrix proteins (Fig. 3c), and normalizes the liver function (Fig. 6).

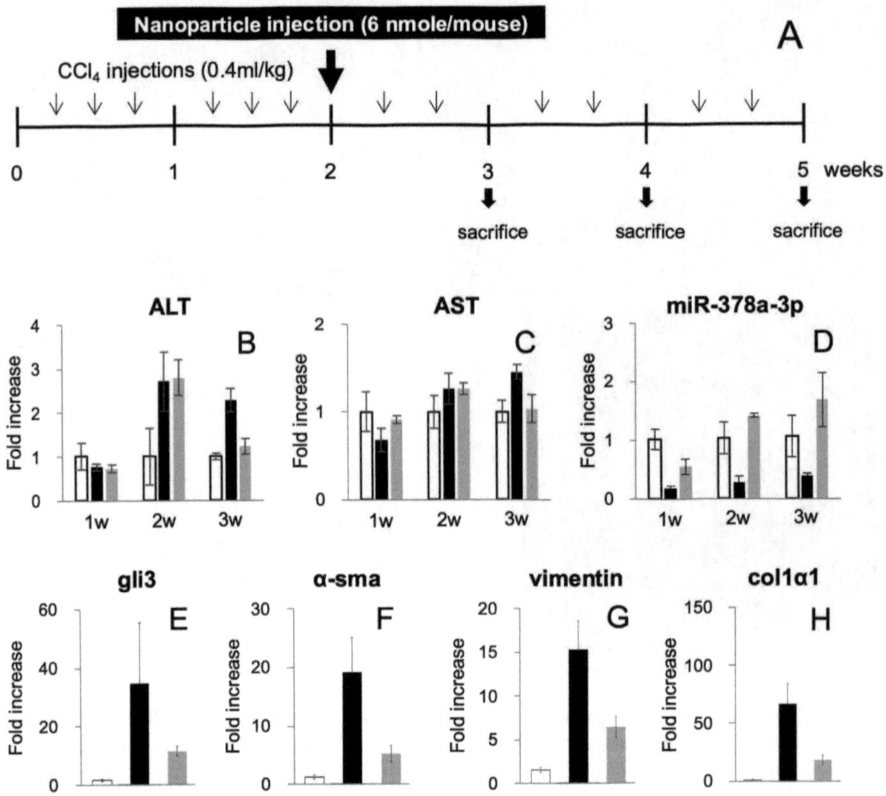

Fig. 6 Injection protocol and results. (**a**) Injection schedule and mRNA levels of (**b**) ALT, (**c**) AST, (**d**) miR-378-3p, (**e**) gli3, (**f**) α-SMA, (**g**) vimentin, and (**h**) collage for control (white), CCl_4 injected (black), and LTU-miRNA NP-treated animals (gray). These images originally appeared in *Nature Communications* [62]

5 Conclusions

Liver fibrosis is a common pathological feature for most chronic liver diseases regardless of the etiology and molecular progression to liver cirrhosis, which has high global morbidity and mortality [4]. A cure currently does not exist for liver fibrosis/cirrhosis, which was once thought to be irreversible, but recent medical evidence suggests that liver fibrosis might be reversed if treated properly. The diagnosis and management of liver diseases are also difficult because patients are asymptomatic until the advanced stages. As a substantial amount of research has linked a broad array of miRNAs to the pathogenesis of liver fibrosis, they also represent potential targets for the diagnosis and therapeutics of liver fibrosis [19, 22], and several clinical trials using miRNA-based treatment are in progress. However, limitations of the in vivo application of miRNAs include unwanted side effects owing to their target multiplicity and redundancy and the development of a suitable delivery system [127, 128].

It has been shown that hepatic levels of miR-378a-3p correlate inversely with gli3 expression, and fibric liver can be rescued by the delivery of this molecule in an animal model using biodegradable NPs. A broad array of miRNAs are closely associated with the pathogenesis of liver fibrosis and represent potential targets for diagnosis and treatment of this disease [19, 22]. For example, Hu et al. used PLGA-based NPs to deliver miR-449b-5p to a rat model of ischemia/reperfusion (I/R) hepatic injury [134]. Although the I/R model is different from the liver fibrosis model, miR-449b-5p-loaded NPs inhibit the expression of high-mobility group box 1 (HMGB1), which is a target of miR-449b-5p, and mitigates I/R-induced hepatic injuries in rats [134]. The results using LTU-miRNA NP are also promising and underscore the valuable role of a delivery system with an intelligent design, but liver fibrosis has been caused by an artificial factor and treated with only one miRNA. Thus, directly translating our results for humans could be problematic because liver fibrosis is a chronic disease with multifactorial causes. A comprehensive strategy that delivers multiple miRNAs should be considered to advance novel therapies for fibrotic liver diseases, as many factors are involved in liver pathogenesis. A panel of miRNAs could even be identified for a specific patient and encapsulated into a cocktail of NPs, as technologies in high-throughput screening and personalized medicine have also shown recent advancements. The miRNAs for these types of therapies can be encapsulated into various formulations of NPs designed for multiple release kinetics to ensure the bioavailability of specific miRNA at the desired times. Table 1 gives a short list of potential targets and the corresponding miRNAs could be used for comprehensive miRNA delivery. Therefore, miRNAs are attractive for medical therapeutics because of their ability to regulate numerous pathways, molecular targets can be easily changed by altering the miRNA's genetic sequence, and a delivery system can be designed for safe, effective, and targeted delivery that minimizes unwanted side effects.

References

1. Boyer, T. D., & Lindor, K. D. (2016). *Zakim and Boyer's hepatology: A textbook of liver disease e-book*. Elsevier Health Sciences.
2. Higgins, G. (1931). Experimental pathology of the liver. *Archives of Pathology, 12*, 186–202.
3. Michalopoulos, G. K., & DeFrances, M. C. (1997). Liver regeneration. *Science, 276*(5309), 60–66.
4. Bataller, R., & Brenner, D. A. (2005). Liver fibrosis. *The Journal of Clinical Investigation, 115*(2), 209–218.
5. Schuppan, D., & Afdhal, N. H. (2008). Liver cirrhosis. *Lancet, 371*(9615), 838–851.
6. Friedman, S. L. (2008a). Hepatic stellate cells: Protean, multifunctional, and enigmatic cells of the liver. *Physiological Reviews, 88*(1), 125–172.
7. Friedman, S. L. (2008b). Mechanisms of hepatic fibrogenesis. *Gastroenterology, 134*(6), 1655–1669.

8. Bosetti, C., Levi, F., Lucchini, F., Zatonski, W. A., Negri, E., & La Vecchia, C. (2007). Worldwide mortality from cirrhosis: An update to 2002. *Journal of Hepatology, 46*(5), 827–839.
9. Moon, A. M., Singal, A. G., & Tapper, E. B. (2020). Contemporary epidemiology of chronic liver disease and cirrhosis. *Clinical Gastroenterology and Hepatology, 18*(12), 2650–2666.
10. Tsuchida, T., & Friedman, S. L. (2017). Mechanisms of hepatic stellate cell activation. *Nature Reviews. Gastroenterology & Hepatology, 14*(7), 397–411.
11. Trautwein, C., Friedman, S. L., Schuppan, D., & Pinzani, M. (2015). Hepatic fibrosis: Concept to treatment. *Journal of Hepatology, 62*(1 Suppl), S15–S24.
12. Kim, V. N. (2005). MicroRNA biogenesis: Coordinated cropping and dicing. *Nature Reviews. Molecular Cell Biology, 6*(5), 376–385.
13. Lee, R. C., Feinbaum, R. L., & Ambros, V. (1993). The C. elegans heterochronic gene lin-4 encodes small RNAs with antisense complementarity to lin-14. *Cell, 75*(5), 843–854.
14. Bartel, D. P. (2009). MicroRNAs: Target recognition and regulatory functions. *Cell, 136*(2), 215–233.
15. Bartel, D. P. (2004). MicroRNAs: Genomics, biogenesis, mechanism, and function. *Cell, 116*(2), 281–297.
16. Lee, H. M., Nguyen, D. T., & Lu, L. F. (2014). Progress and challenge of microRNA research in immunity. *Frontiers in Genetics, 5*, 178.
17. Rupaimoole, R., & Slack, F. J. (2017). MicroRNA therapeutics: Towards a new era for the management of cancer and other diseases. *Nature Reviews. Drug Discovery, 16*(3), 203–222.
18. Shenoy, A., & Blelloch, R. H. (2014). Regulation of microRNA function in somatic stem cell proliferation and differentiation. *Nature Reviews. Molecular Cell Biology, 15*(9), 565–576.
19. Murakami, Y., & Kawada, N. (2017). MicroRNAs in hepatic pathophysiology. *Hepatology Research, 47*(1), 60–69.
20. Szabo, G., & Bala, S. (2013). MicroRNAs in liver disease. *Nature Reviews. Gastroenterology & Hepatology, 10*(9), 542–552.
21. Wang, X. W., Heegaard, N. H., & Orum, H. (2012). MicroRNAs in liver disease. *Gastroenterology, 142*(7), 1431–1443.
22. Mahgoub, A., & Steer, C. J. (2016). MicroRNAs in the evaluation and potential treatment of liver diseases. *Journal of Clinical Medicine, 5*(5), 52.
23. Yu, B., Zhao, X., Lee, L. J., & Lee, R. J. (2009). Targeted delivery systems for oligonucleotide therapeutics. *The AAPS Journal, 11*(1), 195–203.
24. Giannitrapani, L., Soresi, M., Bondì, M. L., Montalto, G., & Cervello, M. (2014). Nanotechnology applications for the therapy of liver fibrosis. *World Journal of Gastroenterology, 20*(23), 7242–7251.
25. Poilil Surendran, S., George Thomas, R., Moon, M. J., & Jeong, Y. Y. (2017). Nanoparticles for the treatment of liver fibrosis. *International Journal of Nanomedicine, 12*, 6997–7006.
26. Ditto, A. J., Shah, P. N., Gump, L. R., & Yun, Y. H. (2009a). Nanospheres formulated from L-tyrosine polyphosphate exhibiting sustained release of polyplexes and in vitro controlled transfection properties. *Molecular Pharmaceutics, 6*(3), 986–995.
27. Ditto, A. J., Shah, P. N., & Yun, Y. H. (2009b). Non-viral gene delivery using nanoparticles. *Expert Opinion on Drug Delivery, 6*(11), 1149–1160.
28. Shah, P. N., & Yun, Y. H. (2013). Cellular interactions with biodegradable polyurethanes formulated from L-tyrosine. *Journal of Biomaterials Applications, 27*(8), 1017–1031.
29. Muthiah, M., Park, I. K., & Cho, C. S. (2013). Nanoparticle-mediated delivery of therapeutic genes: Focus on miRNA therapeutics. *Expert Opinion on Drug Delivery, 10*(9), 1259–1273.
30. Choi, S. S., Omenetti, A., Witek, R. P., Moylan, C. A., Syn, W. K., Jung, Y., Yang, L., Sudan, D. L., Sicklick, J. K., Michelotti, G. A., Rojkind, M., & Diehl, A. M. (2009). Hedgehog pathway activation and epithelial-to-mesenchymal transitions during myofibroblastic transformation of rat hepatic cells in culture and cirrhosis. *American Journal of Physiology. Gastrointestinal and Liver Physiology, 297*(6), G1093–G1106.

31. Iwaisako, K., Brenner, D. A., & Kisseleva, T. (2012). What's new in liver fibrosis? The origin of myofibroblasts in liver fibrosis. *J Gastroenterol Hepatol, 27 Suppl 2*(Suppl 2), 65–68.
32. Mederacke, I., Hsu, C. C., Troeger, J. S., Huebener, P., Mu, X., Dapito, D. H., Pradere, J. P., & Schwabe, R. F. (2013). Fate tracing reveals hepatic stellate cells as dominant contributors to liver fibrosis independent of its aetiology. *Nature Communications, 4*, 2823.
33. Omenetti, A., Choi, S., Michelotti, G., & Diehl, A. M. (2011). Hedgehog signaling in the liver. *Journal of Hepatology, 54*(2), 366–373.
34. Engel, M. E., McDonnell, M. A., Law, B. K., & Moses, H. L. (1999). Interdependent SMAD and JNK signaling in transforming growth factor-beta-mediated transcription. *The Journal of Biological Chemistry, 274*(52), 37413–37420.
35. Hanafusa, H., Ninomiya-Tsuji, J., Masuyama, N., Nishita, M., Fujisawa, J., Shibuya, H., Matsumoto, K., & Nishida, E. (1999). Involvement of the p38 mitogen-activated protein kinase pathway in transforming growth factor-beta-induced gene expression. *The Journal of Biological Chemistry, 274*(38), 27161–27167.
36. Hellerbrand, C., Stefanovic, B., Giordano, F., Burchardt, E. R., & Brenner, D. A. (1999). The role of TGFbeta1 in initiating hepatic stellate cell activation in vivo. *Journal of Hepatology, 30*(1), 77–87.
37. Reif, S., Lang, A., Lindquist, J. N., Yata, Y., Gabele, E., Scanga, A., Brenner, D. A., & Rippe, R. A. (2003). The role of focal adhesion kinase-phosphatidylinositol 3-kinase-akt signaling in hepatic stellate cell proliferation and type I collagen expression. *The Journal of Biological Chemistry, 278*(10), 8083–8090.
38. Michelotti, G. A., Xie, G., Swiderska, M., Choi, S. S., Karaca, G., Krüger, L., Premont, R., Yang, L., Syn, W. K., Metzger, D., & Diehl, A. M. (2013). Smoothened is a master regulator of adult liver repair. *The Journal of Clinical Investigation, 123*(6), 2380–2394.
39. Kendall, T. J., Hennedige, S., Aucott, R. L., Hartland, S. N., Vernon, M. A., Benyon, R. C., & Iredale, J. P. (2009). p75 Neurotrophin receptor signaling regulates hepatic myofibroblast proliferation and apoptosis in recovery from rodent liver fibrosis. *Hepatology, 49*(3), 901–910.
40. Krizhanovsky, V., Yon, M., Dickins, R. A., Hearn, S., Simon, J., Miething, C., Yee, H., Zender, L., & Lowe, S. W. (2008). Senescence of activated stellate cells limits liver fibrosis. *Cell, 134*(4), 657–667.
41. Oh, Y., Park, O., Swierczewska, M., Hamilton, J. P., Park, J. S., Kim, T. H., et al. (2016). Systemic PEGylated TRAIL treatment ameliorates liver cirrhosis in rats by eliminating activated hepatic stellate cells. *Hepatology, 64*(1), 209–223.
42. Bartel, D. P. (2018). Metazoan microRNAs. *Cell, 173*(1), 20–51.
43. Gamazon, E. R., Innocenti, F., Wei, R., Wang, L., Zhang, M., Mirkov, S., Ramírez, J., Huang, R. S., Cox, N. J., Ratain, M. J., & Liu, W. (2013). A genome-wide integrative study of microRNAs in human liver. *BMC Genomics, 14*, 395.
44. Chang, J., Guo, J. T., Jiang, D., Guo, H., Taylor, J. M., & Block, T. M. (2008). Liver-specific microRNA miR-122 enhances the replication of hepatitis C virus in nonhepatic cells. *Journal of Virology, 82*(16), 8215–8223.
45. Chang, J., Nicolas, E., Marks, D., Sander, C., Lerro, A., Buendia, M. A., Xu, C., Mason, W. S., Moloshok, T., Bort, R., Zaret, K. S., & Taylor, J. M. (2004). miR-122, a mammalian liver-specific microRNA, is processed from hcr mRNA and may downregulate the high affinity cationic amino acid transporter CAT-1. *RNA Biology, 1*(2), 106–113.
46. Lagos-Quintana, M., Rauhut, R., Yalcin, A., Meyer, J., Lendeckel, W., & Tuschl, T. (2002). Identification of tissue-specific microRNAs from mouse. *Current Biology, 12*(9), 735–739.
47. Jopling, C. L., Yi, M., Lancaster, A. M., Lemon, S. M., & Sarnow, P. (2005). Modulation of hepatitis C virus RNA abundance by a liver-specific MicroRNA. *Science, 309*(5740), 1577–1581.
48. Tsai, W. C., Hsu, S. D., Hsu, C. S., Lai, T. C., Chen, S. J., Shen, R., Huang, Y., Chen, H. C., Lee, C. H., Tsai, T. F., Hsu, M. T., Wu, J. C., Huang, H. D., Shiao, M. S., Hsiao, M., & Tsou,

A. P. (2012). MicroRNA-122 plays a critical role in liver homeostasis and hepatocarcinogenesis. *The Journal of Clinical Investigation, 122*(8), 2884–2897.

49. Cheung, O., Puri, P., Eicken, C., Contos, M. J., Mirshahi, F., Maher, J. W., Kellum, J. M., Min, H., Luketic, V. A., & Sanyal, A. J. (2008). Nonalcoholic steatohepatitis is associated with altered hepatic MicroRNA expression. *Hepatology, 48*(6), 1810–1820.
50. Halász, T., Horváth, G., Pár, G., Werling, K., Kiss, A., Schaff, Z., & Lendvai, G. (2015). miR-122 negatively correlates with liver fibrosis as detected by histology and FibroScan. *World Journal of Gastroenterology, 21*(25), 7814–7823.
51. Li, J., Ghazwani, M., Zhang, Y., Lu, J., Li, J., Fan, J., Gandhi, C. R., & Li, S. (2013). miR-122 regulates collagen production via targeting hepatic stellate cells and suppressing P4HA1 expression. *Journal of Hepatology, 58*(3), 522–528.
52. Shi, J., Aisaki, K., Ikawa, Y., & Wake, K. (1998). Evidence of hepatocyte apoptosis in rat liver after the administration of carbon tetrachloride. *The American Journal of Pathology, 153*(2), 515–525.
53. Williams, A. T., & Burk, R. F. (1990). Carbon tetrachloride hepatotoxicity: An example of free radical-mediated injury. *Seminars in Liver Disease, 10*(4), 279–284.
54. Hsu, S. H., Wang, B., Kota, J., Yu, J., Costinean, S., Kutay, H., Yu, L., Bai, S., La Perle, K., Chivukula, R. R., Mao, H., Wei, M., Clark, K. R., Mendell, J. R., Caligiuri, M. A., Jacob, S. T., Mendell, J. T., & Ghoshal, K. (2012). Essential metabolic, anti-inflammatory, and anti-tumorigenic functions of miR-122 in liver. *The Journal of Clinical Investigation, 122*(8), 2871–2883.
55. Ahsani, Z., Mohammadi-Yeganeh, S., Kia, V., Karimkhanloo, H., Zarghami, N., & Paryan, M. (2017). WNT1 gene from WNT signaling pathway is a direct target of miR-122 in hepatocellular carcinoma. *Applied Biochemistry and Biotechnology, 181*(3), 884–897.
56. Cao, F., & Yin, L. X. (2019). miR-122 enhances sensitivity of hepatocellular carcinoma to oxaliplatin via inhibiting MDR1 by targeting Wnt/β-catenin pathway. *Experimental and Molecular Pathology, 106*, 34–43.
57. Sun, Y., Wang, H., Li, Y., Liu, S., Chen, J., & Ying, H. (2018). miR-24 and miR-122 negatively regulate the transforming growth factor-β/Smad signaling pathway in skeletal muscle fibrosis. *Mol Ther Nucleic Acids, 11*, 528–537.
58. Yin, S., Fan, Y., Zhang, H., Zhao, Z., Hao, Y., Li, J., Sun, C., Yang, J., Yang, Z., Yang, X., Lu, J., & Xi, J. J. (2016). Differential TGFβ pathway targeting by miR-122 in humans and mice affects liver cancer metastasis. *Nature Communications, 7*, 11012.
59. Schueller, F., Roy, S., Trautwein, C., Luedde, T., & Roderburg, C. (2016). miR-122 expression is not regulated during activation of hepatic stellate cells. *Journal of Hepatology, 65*(4), 865–867.
60. Guo, C. J., Pan, Q., Cheng, T., Jiang, B., Chen, G. Y., & Li, D. G. (2009). Changes in microRNAs associated with hepatic stellate cell activation status identify signaling pathways. *The FEBS Journal, 276*(18), 5163–5176.
61. Hyun, J., Park, J., Wang, S., Kim, J., Lee, H. H., Seo, Y. S., & Jung, Y. (2016a). MicroRNA expression profiling in CCl_4-induced liver fibrosis of Mus musculus. *International Journal of Molecular Sciences, 17*(6), 691.
62. Hyun, J., Wang, S., Kim, J., Rao, K. M., Park, S. Y., Chung, I., Ha, C. S., Kim, S. W., Yun, Y. H., & Jung, Y. (2016b). MicroRNA-378 limits activation of hepatic stellate cells and liver fibrosis by suppressing Gli3 expression. *Nature Communications, 7*, 10993.
63. Lakner, A. M., Steuerwald, N. M., Walling, T. L., Ghosh, S., Li, T., McKillop, I. H., Russo, M. W., Bonkovsky, H. L., & Schrum, L. W. (2012). Inhibitory effects of microRNA 19b in hepatic stellate cell-mediated fibrogenesis. *Hepatology, 56*(1), 300–310.
64. Roderburg, C., Urban, G. W., Bettermann, K., Vucur, M., Zimmermann, H., Schmidt, S., Janssen, J., Koppe, C., Knolle, P., Castoldi, M., Tacke, F., Trautwein, C., & Luedde, T. (2011). Micro-RNA profiling reveals a role for miR-29 in human and murine liver fibrosis. *Hepatology, 53*(1), 209–218.

65. McDaniel, K., Huang, L., Sato, K., Wu, N., Annable, T., Zhou, T., Ramos-Lorenzo, S., Wan, Y., Huang, Q., Francis, H., Glaser, S., Tsukamoto, H., Alpini, G., & Meng, F. (2017). The let-7/Lin28 axis regulates activation of hepatic stellate cells in alcoholic liver injury. *The Journal of Biological Chemistry, 292*(27), 11336–11347.
66. Ma, L., Liu, J., Xiao, E., Ning, H., Li, K., Shang, J., & Kang, Y. (2021). MiR-15b and miR-16 suppress TGF-β1-induced proliferation and fibrogenesis by regulating LOXL1 in hepatic stellate cells. *Life Sciences, 270*, 119144.
67. Kim, K. M., Han, C. Y., Kim, J. Y., Cho, S. S., Kim, Y. S., Koo, J. H., Lee, J. M., Lim, S. C., Kang, K. W., Kim, J. S., Hwang, S. J., Ki, S. H., & Kim, S. G. (2018). Gα(12) overexpression induced by miR-16 dysregulation contributes to liver fibrosis by promoting autophagy in hepatic stellate cells. *Journal of Hepatology, 68*(3), 493–504.
68. Lan, T., Li, C., Yang, G., Sun, Y., Zhuang, L., Ou, Y., Li, H., Wang, G., Kisseleva, T., Brenner, D., & Guo, J. (2018). Sphingosine kinase 1 promotes liver fibrosis by preventing miR-19b-3p-mediated inhibition of CCR2. *Hepatology, 68*(3), 1070–1086.
69. Brandon-Warner, E., Benbow, J. H., Swet, J. H., Feilen, N. A., Culberson, C. R., McKillop, I. H., deLemos, A. S., Russo, M. W., & Schrum, L. W. (2018). Adeno-associated virus serotype 2 vector-mediated reintroduction of microRNA-19b attenuates hepatic fibrosis. *Human Gene Therapy, 29*(6), 674–686.
70. Zhang, Z., Zha, Y., Hu, W., Huang, Z., Gao, Z., Zang, Y., Chen, J., Dong, L., & Zhang, J. (2013). The autoregulatory feedback loop of microRNA-21/programmed cell death protein 4/activation protein-1 (MiR-21/PDCD4/AP-1) as a driving force for hepatic fibrosis development. *The Journal of Biological Chemistry, 288*(52), 37082–37093.
71. Ning, B. F., Ding, J., Liu, J., Yin, C., Xu, W. P., Cong, W. M., Zhang, Q., Chen, F., Han, T., Deng, X., Wang, P. Q., Jiang, C. F., Zhang, J. P., Zhang, X., Wang, H. Y., & Xie, W. F. (2014). Hepatocyte nuclear factor 4α-nuclear factor-κB feedback circuit modulates liver cancer progression. *Hepatology, 60*(5), 1607–1619.
72. Kennedy, L. L., Meng, F., Venter, J. K., Zhou, T., Karstens, W. A., Hargrove, L. A., Wu, N., Kyritsi, K., Greene, J., Invernizzi, P., Bernuzzi, F., Glaser, S. S., Francis, H. L., & Alpini, G. (2016). Knockout of microRNA-21 reduces biliary hyperplasia and liver fibrosis in cholestatic bile duct ligated mice. *Laboratory Investigation, 96*(12), 1256–1267.
73. Wu, K., Ye, C., Lin, L., Chu, Y., Ji, M., Dai, W., Zeng, X., & Lin, Y. (2016). Inhibiting miR-21 attenuates experimental hepatic fibrosis by suppressing both the ERK1 pathway in HSC and hepatocyte EMT. *Clinical Science (London, England), 130*(16), 1469–1480.
74. Genz, B., Coleman, M. A., Irvine, K. M., Kutasovic, J. R., Miranda, M., Gratte, F. D., Tirnitz-Parker, J. E. E., Olynyk, J. K., Calvopina, D. A., Weis, A., Cloonan, N., Robinson, H., Hill, M. M., Al-Ejeh, F., & Ramm, G. A. (2019). Overexpression of miRNA-25-3p inhibits Notch1 signaling and TGF-β-induced collagen expression in hepatic stellate cells. *Scientific Reports, 9*(1), 8541.
75. Li, Z., Ji, L., Su, S., Zhu, X., Cheng, F., Jia, X., Zhou, Q., & Zhou, Y. (2018b). Leptin up-regulates microRNA-27a/b-3p level in hepatic stellate cells. *Experimental Cell Research, 366*(1), 63–70.
76. Matsumoto, Y., Itami, S., Kuroda, M., Yoshizato, K., Kawada, N., & Murakami, Y. (2016). MiR-29a assists in preventing the activation of human stellate cells and promotes recovery from liver fibrosis in mice. *Molecular Therapy, 24*(10), 1848–1859.
77. Huang, Y. H., Tiao, M. M., Huang, L. T., Chuang, J. H., Kuo, K. C., Yang, Y. L., & Wang, F. S. (2015). Activation of Mir-29a in activated hepatic stellate cells modulates its profibrogenic phenotype through inhibition of histone deacetylases 4. *PLoS One, 10*(8), e0136453.
78. Liang, C., Bu, S., & Fan, X. (2016). Suppressive effect of microRNA-29b on hepatic stellate cell activation and its crosstalk with TGF-β1/Smad3. *Cell Biochemistry and Function, 34*(5), 326–333.
79. Lin, H. Y., Wang, F. S., Yang, Y. L., & Huang, Y. H. (2019). MicroRNA-29a suppresses CD36 to ameliorate high fat diet-induced steatohepatitis and liver fibrosis in mice. *Cell, 8*(10), 1298.

80. Zheng, J., Wang, W., Yu, F., Dong, P., Chen, B., & Zhou, M. T. (2018). MicroRNA-30a suppresses the activation of hepatic stellate cells by inhibiting epithelial-to-mesenchymal transition. *Cellular Physiology and Biochemistry, 46*(1), 82–92.
81. Tu, X., Zheng, X., Li, H., Cao, Z., Chang, H., Luan, S., Zhu, J., Chen, J., Zang, Y., & Zhang, J. (2015). MicroRNA-30 protects against carbon tetrachloride-induced liver fibrosis by attenuating transforming growth factor beta signaling in hepatic stellate cells. *Toxicological Sciences, 146*(1), 157–169.
82. Chen, J., Yu, Y., Li, S., Liu, Y., Zhou, S., Cao, S., Yin, J., & Li, G. (2017). MicroRNA-30a ameliorates hepatic fibrosis by inhibiting Beclin1-mediated autophagy. *Journal of Cellular and Molecular Medicine, 21*(12), 3679–3692.
83. Li, X., Chen, Y., Wu, S., He, J., Lou, L., Ye, W., & Wang, J. (2015). microRNA-34a and microRNA-34c promote the activation of human hepatic stellate cells by targeting peroxisome proliferator-activated receptor γ. *Molecular Medicine Reports, 11*(2), 1017–1024.
84. Li, W. Q., Chen, C., Xu, M. D., Guo, J., Li, Y. M., Xia, Q. M., Liu, H. M., He, J., Yu, H. Y., & Zhu, L. (2011). The rno-miR-34 family is upregulated and targets ACSL1 in dimethylnitrosamine-induced hepatic fibrosis in rats. *The FEBS Journal, 278*(9), 1522–1532.
85. Lei, Y., Wang, Q. L., Shen, L., Tao, Y. Y., & Liu, C. H. (2019). MicroRNA-101 suppresses liver fibrosis by downregulating PI3K/Akt/mTOR signaling pathway. *Clinics and Research in Hepatology and Gastroenterology, 43*(5), 575–584.
86. Tu, X., Zhang, H., Zhang, J., Zhao, S., Zheng, X., Zhang, Z., Zhu, J., Chen, J., Dong, L., Zang, Y., & Zhang, J. (2014). MicroRNA-101 suppresses liver fibrosis by targeting the TGFβ signalling pathway. *The Journal of Pathology, 234*(1), 46–59.
87. Nakamura, M., Kanda, T., Sasaki, R., Haga, Y., Jiang, X., Wu, S., Nakamoto, S., & Yokosuka, O. (2015). MicroRNA-122 inhibits the production of inflammatory cytokines by targeting the PKR activator PACT in human hepatic stellate cells. *PLoS One, 10*(12), e0144295.
88. Teng, K. Y., Barajas, J. M., Hu, P., Jacob, S. T., & Ghoshal, K. (2020). Role of B cell lymphoma 2 in the regulation of liver fibrosis in miR-122 knockout mice. *Biology (Basel), 9*(7), 157.
89. Hyun, J., Wang, S., Kim, J., Kim, G. J., & Jung, Y. (2015). MicroRNA125b-mediated Hedgehog signaling influences liver regeneration by chorionic plate-derived mesenchymal stem cells. *Scientific Reports, 5*, 14135.
90. Wang, Y., Du, J., Niu, X., Fu, N., Wang, R., Zhang, Y., Zhao, S., Sun, D., & Nan, Y. (2017). MiR-130a-3p attenuates activation and induces apoptosis of hepatic stellate cells in nonalcoholic fibrosing steatohepatitis by directly targeting TGFBR1 and TGFBR2. *Cell Death & Disease, 8*(5), e2792.
91. Roderburg, C., Luedde, M., Vargas Cardenas, D., Vucur, M., Mollnow, T., Zimmermann, H. W., Koch, A., Hellerbrand, C., Weiskirchen, R., Frey, N., Tacke, F., Trautwein, C., & Luedde, T. (2013). miR-133a mediates TGF-β-dependent derepression of collagen synthesis in hepatic stellate cells during liver fibrosis. *Journal of Hepatology, 58*(4), 736–742.
92. Yang, X., Dan, X., Men, R., Ma, L., Wen, M., Peng, Y., & Yang, L. (2017b). MiR-142-3p blocks TGF-β-induced activation of hepatic stellate cells through targeting TGFβRI. *Life Sciences, 187*, 22–30.
93. Yang, J., Lu, Y., Yang, P., Chen, Q., Wang, Y., Ding, Q., Xu, T., Li, X., Li, C., Huang, C., Meng, X., Li, J., Zhang, L., & Wang, X. (2019). MicroRNA-145 induces the senescence of activated hepatic stellate cells through the activation of p53 pathway by ZEB2. *Journal of Cellular Physiology, 234*(5), 7587–7599.
94. Du, J., Niu, X., Wang, Y., Kong, L., Wang, R., Zhang, Y., Zhao, S., & Nan, Y. (2015). MiR-146a-5p suppresses activation and proliferation of hepatic stellate cells in nonalcoholic fibrosing steatohepatitis through directly targeting Wnt1 and Wnt5a. *Scientific Reports, 5*, 16163.
95. Zou, Y., Cai, Y., Lu, D., Zhou, Y., Yao, Q., & Zhang, S. (2017). MicroRNA-146a-5p attenuates liver fibrosis by suppressing profibrogenic effects of TGFβ1 and lipopolysaccharide. *Cellular Signalling, 39*, 1–8.

96. Zhou, L., Liu, S., Han, M., Ma, Y., Feng, S., Zhao, J., Lu, H., Yuan, X., & Cheng, J. (2018). miR-185 inhibits fibrogenic activation of hepatic stellate cells and prevents liver fibrosis. *Mol Ther Nucleic Acids, 10*, 91–102.
97. Ju, B., Nie, Y., Yang, X., Wang, X., Li, F., Wang, M., Wang, C., & Zhang, H. (2019). miR-193a/b-3p relieves hepatic fibrosis and restrains proliferation and activation of hepatic stellate cells. *Journal of Cellular and Molecular Medicine, 23*(6), 3824–3832.
98. Song, L. Y., Ma, Y. T., Wu, C. F., Wang, C. J., Fang, W. J., & Liu, S. K. (2017). MicroRNA-195 activates hepatic stellate cells in vitro by targeting Smad7. *BioMed Research International, 2017*, 1945631.
99. Bi, Z. M., Zhou, Q. F., Geng, Y., & Zhang, H. M. (2016). Human umbilical cord mesenchymal stem cells ameliorate experimental cirrhosis through activation of keratinocyte growth factor by suppressing microRNA-199. *European Review for Medical and Pharmacological Sciences, 20*(23), 4905–4912.
100. Li, L., Ran, J., Li, L., Chen, G., Zhang, S., & Wang, Y. (2020). Gli3 is a novel downstream target of miR-200a with an anti-fibrotic role for progression of liver fibrosis in vivo and in vitro. *Molecular Medicine Reports, 21*(4), 1861–1871.
101. Yang, J. J., Tao, H., Liu, L. P., Hu, W., Deng, Z. Y., & Li, J. (2017a). miR-200a controls hepatic stellate cell activation and fibrosis via SIRT1/Notch1 signal pathway. *Inflammation Research, 66*(4), 341–352.
102. Yu, F., Zheng, Y., Hong, W., Chen, B., Dong, P., & Zheng, J. (2015). MicroRNA-200a suppresses epithelial-to-mesenchymal transition in rat hepatic stellate cells via GLI family zinc finger 2. *Molecular Medicine Reports, 12*(6), 8121–8128.
103. Ma, T., Cai, X., Wang, Z., Huang, L., Wang, C., Jiang, S., & Hua, Y. (2017). Liu Q (2017) miR-200c accelerates hepatic stellate cell-induced liver fibrosis via targeting the FOG2/PI3K pathway. *BioMed Research International*, 2670658. https://doi.org/10.1155/2017/2670658
104. Ma, L., Yang, X., Wei, R., Ye, T., Zhou, J. K., Wen, M., Men, R., Li, P., Dong, B., Liu, L., Fu, X., Xu, H., Aqeilan, R. I., Wei, Y. Q., Yang, L., & Peng, Y. (2018). MicroRNA-214 promotes hepatic stellate cell activation and liver fibrosis by suppressing Sufu expression. *Cell Death & Disease, 9*(7), 718.
105. Okada, H., Honda, M., Campbell, J. S., Takegoshi, K., Sakai, Y., Yamashita, T., Shirasaki, T., Takabatake, R., Nakamura, M., Tanaka, T., & Kaneko, S. (2015). Inhibition of microRNA-214 ameliorates hepatic fibrosis and tumor incidence in platelet-derived growth factor C transgenic mice. *Cancer Science, 106*(9), 1143–1152.
106. Dong, R., Zheng, Y., Chen, G., Zhao, R., Zhou, Z., & Zheng, S. (2015). miR-222 overexpression may contribute to liver fibrosis in biliary atresia by targeting PPP2R2A. *Journal of Pediatric Gastroenterology and Nutrition, 60*(1), 84–90.
107. Jiang, X., Jiang, L., Shan, A., Su, Y., Cheng, Y., Song, D., Ji, H., Ning, G., Wang, W., & Cao, Y. (2018). Targeting hepatic miR-221/222 for therapeutic intervention of nonalcoholic steatohepatitis in mice. *eBioMedicine, 37*, 307–321.
108. Ogawa, T., Enomoto, M., Fujii, H., Sekiya, Y., Yoshizato, K., Ikeda, K., & Kawada, N. (2012). MicroRNA-221/222 upregulation indicates the activation of stellate cells and the progression of liver fibrosis. *Gut, 61*(11), 1600–1609.
109. Yu, F., Fan, X., Chen, B., Dong, P., & Zheng, J. (2016). Activation of hepatic stellate cells is inhibited by microRNA-378a-3p via Wnt10a. *Cellular Physiology and Biochemistry, 39*(6), 2409–2420.
110. Kim, J., Lee, C., Shin, Y., Wang, S., Han, J., Kim, M., Kim, J. M., Shin, S. C., Lee, B. J., Kim, T. J., & Jung, Y. (2020). sEVs from tonsil-derived mesenchymal stromal cells alleviate activation of hepatic stellate cells and liver fibrosis through miR-486-5p. *Molecular Therapy, 29*(4), 1471–1486.Published online December 19. https://doi.org/10.1016/j.ymthe.2020.12.025
111. Ji, F., Wang, K., Zhang, Y., Mao, X. L., Huang, Q., Wang, J., Ye, L., & Li, Y. (2019). MiR-542-3p controls hepatic stellate cell activation and fibrosis via targeting BMP-7. *Journal of Cellular Biochemistry, 120*(3), 4573–4581.

112. Tao, L., Xue, D., Shen, D., Ma, W., Zhang, J., Wang, X., Zhang, W., Wu, L., Pan, K., Yang, Y., Nwosu, Z. C., Dooley, S., Seki, E., & Liu, C. (2018). MicroRNA-942 mediates hepatic stellate cell activation by regulating BAMBI expression in human liver fibrosis. *Archives of Toxicology, 92*(9), 2935–2946.
113. Tao, L., Wu, L., Zhang, W., Ma, W. T., Yang, G. Y., Zhang, J., Xue, D. Y., Chen, B., & Liu, C. (2020). Peroxisome proliferator-activated receptor γ inhibits hepatic stellate cell activation regulated by miR-942 in chronic hepatitis B liver fibrosis. *Life Sciences, 253*, 117572.
114. Noetel, A., Kwiecinski, M., Elfimova, N., Huang, J., & Odenthal, M. (2012). microRNA are central players in anti- and profibrotic gene regulation during liver fibrosis. *Front Physiol, 3*, 49.
115. Huang, Y. H., Yang, Y. L., & Wang, F. S. (2018). The role of miR-29a in the regulation, function, and signaling of liver fibrosis. *International Journal of Molecular Sciences, 19*(7), 1889.
116. Sekiya, Y., Ogawa, T., Yoshizato, K., Ikeda, K., & Kawada, N. (2011). Suppression of hepatic stellate cell activation by microRNA-29b. *Biochemical and Biophysical Research Communications, 412*(1), 74–79.
117. Knabel, M. K., Ramachandran, K., Karhadkar, S., Hwang, H. W., Creamer, T. J., Chivukula, R. R., Sheikh, F., Clark, K. R., Torbenson, M., Montgomery, R. A., Cameron, A. M., Mendell, J. T., & Warren, D. S. (2015). Systemic delivery of scAAV8-encoded MiR-29a ameliorates hepatic fibrosis in carbon tetrachloride-treated mice. *PLoS One, 10*(4), e0124411.
118. Wang, J., Chu, E. S., Chen, H. Y., Man, K., Go, M. Y., Huang, X. R., Lan, H. Y., Sung, J. J., & Yu, J. (2015). microRNA-29b prevents liver fibrosis by attenuating hepatic stellate cell activation and inducing apoptosis through targeting PI3K/AKT pathway. *Oncotarget, 6*(9), 7325–7338.
119. Zhang, Y., Ghazwani, M., Li, J., Sun, M., Stolz, D. B., He, F., Fan, J., Xie, W., & Li, S. (2014). MiR-29b inhibits collagen maturation in hepatic stellate cells through down-regulating the expression of HSP47 and lysyl oxidase. *Biochemical and Biophysical Research Communications, 446*(4), 940–944.
120. Kwiecinski, M., Elfimova, N., Noetel, A., Töx, U., Steffen, H. M., Hacker, U., Nischt, R., Dienes, H. P., & Odenthal, M. (2012). Expression of platelet-derived growth factor-C and insulin-like growth factor I in hepatic stellate cells is inhibited by miR-29. *Laboratory Investigation, 92*(7), 978–987.
121. Kwiecinski, M., Noetel, A., Elfimova, N., Trebicka, J., Schievenbusch, S., Strack, I., Molnar, L., von Brandenstein, M., Töx, U., Nischt, R., Coutelle, O., Dienes, H. P., & Odenthal, M. (2011). Hepatocyte growth factor (HGF) inhibits collagen I and IV synthesis in hepatic stellate cells by miRNA-29 induction. *PLoS One, 6*(9), e24568.
122. Roy, S., Benz, F., Vargas Cardenas, D., Vucur, M., Gautheron, J., Schneider, A., Hellerbrand, C., Pottier, N., Alder, J., Tacke, F., Trautwein, C., Roderburg, C., & Luedde, T. (2015). miR-30c and miR-193 are a part of the TGF-β-dependent regulatory network controlling extracellular matrix genes in liver fibrosis. *Journal of Digestive Diseases, 16*(9), 513–524.
123. Duisters, R. F., Tijsen, A. J., Schroen, B., Leenders, J. J., Lentink, V., van der Made, I., Herias, V., van Leeuwen, R. E., Schellings, M. W., Barenbrug, P., Maessen, J. G., Heymans, S., Pinto, Y. M., & Creemers, E. E. (2009). miR-133 and miR-30 regulate connective tissue growth factor: Implications for a role of microRNAs in myocardial matrix remodeling. *Circulation Research, 104*(2), 170–178.
124. Fornari, F., Gramantieri, L., Ferracin, M., Veronese, A., Sabbioni, S., Calin, G. A., Grazi, G. L., Giovannini, C., Croce, C. M., Bolondi, L., & Negrini, M. (2008). MiR-221 controls CDKN1C/p57 and CDKN1B/p27 expression in human hepatocellular carcinoma. *Oncogene, 27*(43), 5651–5661.
125. Iizuka, M., Ogawa, T., Enomoto, M., Motoyama, H., Yoshizato, K., Ikeda, K., & Kawada, N. (2012). Induction of microRNA-214-5p in human and rodent liver fibrosis. *Fibrogenesis & Tissue Repair, 5*(1), 12.

126. Chen, X., Mangala, L. S., Rodriguez-Aguayo, C., Kong, X., Lopez-Berestein, G., & Sood, A. K. (2018). RNA interference-based therapy and its delivery systems. *Cancer Metastasis Reviews, 37*(1), 107–124.
127. Dowdy, S. F. (2017). Overcoming cellular barriers for RNA therapeutics. *Nature Biotechnology, 35*(3), 222–229.
128. Lee, S. W. L., Paoletti, C., Campisi, M., Osaki, T., Adriani, G., Kamm, R. D., Mattu, C., & Chiono, V. (2019). MicroRNA delivery through nanoparticles. *Journal of Controlled Release, 313*, 80–95.
129. Raemdonck, K., Vandenbroucke, R. E., Demeester, J., Sanders, N. N., & De Smedt, S. C. (2008). Maintaining the silence: Reflections on long-term RNAi. *Drug Discovery Today, 13*(21–22), 917–931.
130. Giacca, M., & Zacchigna, S. (2012). Virus-mediated gene delivery for human gene therapy. *Journal of Controlled Release, 161*(2), 377–388.
131. Kasuya, T., & Kuroda, S. (2009). Nanoparticles for human liver-specific drug and gene delivery systems: In vitro and in vivo advances. *Expert Opinion on Drug Delivery, 6*(1), 39–52.
132. Sen Gupta, A., & Lopina, S. T. (2004). Synthesis and characterization of l-tyrosine based novel polyphosphates for potential biomaterial applications. *Polymer, 45*(14), 4653–4662.
133. Sen Gupta, A., & Lopina, S. T. (2005). Properties of l-tyrosine based polyphosphates pertinent to potential biomaterial applications. *Polymer, 46*(7), 2133–2140.
134. Hu, F., Yang, D., Qian, B., Fan, S., Zhu, Q., Ren, H., Li, X., & Zhai, B. (2019). The exogenous delivery of microRNA-449b-5p using spermidine-PLGA nanoparticles efficiently decreases hepatic injury. *RSC Advances, 9*(60), 35135–35144.

MicroRNA as a Versatile Regulator of Wnt the Signaling Pathway in Colorectal Cancer

Jehyun Nam and Sang-Woo Kim

Abstract Colorectal cancer (CRC) is one of the leading causes of cancer-related death that accounted for about one million deaths in 2018 globally. CRC incidence and deaths have steadily increased, especially in developing nations, where people are adopting the Western diet and lifestyle. Obesity, red meat consumption, alcohol, and tobacco have been recognized as important risk factors. Wingless-related integration site (Wnt) signaling orchestrates various biological processes during early development and adulthood, and its aberrant regulation has been causally associated with human diseases, including cancer. Hyperactivation of Wnt signaling is believed to be involved in the initiation and progression of CRC, and found in almost all cases. As such, interventions in Wnt signaling using microRNAs (miRNAs) as a novel therapy have been actively pursued. MiRNAs are endogenous small noncoding RNAs implicated in the regulation of normal cell physiology. Additionally, miRNAs influence numerous cancer-relevant processes, such as cell proliferation, stemness, epithelial-to-mesenchymal transition, and migration, and can act as either oncogenes or tumor suppressors. In agreement with this, many types of human cancers demonstrate abnormal up- or down-regulation of miRNA levels, highlighting therapeutic opportunities in cancer by restoring normal miRNA expression and function. The high mutation rates of Wnt pathway components and frequent dysregulation of miRNAs observed in CRC prompted us to review the interrelationship between Wnt signaling and miRNAs. In this chapter, we summarize the current knowledge concerning their interplay in CRC and discuss the therapeutic potential of miRNAs in the treatment of CRC.

J. Nam
Department of Integrated Biological Science, Pusan National University, Pusan, Republic of Korea

S.-W. Kim (✉)
Department of Integrated Biological Science, Pusan National University, Pusan, Republic of Korea

Department of Biological Sciences, College of Natural Science, Pusan National University, Pusan, Republic of Korea
e-mail: kimsw@psuan.ac.kr

Y. H. Yun, K. E. Yoder (eds.), *Biotechnologies for Gene Therapy*,
https://doi.org/10.1007/978-3-030-93333-3_2

1 Introduction

Colorectal cancer (CRC) is the third most commonly diagnosed cancer and the second leading cause of cancer-related deaths [1]. Although significant progress has been made in the development of diagnostic methods and therapeutic agents for CRC, the death rate from CRC remains high [2]. During the initiation and development of the CRC from normal cells or tissues, accumulated mutations are observed in genes with important functions in regulating cell proliferation or DMA damage response, such as proto-oncogenes and tumor suppressor genes [3]. These genetic instabilities cause abnormal regulation of multiple intracellular signaling pathways including Wnt, epidermal growth factor receptor (EGFR), phosphatidylinositol-3 kinase/Protein kinase B (PI3K/AKT), and mitogen-activated protein kinase (MAPK) pathways [3]. A genome scale analysis has identified more than 90% of CRC patients carried mutations in one or more of the downstream components of the Wnt signaling pathway, e.g., the loss-of-function mutations of adenomatous polyposis coli (APC) or the activating gain-of-function mutations of β-catenin [4]. Dysregulation of Wnt signaling by these and other mutations could disturb the normal tissue homeostasis and signaling networks, eventually leading to the malignant transformation of cells and the occurrence of CRC [5]. Therefore, Wnt signaling has attracted much attention as a potential target for developing CRC therapies.

MicroRNA is a small noncoding RNA that consists of 20–25 nucleotides and regulates target genes by binding to 3'untranslated region (3′UTR) of mRNAs [6]. MicroRNAs mediate target gene silencing via miRNA-induced silencing complex (miRISC), which minimally consists of the guide strand and Argonaute (AGO). The target specificity of miRISC is determined by the interaction between miRNA and the complementary sequences on target mRNA, called miRNA response elements (MREs). The degree of complementarity between miRNA and MRE generally determines whether the target mRNA is degraded or translation is repressed. Once complementarity between miRNA and MRE has been accomplished, the target mRNA is cleaved with AGO2 endonuclease activity and/or translation of the target mRNA is arrested, ameliorating gene expression [7]. Although most studies have reported inhibition of gene expression, miRNA can also promote translational activation under specific conditions [8, 9].

MicroRNAs can have multiple target genes and are involved in the regulation of a variety of physiological and pathological processes [10]. It has become ever more apparent that miRNAs are dysregulated in most cancers [11]. Many of these miRNAs either contribute to or repress the cancer phenotype by inhibiting the expression of tumor suppressors or oncogenes respectively [12]. The oncogenic miRNAs (oncomiRs) or tumor suppressor miRNAs can be used as diagnostic, prognostic, and/or predictive biomarkers of cancer [13]. Conceptually, if the oncomiRs or tumor suppressor miRNAs are to be inhibited or stimulated respectively, it could suppress cell proliferation, cell survival, angiogenesis, invasion, and metastasis of tumor cells to restore normal cell physiology, making them an attractive therapeutic target [14]. CRC is not an exception; therefore, increasing evidence suggests dysregulation of miRNAs are implicated in the pathogenesis of CRC, and devising ways of controlling their expression for therapeutic purpose has been considered as a novel strategy for the treatment of CRC [15]. As Wnt pathway are frequently dysregulated in this type of cancer, miRNAs regulate the signaling components of Wnt have attracted much attention [16, 17]. In this chapter, we summarize the current knowledge of CRC pathogenesis and the role of the Wnt signaling pathway during CRC development, illuminate the regulatory actions of miRNAs associated with the Wnt signaling pathway, and the possibility of harnessing miRNAs therapeutically to treat CRC.

2 General Pathogenesis of Colorectal Cancer

Colorectal cancer typically results from focal changes within benign and precancerous polyps [18]. These polyps are condensed growths or aggregations of abnormal cells within the intestinal mucosa that infiltrate into the intestinal lumen. The proliferating cells within these polyps accumulate genetic changes and acquire the ability to invade the bowel wall [19]. They eventually become more aggressive, spread to local lymph nodes, and finally migrate to distant organs to form metastases mostly in the liver and lungs. Most CRC develops through a multi-step process involving a series of histological, morphological, and genetic changes that accumulate over time [20]. These processes are not individual events but related to each other. For example, recurrent driver mutations in tumor suppressor genes or oncogenes, such as APC, TP53, SMAD4, and KRAS, play an important role during the adenoma–carcinoma sequence and accelerate the progression of CRCs [21].

Three distinct pathways of genetic instability contributing to CRC development are chromosomal instability (CIN), CpG island methylator phenotype (CIMP), and microsatellite instability (MSI) [22]. CIN is observed in about 70% of CRC patients and results from a cascade of accumulating mutations including β-catenin and APC. CIMP is characterized by hypermethylation of the aberrant gene promoter region, leading to transcriptional repression of tumor suppressor and DNA repair genes. Changes in microsatellite repeats are demonstrated in MSI, which is caused by the disruption of DNA repair genes, such as mismatch repair (MMR) genes [3]. MSI can induce uncertain replication of repetitive DNA sequences in short-noncoding regions and increase the possibility of additional genetic mutations. Each of these mechanisms significantly differs in specific pathological features, mechanisms of carcinogenesis, and process of tumor development [23]. The molecular aspects of these pathways have been used clinically in the diagnosis, screening, and management of patients with CRC [24]. The relative contributions of these mechanisms to the development of individual CRC cases differ but are not mutually exclusive.

Thanks to completion of The Cancer Genome Atlas project on clinical CRC samples, CRC can also be categorized into hypermutated (16%, harboring over 12 nonsilent exonic mutations per million bases) or nonhypermutated (84%, harboring less than 12 nonsilent exonic mutations per million bases) tumors [25]. Hypermutated

CRC samples are characterized by a high frequency of transcriptional silencing of the MLH1 gene, and the genes with recurring mutations in the nonhypermutated CRCs include APC, KRAS, and p53 [25, 26]. Each category exhibits several mutations in proto-oncogenes and tumor suppressor genes, but the most prominent genes mutated are the components of the Wnt signaling pathway. Therefore, it is important to understand the role of the Wnt signaling pathway in CRC.

3 Overview of Wnt Signaling Pathways

Wnt signaling is a highly conserved pathway that is critical for both embryonic development and homeostasis of adult tissue, including the intestine [27]. The signaling components of the Wnt pathway are differentially expressed during intestinal epithelial development, regulating self-renewal of stem/progenitor cells, proliferation, apoptosis, differentiation, and migration [28]. Consistently, aberration of the Wnt pathway has been causally associated with every stage of tumor development, including cell proliferation, cell survival, migration, invasion, and metastasis [29]. Wnt signaling is classified into two distinct pathways: canonical (β-catenin dependent) and noncanonical (β-catenin independent) pathways (Fig. 1) [30].

3.1 Canonical Pathway

β-catenin is the key effector in the canonical pathway, and its expression levels are regulated by committed protein complexes [31]. In the absence of a Wnt ligand, cytoplasmic β-catenin protein is continually degraded by the Axin complex, which consists of Axin, APC, casein kinase (CK1), and glycogen synthase kinase 3β (GSK3β) [32]. When the amino terminal region of β-catenin is sequentially

Fig. 1 (continued) In the Wnt/Ca^{2+} signaling pathway (bottom right panel), the interaction of Wnt ligands with the ROR/RYK-Frizzled receptor complex leads to phosphorylation and activation of phospholipase C (PLC). Activated PLC induces intracellular Ca^{2+} fluxes, which in turn trigger activation of the PKC/CDC42 axis and regulate transcription of target genes through calcineurin and NFAT. APC, adenomatous polyposis coli; BCL9L, B-cell CLL/lymphoma 9 protein like; β-TrCP, β-Transducin repeat containing protein; CK1, casein kinase 1; CDC42, cell division control protein 42; DAAM1, Dvl-associated activator of morphogenesis; DKK, dickkopf Wnt signaling pathway inhibitor 1; Dvl, Disheveled; GSK3β, glycogen synthase kinase 3 β; LRP5/6, low-density lipoprotein receptor-related protein5/6; NFAT, nuclear factor of activated T cell; PCP, planar cell polarity; PKC, protein kinase C; ROR, bind tyrosine kinase-like orphan receptor; RNF43, ring finger protein 43; RYK, receptor-like tyrosine kinase; sFRP, secreted frizzled-related protein 1; TCF/LEF, T-cell factor/lymphoid enhancer factor

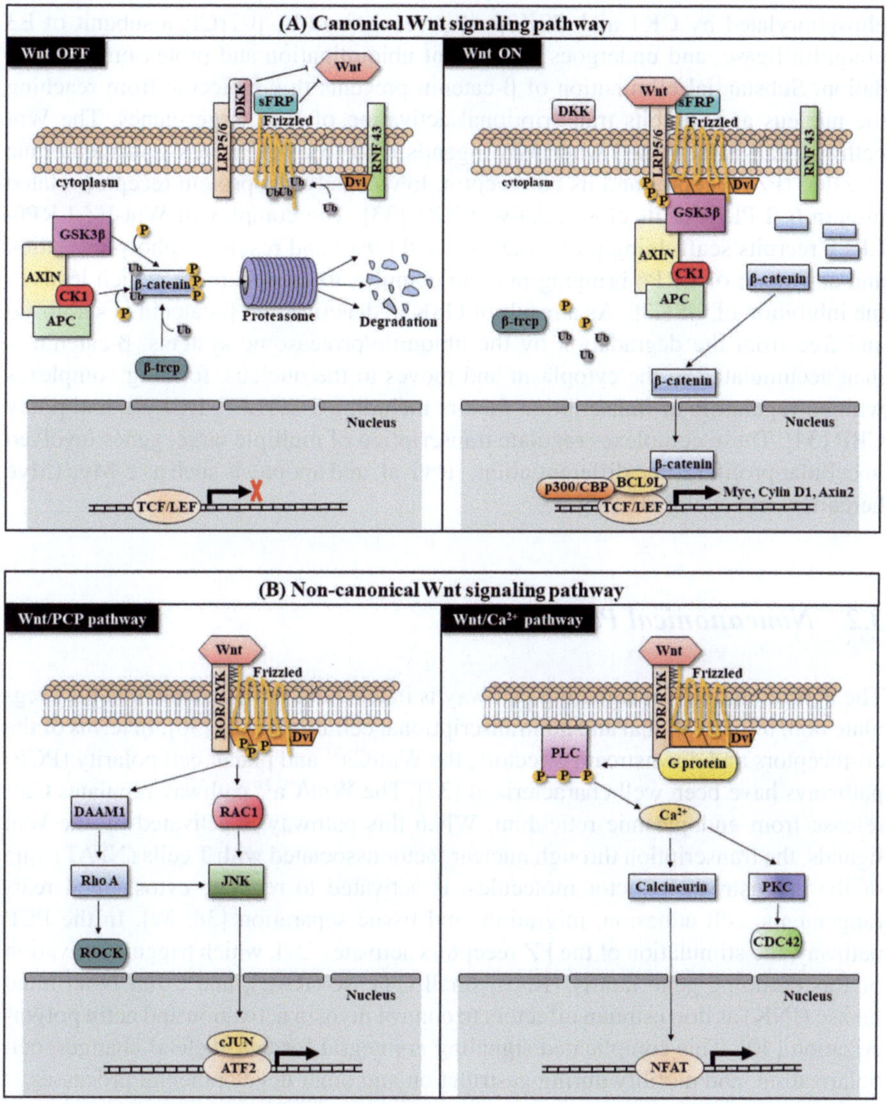

Fig. 1 A schematic illustration of canonical and noncanonical Wnt signaling pathways. (**a**) Canonical Wnt signaling pathway. In Wnt off state (top left panel), DKK and sFRP act as antagonists of Wnt ligand, which prevents Wnt ligands from binding to the receptor, and RNF43 promote ubiquitination of Frizzled for destruction. β-catenin is phosphorylated by the destruction complex consisting of APC, AXIN, CK1, and GSK3β. This complex leads to the ubiquitination of β-catenin by β-TrCP and subsequent proteasomal degradation. Once Wnt ligands bind to the Frizzled-LRP5/6 co-receptor complex (top right panel), Frizzled interacts with cytoplasmic protein Dvl, an upstream regulator of GSK3β, which disrupts the function of β-catenin destruction complex. β-catenin has now stabilized and accumulated in the cytoplasm. Then, it migrates into the nucleus and regulates target gene expression, such as Myc, cyclin D1, and Axin, by interacting with TCF/LEF and other co-activators including P300/CBP and BCL9L. (**b**) Noncanonical Wnt signaling pathway. In Wnt/PCP signaling (bottom left panel), Wnt ligands bind to the ROR-Frizzled receptor complex, activating Dvl. Then, Dvl mediates activation of Rho via DAAM1, which in turn activates Rho kinase (ROCK). Dvl also activates Rac1. Activated Rac1 results in transcription of the target gene by stimulating cJUN via JNK.

phosphorylated by CK1 and GSK3β, it is recognized by β-TrCP, a subunit of E3 ubiquitin ligase, and undergoes subsequent ubiquitination and proteasomal degradation. Substantial elimination of β-catenin prevents this molecule from reaching the nucleus and inhibits transcriptional activation of Wnt target genes. The Wnt pathway can be activated when Wnt ligands bind to a seven-pass transmembrane Frizzled (FZ) receptor and its co-receptor, low-density lipoprotein receptor related protein 6 (LPR6) or its close relative LRP5 [33]. The complex of Wnt-FZ-LRP6/LRP5 recruits scaffolding protein Disheveled (Dsh) and results in phosphorylation and activation of LRP6 bringing in Axin complex to the receptors, which leads to the inhibition of GSK3β. As a result of GSK3β deactivation, β-catenin is stabilized and free from the degradation by the ubiquitin/proteasome systems. β-catenin is then accumulated in the cytoplasm and moves to the nucleus, forming complexes with co-regulators of transcription factors including TCF/LEF, BCL9L, and p300/CBP [34]. These complexes regulate transcription of multiple target genes involved in cellular proliferation, differentiation, survival, and apoptosis such as c-Myc (Myc hereafter) and cyclin D1 [35].

3.2 *Noncanonical Pathway*

The noncanonical Wnt signaling pathway is independent of β-catenin and may regulate both transcriptional and nontranscriptional cellular factors [36]. In terms of the co-receptors and downstream effectors, the Wnt/Ca^{2+} and planar cell polarity (PCP) pathways have been well characterized [37]. The Wnt/Ca^{2+} pathway regulates Ca^{2+} release from endoplasmic reticulum. When this pathway is activated by the Wnt ligands, the transcription through nuclear factor associated with T cells (NFAT), one of the downstream effector molecules, is activated to regulate cytoskeletal rearrangements, cell adhesion, migration, and tissue separation [38, 39]. In the PCP pathway, the stimulation of the FZ receptors activates Dvl, which triggers activation of Ras homolog gene family (Rho), small GTPase (RAC), and c-Jun N-terminal kinase (JNK) as downstream effectors to control myosin activation and actin polymerization [40]. This complicated signaling is integral for cytoskeletal changes, cell polarization, and motility during gastrulation and other developmental processes.

3.3 *Dysregulation of Wnt Signaling Pathway in CRC*

Given the essential role of the Wnt signaling pathway in the homeostasis of embryonic and adult tissues, the hyperactivation of this pathway is regarded as an essential mechanism of CRC carcinogenesis [41]. Dysregulation of Wnt canonical signaling can affect various steps of tumorigenesis such as proliferation, growth, inhibition of cell death, and metastasis, resulting in polyps and carcinoma of the upper and lower gastrointestinal system [42]. Aberrant regulation of this signaling is mainly caused by frequent mutations in the genes encoding for the components in the Wnt signaling pathway.

Wnt signaling pathways are known to be activated in 93% of nonhypermutated CRCs and 97% of hypermutated CRCs [43]. Nonhypermutated CRCs are enriched for somatic mutations in the APC (81%), TP53 (60%), KRAS (43%), PIK3CA (18%), FBXW7 (11%), SMAD4 (10%), TCF7L2 (9%), NRAS (9%), FAM123B (7%), CTNNB1 (β-catenin) (5%), ACVR1B (4%), and SOX9 (4%) genes. FAM123B (also known as WTX) is an X-linked negative regulator of Wnt signaling, and most mutations involve the loss of function, which is found in about 10% of CRC tumors. In hypermutated CRCs, ACVR2A (63%), APC (51%), TGFBR2 (51%), BRAF (46%), MSH3 (40%), and MSH6 (40%) genes are frequently mutated [44]. The majority of these mutated genes are characterized by direct or indirect associations with Wnt signaling pathways [44].

Although APC mutations are initial and critical events during CRC development, the acquisition of further genetic and epigenetic changes, including dysregulation of miRNAs, is needed to evolve into a lethal metastatic carcinoma [45, 46]. In other words, the activation of the Wnt pathway involves either APC inactivation or β-catenin activation cooperating with stepwise changes in additional genes and miRNAs, such as KRAS, SMAD4, p53, mismatch repair genes, let-7, miR-17-92 cluster, miR-34a, and miR-155 [15, 19, 47]. These multiple genetic and epigenetic aberrations involved in CRC carcinogenesis make developing targeted therapies difficult and necessitate alternative strategies for developing novel therapeutics for improving outcomes for CRC patients.

In 2018, patisiran (brand name, Onpattro; Alnylam Pharmaceuticals Inc.) has been announced as the first small interfering RNA (siRNA)-based drug approved by the United States Food and Drug Administration (FDA) and the European Commission (EC) for the treatment of polyneuropathy in people with hereditary variant transthyretin amyloidosis (AATRv) [48]. Patisiran is a double-stranded oligonucleotide that reduces the accumulation of misfolded mutant transthyretin (TTR) protein fibrils. Ongoing efforts to develop miRNA drugs as therapies for a variety of human diseases, including cancer, continue to expand and several miRNAs are undergoing clinical trials (Table 1) [48–50].

4 MicroRNA as an Important Regulator of Wnt Signaling Pathway in CRC Pathogenesis

MiRNAs are small noncoding RNAs consisting of 20–25 nucleotides and attenuate gene expression by binding to the 3′UTR of target mRNA. They play a critical role in the Wnt signaling pathway in normal cells by targeting the components of this pathway. Consistently, dysregulation of miRNAs can hyperactivate the Wnt signaling pathway, which can eventually lead to colorectal carcinogenesis [47, 51]. The

Table 1 Representative miRNA-based clinical trials in multiple human diseases

Target miRNA	Therapeutic molecules	Disease	Phase status
miR-16	MesomiR-1	Mesothelioma Lung cancer	I
miR-17	RGLS4326	Polycystic kidney disease	I
miR-21	RG-012	Alport nephropathy	Preclinical trial
miR-29	MGN-4220 MRG-201	Keloid, fibrous scar tissue formation	I
miR-34	MRX34	Liver cancer, lymphoma, melanoma	I
miR-92	MRG-110	Wound healing, heart failure Mesothelioma Lung cancer	I
miR-122	Miravirsen RG-101	Hepatitis C virus Viral effect	II IB
miR-143/145	MGN-2677	Vascular disease	Preclinical trial
miR-155	MRG-106 (Cobomarsen)	T cell lymphoma/ Mycosis fungoides Cutaneous T-cell lymphoma	II
miR-208	MGN-9103	Chronic heart failure	Preclinical trial
miR-378	MGN-5804	Cardiometabolic disease	Preclinical trial
miR-451	MGN-4893	Disorders of abnormal red blood cell production	Preclinical trial

components of the Wnt pathway targeted by miRNAs include Wnt ligands/receptors (Wnt, Frizzled, Dvl, and LRP5/6), β-catenin, Wnt antagonists (DKK and sFRP), β-catenin destruction complexes (APC, GSK3β, Axin, CK1, and β-TrCP), and transcriptional co-activators (TCF/LEF, BCL9L, and p300/CBP). Components in other signaling pathways (TP53, TGFβ2, NUMB, and YY1) that promote CRC development when dysregulated can also be targeted by miRNAs [17, 51]. A list of miRNAs and their target genes associated with Wnt signaling and CRC is shown in Fig. 2 and Table 2.

4.1 MiRNAs Regulating Oncogenic Potential in CRC Via Wnt Signaling

It has been reported miRNAs regulate various steps of CRC pathogenesis, such as stemness, epithelial to mesenchymal transition (EMT), cell proliferation, and invasion via Wnt signaling (Table 3) [51–54]. With regard to stemness, miRNAs associated with the canonical Wnt signaling pathway are important regulators of the

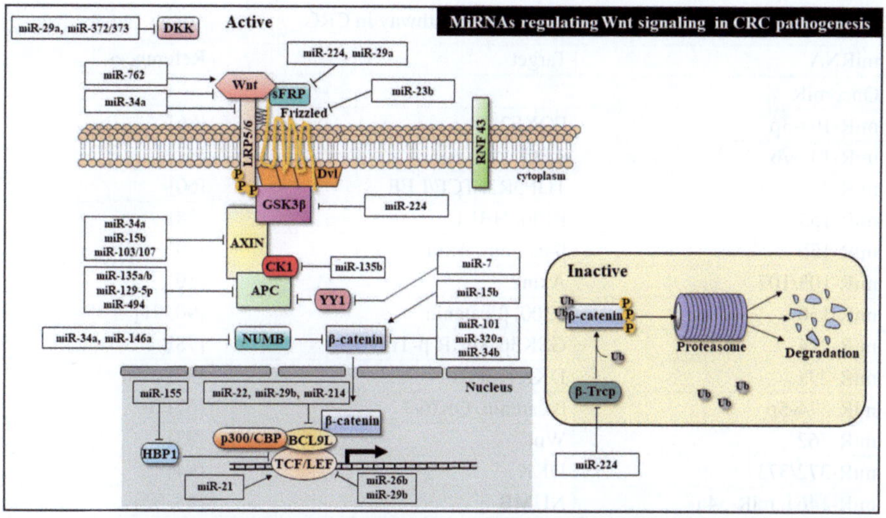

Fig. 2 MicroRNAs regulating the Wnt signaling pathway in CRC pathogenesis. MiRNAs regulate Wnt signaling pathway in multiple steps by targeting Wnt ligands/receptors and downstream components of Wnt in CRC pathogenesis: *sFRP,* secreted frizzled-related protein 1; *APC,* adenomatous polyposis coli; *BCL9L,* B-cell CLL/lymphoma 9 protein like; *β-TrCP,* β-Transducin repeat containing protein; *CK1,* casein kinase 1; *DKK,* dickkopf Wnt signaling pathway inhibitor 1; *Dvl,* Disheveled; *GSK3β,* glycogen synthase kinase 3 β; *HBP,* HMG-Box transcription factor 1; *LRP5/6,* low-density lipoprotein receptor-related protein5/6; *TCF/LEF,* T-cell factor/lymphoid enhancer factor; and *YY1,* Yin Yang 1

function of colorectal cancer stem cells (CRCSCs) [55]. For instance, epigenetically silenced miR-34a activates the asymmetric division process of CRCSCs, enhances self-renewal ability, and increases the expression of stem-cell markers by promoting the Wnt signaling pathway [56]. Moreover, the activated Wnt–β-catenin–TCF4–Rab27B pathway, which is involved in vesicular fusion and trafficking, enhances the secretion of miR-146a-enveloped exosomes that increases the expression of stem cell markers and the ability to self renew in recipient CRC cells [57]. In addition, miR-146a has been shown to be transactivated by the β-catenin–TCF4 complex in CRC cells in which miR-146a contributes to the symmetric division of CRCSCs by targeting NUMB, a protein that controls β-catenin stability through polyubiquitylation [58]. MiR-103/107 downregulate Axin2 to prolong the duration of Wnt/β-catenin signaling, thereby enhancing the expression of stem-cell markers, such as CD133 and Sox2, sphere formation, and cell proliferation [59]. MiR-21 induces stemness through downregulation of TGFβR2 and activation of the β-catenin TCF/LEF signaling pathway [60]. MiR-92a upregulates the Wnt signaling pathway activity by directly targeting KLF4, GSK3β, and DKK3, negative regulators of the Wnt signaling pathway, which promotes stem-cell-like properties in CRC such as sphere formation and the expression of stem-cell markers [61].

MicroRNAs are also involved in the process of EMT through the Wnt signaling pathway in CRC. EMT is enhanced by miR-582, miR-183, miR-496, miR-431, and

Table 2 MicroRNAs regulating Wnt signaling pathway in CRC

miRNA	Target	References
OncomiR		
miR-19a-3p	FOXF2	[86]
miR-135a/b	APC	[87]
miR-21	TGFβR2, TCF/LEF	[60]
miR-155	P130, HBP1	[88]
miR-15b	β-catenin, Axin2	[89]
miR-103/107	Axin2	[59]
miR-150	P300, β-catenin	[90, 91]
miR-224	GSK3β, sFRP, β-TrCP	[78]
miR-29a	DKK1, sFRP	[92]
miR-574-5p	β-catenin, QKI6/7	[67]
miR-762	Wnt	[93]
miR-372/373	DKK	[94]
miR-146a, miR-34a	NUMB	[58, 95]
Tumor suppressor miR		
miR-23b	Frizzled (FZD)	[96]
miR-7	Ying Yang 1 (YY1)	[97]
miR-26b	LEF1	[98]
miR-93	SMAD	[99]
miR-320b	c-myc, cyclin D1	[100]
miR-34b	β-catenin	[66]
miR-29b,	BCL9L, TCF7L, and SNAI1	[101]
miR-22, miR-214	BCL9L	[64]

APC adenomatous polyposis coli; *BCL9L* B-cell CLL/lymphoma 9-like protein like; *β-TrCP* beta-transducin repeat containing protein; *DKK1* Dickkopf Wnt signaling pathway inhibitor 1; *FOXF2* Forkhead box F2; *GSK3β* glycogen synthase kinase 3 bet; *HBP1* HMG-Box transcription factor 1; *sFRP* secreted frizzled-related protein; *TGFβR2* transforming growth factor beta receptor 2; *TCF/LEF* T cell-specific transcription factor/lymphoid enhancer binding factor

Table 3 Roles of microRNAs in oncogenic potential via Wnt signaling

Function	miRNA	References
Stemness	miR-34a, miR-146a, miR-103/107, miR-21, miR-92a miR-3622a	[58–61, 95, 102]
EMT	miR-582, miR-496, miR-431, miR-452, miR-200c, miR-206, miR139-5p, miR-22, miR-214	[56, 62–64]
Cell proliferation/ invasion	miR-452, miR34b, miR-552, miR-574-5p, miR-590-3p, miR532-3p, miR-27a	[65–68, 98, 103–105]
Metastasis	miR-452, miR-103/107, miR-532-3p, miR-26b, miR-762, miR-22, miR-214, miR-27a	[59, 64, 65, 93, 98, 104, 105]

EMT epithelial to mesenchymal transition

miR-452 directly or indirectly, whereas miR-200c, miR-206, and miR-139-5p suppress EMT by targeting GSK3β [56, 62, 63]. MiR-22 and miR-214 regulate the EMT by directly targeting the BCL9L gene in CRC [64]. Consistent with these findings, several studies have demonstrated aberrant expression of miRNAs contributes to cell proliferation and invasiveness of CRC. MiR-452 acts as a stimulator of Wnt signaling and subsequently induces CRC cell proliferation and invasion by directly targeting GSK3β [65]. In addition, the miR-452–GSK3β–TCF4/LEF1 positive feedback loop plays a central role in regulating CRC initiation and progression [65]. The expression level of miR-34b is significantly down-regulated in human colon cancer samples, and overexpression of miR-34b effectively inhibits the proliferation and apoptosis in colon cancer cells, possibly through the regulation of the Wnt signaling pathway [66]. MiR-574-5p also increases CRC cell proliferation and invasion by regulating expression of QKI6/7 [67], and miR-590-3p positively regulates cell proliferation via targeting WIF1 and DKK1 [68]. Considering the critical roles of miRNAs in the regulation of stemness, EMT and other oncogenic factors in CRC, restoration of the normal expression levels of these dysregulated miRNAs could be an effective therapeutic strategy for CRC.

4.2 MiRNAs Regulating Drug Resistance of CRC Via Wnt Signaling

5-fluorouracil (5-FU) is one of the most widely applied anticancer drugs for treating CRC, and novel anti-CRC drugs, such as cetuximab (monoclonal antibody) and oxaliplatin, are also commonly used in clinical practice [69]. However, acquired resistance to these agents presents a major obstacle. Recent studies suggest that miRNAs might be associated with resistance to anti-cancer agents [70]. For instance, ectopic expression of miR-92a causes chemoresistance to 5-FU-induced apoptosis in vitro [61]. In addition, miR-30-5p expression is significantly reduced in 5-FU-resistant CRC cells, suggesting that this miRNA might associate with sensitivity to 5-FU [71]. It has been shown miR-125b regulates the CXCL12/CXCR4 axis by suppressing APC expression in vitro and induces 5-FU resistance in a xenograft model [72]. In contrast, miR-149 and miR-320 increase 5-FU sensitivity by decreasing forkhead box protein M1 (FOXM1) expression and subsequent inhibition of β-catenin localization in the nucleus [73, 74]. Similarly, miR-103/107 promotes the Wnt/β-catenin signaling pathway by directly inhibiting Axin2 expression, which contributes to oxaliplatin resistance and tumor recurrence in xenograft models of CRC [59]. In the case of cetuximab-resistant CRC, miR-100 and miR-125b upregulate Wnt signaling by repressing the expression of negative regulators of this signaling pathway [75, 76]. These studies demonstrate that aberrant miRNAs in CRC have been recognized as effective therapeutic targets for overcoming drug resistance.

Table 4 MicroRNAs modulating drug resistance via Wnt signaling pathway in CRC

miRNA	Target	Drug resistance	Cell line/in vivo	References
miR-149	FOXM1	5-fluorouracil (5-FU)	HCT-8, LoVo	[73]
miR-320	FOXM1	5-fluorouracil (5-FU)	HT-29, HCT116	[74]
miR-125b	APC	5-fluorouracil (5-FU)	HCT116, SW620 Xenograft	[72]
miR-92a	GSK3β, DKK	5-fluorouracil (5-FU)	HT-29, HCT116 Xenograft	[61]
miR-30-5p	USP22	5-fluorouracil (5-FU)	Caco2	[71]
miR-100/125b	Wnt negative regulator	Cetuximab	Caco2, SW403, SW948	[75]
miR-506	β-catenin	Oxaliplatin	HCT116	[106]
miR-103/107	Axin2	Oxaliplatin	HCT116	[59]

APC adenomatous polyposis coli; *DKK* dickkopf–Wnt signaling pathway; *FOXM1* Forkhead box protein M1 inhibitor; *USP22* ubiquitin-specific peptidase 22

Multiple studies have been recently conducted based on the assumption that restoring expression of miRNAs involved in sensitivity to anti-cancer agents may improve therapeutic and clinical outcomes for patients with CRC by overcoming drug resistance [77]. It has been reported that overexpression of miR-30-5p in 5-FU-resistant cells reduces tumor sphere formation and cell viability, whereas the inhibition of miR-30-5p reverses the process [71]. Moreover, ectopic expression of miR-30-5p attenuates the expression of Wnt signaling target genes (Axin2 and Myc), but antagomiR-92a significantly enhances chemosensitivity in vivo by upregulating KLF4, GSK3β, and DKK3, negative regulators of the Wnt pathway [61]. In addition, inhibition of miR-100 and miR-125b has been shown overcome resistance to cetuximab through regulation of the Wnt signaling pathway [75], whereas the knockdown of miR-224 restores the expression of GSK3β and SFRP2 and inhibits Wnt/β-catenin-mediated cell metastasis and proliferation [78]. Together, using miRNAs as therapeutic agents for drug resistance CRC warrants further investigation. We listed miRNAs involved in chemoresistance via the Wnt signaling pathway in CRC in Table 4.

4.3 *MiRNAs Regulated by the Wnt/Myc Axis and their Role in CRC*

Upon activation of Wnt signaling, expression of a large number of target genes is up-regulated, such as *Myc*, *Cyclin D*, *VEGF*, *BMP4*, and *Survivin* [79]. A recent work has shown phenotype of *APC* deletion, which hyperactivates Wnt signaling, can be rescued by *Myc* deletion in the intestine suggesting *Myc* might be a critical target of the Wnt pathway [80]. *Myc* is a well-established oncogene in CRC and regulates transcription of *p53*, *p27*, *ARF*, *Cyclin A*, *Cyclin E*, and many other genes [81]. It also can exert its effect by regulating the expression of miRNAs via Myc/miRNA axes. For example, the Myc/miR-27b-3p/ATG10 (Autophagy Related 10) regulatory axis has been shown to be involved in oxaliplatin resistance by regulating autophagy in CRC [82]. Given autophagy can protect cancer cells from the cytotoxic effect of chemotherapeutic drugs, *Myc* upregulates ATG10 and autophagy by inhibiting expression of miR-27b-3p, rendering CRC cells resistant to oxaliplatin [82]. Tumor suppressor p53 is also frequently dysregulated in CRC, which contributes to CRC pathogenesis. Our work demonstrates dysregulation of Wnt/Myc and p53 is interrelated through miRNAs in CRC, and hyperactivation of the Wnt/Myc pathway downregulates p53 expression by augmenting p53-targeting miR-552 levels [83]. Expression of miR-552 is upregulated in CRC patient tumor tissues compared with adjacent normal tissues, and CRC cells ectopically expressing this miRNA are resistant to doxorubicin-induced inhibition of cell proliferation, suggesting miR-552 might be an oncogene in CRC (Fig. 3) [83]. A Myc/miRNA axis can also regulate the EMT and metastasis [84]. Activation of Wnt/Myc signaling downregulates miR-29a-3p and miR-200a, which enhances the EMT by upregulating MMP2 (Matrix Metallopeptidase 2) and ZEB1 (Zinc Finger E-Box Binding Homeobox 1). Astaxanthin (AXT) represses the EMT by reversing the effect of Myc on these two miRNAs [85]. Together, these studies suggest dysregulation of miRNAs by the hyperactivated Wnt/Myc signaling might be promising therapeutic targets in CRC.

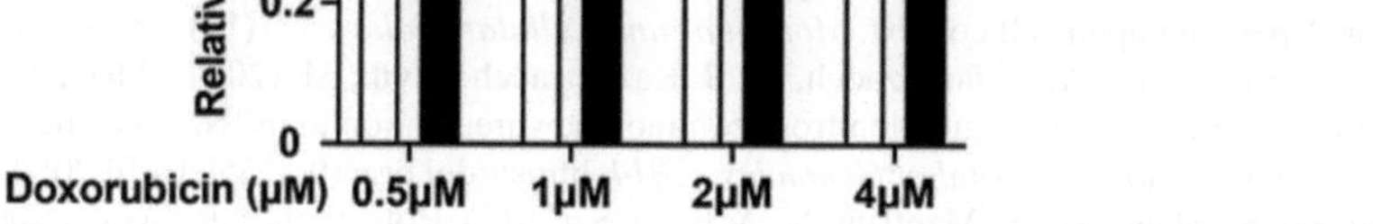

Fig. 3 Wnt/Myc/miR-552 axis in resistance to doxorubicin. HCT116 CRC cells were stably transduced with the murine stem cell virus (MSCV)-puro retroviral control vector or MSCV-puro-miR-552 vector, followed by selection with puromycin (1 mg/ml). Cells ectopically expressing control vector or miR-552 were exposed to doxorubicin (0, 0.5, 1, 2, and 4 μM) and counted 24 h later. Overexpression of miR-552 rendered HCT116 cells resistant to doxorubicin-induced inhibition of cell proliferation (*$p < 0.05$). Cell number after treatment with DMSO for 24 h was set at 1, and the depicted cell proliferation rates after exposure to doxorubicin were normalized to DMSO treatment. The experiments were performed at least three times independently to confirm reproducibility

5 Conclusion and Future Direction

Two siRNA-based drugs, patisiran and givosiran, were approved by the FDA in 2018 and 2019 respectively. Unfortunately, several obstacles currently exist before miRNA therapeutics can be effective in clinical applications. For example, more efficient tools for miRNA delivery need to be developed, and targeted nanoparticles can enhance cell/tissue specificity, avoiding unnecessary side effects caused by needless interactions with normal cells/tissues. Another concern for miRNA therapeutics is the likelihood of miRNAs interacting with multiple target genes, sometimes up to several hundred genes, which may cause a series of unpredictable consequences. However, ongoing research into miRNA therapeutics may provide novel strategies to improve the outcomes of patients with CRC.

Acknowledgement Dr. Yang Hyun Yun (Department of Biomedical Engineering, The University of Akron, Akron, OH 44325, USA) assisted with revision of the manuscript.

References

1. Brenner, H., Kloor, M., & Pox, C. P. (2014). Colorectal cancer. *Lancet, 383*(9927), 1490–1502.
2. Haraldsdottir, S., Einarsdottir, H. M., Smaradottir, A., Gunnlaugsson, A., & Halfdanarson, T. R. (2014). Colorectal cancer-review. *Læknablaðið, 100*(2), 75–82.
3. Armaghany, T., Wilson, J. D., Chu, Q., & Mills, G. (2012). Genetic alterations in colorectal cancer. *Gastrointestinal Cancer Research: GCR., 5*(1), 19.
4. Clevers, H., & Nusse, R. (2012). Wnt/β-catenin signaling and disease. *Cell, 149*(6), 1192–1205.
5. Schatoff, E. M., Leach, B. I., & Dow, L. E. (2017). Wnt signaling and colorectal cancer. *Current Colorectal Cancer Reports., 13*(2), 101–110.
6. O'Brien, J., Hayder, H., Zayed, Y., & Peng, C. (2018). Overview of microRNA biogenesis, mechanisms of actions, and circulation. *Frontiers in Endocrinology., 9*, 402.
7. Catalanotto, C., Cogoni, C., & Zardo, G. (2016). MicroRNA in control of gene expression: An overview of nuclear functions. *International Journal of Molecular Sciences., 17*(10), 1712.
8. Lin, C.-C., Liu, L.-Z., Addison, J. B., Wonderlin, W. F., Ivanov, A. V., & Ruppert, J. M. (2011). A KLF4–miRNA-206 autoregulatory feedback loop can promote or inhibit protein translation depending upon cell context. *Molecular and Cellular Biology., 31*(12), 2513–2527.
9. Valinezhad Orang, A., Safaralizadeh, R., & Kazemzadeh-Bavili, M. (2014). Mechanisms of miRNA-mediated gene regulation from common downregulation to mRNA-specific upregulation. *International Journal of Genomics., 2014*. https://doi.org/10.1155/2014/970607
10. Oliveto, S., Mancino, M., Manfrini, N., & Biffo, S. (2017). Role of microRNAs in translation regulation and cancer. *World Journal of Biological Chemistry., 8*(1), 45.
11. Croce, C. M. (2009). Causes and consequences of microRNA dysregulation in cancer. *Nature Reviews Genetics., 10*(10), 704–714.
12. Svoronos, A. A., Engelman, D. M., & Slack, F. J. (2016). OncomiR or tumor suppressor? The duplicity of microRNAs in cancer. *Cancer Research., 76*(13), 3666–3670.
13. Iorio, M. V., & Croce, C. M. (2012). MicroRNA dysregulation in cancer: Diagnostics, monitoring and therapeutics. A comprehensive review. *EMBO Molecular Medicine., 4*(3), 143–159.

14. Rupaimoole, R., & Slack, F. J. (2017). MicroRNA therapeutics: Towards a new era for the management of cancer and other diseases. *Nature Reviews Drug Discovery., 16*(3), 203.
15. Moridikia, A., Mirzaei, H., Sahebkar, A., & Salimian, J. (2018). MicroRNAs: Potential candidates for diagnosis and treatment of colorectal cancer. *Journal of Cellular Physiology., 233*(2), 901–913.
16. Onyido, E. K., Sweeney, E., & Nateri, A. S. (2016). Wnt-signalling pathways and microRNAs network in carcinogenesis: Experimental and bioinformatics approaches. *Molecular Cancer., 15*(1), 1–17.
17. Rahmani, F., Avan, A., Hashemy, S. I., & Hassanian, S. M. (2018). Role of Wnt/β-catenin signaling regulatory microRNAs in the pathogenesis of colorectal cancer. *Journal of Cellular Physiology., 233*(2), 811–817.
18. Simon, K. (2016). Colorectal cancer development and advances in screening. *Clinical Interventions in Aging., 11*, 967.
19. Testa, U., Pelosi, E., & Castelli, G. (2018). Colorectal cancer: Genetic abnormalities, tumor progression, tumor heterogeneity, clonal evolution and tumor-initiating cells. *Medical Science, 6*(2), 31.
20. Hong, S. N. (2018). Genetic and epigenetic alterations of colorectal cancer. *Intestinal Research., 16*(3), 327.
21. Pino, M. S., & Chung, D. C. (2010). The chromosomal instability pathway in colon cancer. *Gastroenterology, 138*(6), 2059–2072.
22. Worthley, D. L., & Leggett, B. A. (2010). Colorectal cancer: Molecular features and clinical opportunities. *The Clinical Biochemist Reviews., 31*(2), 31.
23. Boland, C. R., & Goel, A. (2010). Microsatellite instability in colorectal cancer. *Gastroenterology., 138*(6), 2073–87.e3.
24. Nguyen, H. T., & Duong, H. Q. (2018). The molecular characteristics of colorectal cancer: Implications for diagnosis and therapy. *Oncology Letters., 16*(1), 9–18.
25. Müller, M. F., Ibrahim, A. E., & Arends, M. J. (2016). Molecular pathological classification of colorectal cancer. *Virchows Archiv., 469*(2), 125–134.
26. Donehower, L. A., Creighton, C. J., Schultz, N., Shinbrot, E., Chang, K., Gunaratne, P. H., et al. (2013). MLH1-silenced and non-silenced subgroups of hypermutated colorectal carcinomas have distinct mutational landscapes. *The Journal of Pathology., 229*(1), 99–110.
27. Clevers, H. (2006). Wnt/β-catenin signaling in development and disease. *Cell, 127*(3), 469–480.
28. Czerwinski, M., Shroyer, N. F., & Spence, J. R. (2018). WNT Signaling in the intestine: Development, homeostasis, disease. In *Physiology of the gastrointestinal tract* (pp. 185–196). Elsevier.
29. Polakis, P. (2012). Wnt signaling in cancer. *Cold Spring Harbor Perspectives in Biology., 4*(5), a008052.
30. Komiya, Y., & Habas, R. (2008). Wnt signal transduction pathways. *Organogenesis, 4*(2), 68–75.
31. MacDonald, B. T., Tamai, K., & He, X. (2009). Wnt/β-catenin signaling: Components, mechanisms, and diseases. *Developmental Cell., 17*(1), 9–26.
32. Stamos, J. L., & Weis, W. I. (2013). The β-catenin destruction complex. *Cold Spring Harbor Perspectives in Biology., 5*(1), a007898.
33. MacDonald, B. T., & He, X. (2012). Frizzled and LRP5/6 receptors for Wnt/β-catenin signaling. *Cold Spring Harbor Perspectives in Biology., 4*(12), a007880.
34. Söderholm, S., & Cantù, C. (2020). The WNT/β-catenin dependent transcription: A tissue-specific business. *Wiley Interdisciplinary Reviews: Systems Biology and Medicine., 13*, e1511.
35. Lecarpentier, Y., Schussler, O., Hébert, J.-L., & Vallée, A. (2019). Multiple targets of the canonical WNT/β-catenin signaling in cancers. *Frontiers in Oncology., 9*, 1248.
36. Flores-Hernández, E., Velázquez, D. M., Castañeda-Patlán, M. C., Fuentes-García, G., Fonseca-Camarillo, G., Yamamoto-Furusho, J. K., et al. (2020). Canonical and non-canonical Wnt signaling are simultaneously activated by Wnts in colon cancer cells. *Cellular Signalling., 72*, 109636.

37. Van Amerongen, R. (2012). Alternative Wnt pathways and receptors. *Cold Spring Harbor Perspectives in Biology., 4*(10), a007914.
38. Duchartre, Y., Kim, Y.-M., & Kahn, M. (2016). The Wnt signaling pathway in cancer. *Critical Reviews in Oncology/Hematology., 99*, 141–149.
39. De, A. (2011). Wnt/Ca2+ signaling pathway: A brief overview. *Acta biochimica et biophysica Sinica., 43*(10), 745–756.
40. Yang, Y., & Mlodzik, M. (2015). Wnt-Frizzled/planar cell polarity signaling: Cellular orientation by facing the wind (Wnt). *Annual Review of Cell and Developmental Biology., 31*, 623–646.
41. Zou, Y., Cai, Y., Lu, D., Zhou, Y., Yao, Q., & Zhang, S. (2017). MicroRNA-146a-5p attenuates liver fibrosis by suppressing profibrogenic effects of TGFβ1 and lipopolysaccharide. *Cellular Signalling, 39*, 1–8.
42. White, B. D., Chien, A. J., & Dawson, D. W. (2012). Dysregulation of Wnt/β-catenin signaling in gastrointestinal cancers. *Gastroenterology, 142*(2), 219–232.
43. Lartigue, J. (2016). Genomic complexity stifles targeted advances in colorectal cancer. *OncologyLive, 2016*, 17.
44. Inamura, K. (2018). Colorectal cancers: An update on their molecular pathology. *Cancers, 10*(1), 26.
45. Aitchison, A., Hakkaart, C., Day, R. C., Morrin, H. R., Frizelle, F. A., & Keenan, J. I. (2020). APC mutations are not confined to hotspot regions in early-onset colorectal cancer. *Cancers, 12*(12), 3829.
46. Kwong, L. N., & Dove, W. F. (2009). APC and its modifiers in colon cancer. *Adv Exp Med Biol, 656*, 85–106.
47. Strubberg, A. M., & Madison, B. B. (2017). MicroRNAs in the etiology of colorectal cancer: Pathways and clinical implications. *Disease Models & Mechanisms., 10*(3), 197–214.
48. Hanna, J., Hossain, G. S., & Kocerha, J. (2019). The potential for microRNA therapeutics and clinical research. *Frontiers in Genetics., 10*, 478.
49. Chakraborty, C., Sharma, A. R., & Sharma, G. (2020). Therapeutic advances of miRNAs: A preclinical and clinical update. *Journal of Advanced Research, 28*, 127–138.
50. Bonneau, E., Neveu, B., Kostantin, E., Tsongalis, G., & De Guire, V. (2019). How close are miRNAs from clinical practice? A perspective on the diagnostic and therapeutic market. *EJIFCC., 30*(2), 114.
51. Peng, Y., Zhang, X., Feng, X., Fan, X., & Jin, Z. (2017). The crosstalk between microRNAs and the Wnt/β-catenin signaling pathway in cancer. *Oncotarget, 8*, 14089–14106.
52. Khan, A. Q., Ahmed, E. I., Elareer, N. R., Junejo, K., Steinhoff, M., & Uddin, S. (2019). Role of miRNA-regulated cancer stem cells in the pathogenesis of human malignancies. *Cell, 8*(8), 840.
53. Wang, J., Du, Y., Liu, X., Cho, W. C., & Yang, Y. (2015). MicroRNAs as regulator of signaling networks in metastatic colon cancer. *BioMed Research International., 2015*, 823620.
54. Jin, D., Fang, Y., Li, Z., Chen, Z., & Xiang, J. (2015). Epithelial-mesenchymal transition-associated microRNAs in colorectal cancer and drug-targeted therapies. *Oncology Reports., 33*(2), 515–525.
55. Caruso, S., Bazan, V., Rolfo, C., Insalaco, L., Fanale, D., Bronte, G., et al. (2012). MicroRNAs in colorectal cancer stem cells: New regulators of cancer stemness? *Oncogene, 1*(11), e32.
56. Wen, X.-Q., Qian, X.-L., Sun, H.-K., Zheng, L.-L., Zhu, W.-Q., Li, T.-Y., et al. (2020). MicroRNAs: Multifaceted regulators of colorectal cancer metastasis and clinical applications. *OncoTargets and Therapy., 13*, 10851.
57. Bruno, S., Chiabotto, G., Favaro, E., Deregibus, M. C., & Camussi, G. (2019). Role of extracellular vesicles in stem cell biology. *American Journal of Physiology-Cell Physiology., 317*(2), C303–C313.
58. Hwang, W.-L., Jiang, J.-K., Yang, S.-H., Huang, T.-S., Lan, H.-Y., Teng, H.-W., et al. (2014). MicroRNA-146a directs the symmetric division of Snail-dominant colorectal cancer stem cells. *Nature Cell Biology., 16*(3), 268–280.

59. Chen, H.-Y., Lang, Y.-D., Lin, H.-N., Liu, Y.-R., Liao, C.-C., Nana, A. W., et al. (2019). miR-103/107 prolong Wnt/β-catenin signaling and colorectal cancer stemness by targeting Axin2. *Scientific Reports, 9*(1), 1–13.
60. Yu, Y., Kanwar, S. S., Patel, B. B., Oh, P.-S., Nautiyal, J., Sarkar, F. H., et al. (2012). MicroRNA-21 induces stemness by downregulating transforming growth factor beta receptor 2 (TGFβR2) in colon cancer cells. *Carcinogenesis, 33*(1), 68–76.
61. Zhang, G.-J., Li, L.-F., Yang, G.-D., Xia, S.-S., Wang, R., Leng, Z.-W., et al. (2017). MiR-92a promotes stem cell-like properties by activating Wnt/β-catenin signaling in colorectal cancer. *Oncotarget, 8*(60), 101760.
62. Su, W., & Liu, Z. (2018). MiR-431 inhibits colorectal cancer cell invasion via repressing CUL4B. *European Review for Medical and Pharmacological Sciences., 22*(10), 3047–3052.
63. Du, F., Cao, T., Xie, H., Li, T., Sun, L., Liu, H., et al. (2020). KRAS mutation-responsive miR-139-5p inhibits colorectal cancer progression and is repressed by Wnt signaling. *Theranostics., 10*(16), 7335.
64. Sun, R., Liu, Z., Han, L., Yang, Y., Wu, F., Jiang, Q., et al. (2019). miR-22 and miR-214 targeting BCL9L inhibit proliferation, metastasis, and epithelial-mesenchymal transition by down-regulating Wnt signaling in colon cancer. *The FASEB Journal., 33*(4), 5411–5424.
65. Li, T., Jian, X., He, H., Lai, Q., Li, X., Deng, D., et al. (2018). MiR-452 promotes an aggressive colorectal cancer phenotype by regulating a Wnt/β-catenin positive feedback loop. *Journal of Experimental & Clinical Cancer Research., 37*(1), 1–15.
66. Ye, K., Xu, C., & Hui, T. (2019). MiR-34b inhibits the proliferation and promotes apoptosis in colon cancer cells by targeting Wnt/β-catenin signaling pathway. *Bioscience Reports, 39*(10), BSR20191799.
67. Ji, S., Ye, G., Zhang, J., Wang, L., Wang, T., Wang, Z., et al. (2013). miR-574-5p negatively regulates Qki6/7 to impact β-catenin/Wnt signalling and the development of colorectal cancer. *Gut, 62*(5), 716–726.
68. Feng, Z., Xu, X., Cen, D., Luo, C., & Wu, S. (2017). miR-590-3p promotes colon cancer cell proliferation via Wnt/beta-catenin signaling pathway by inhibiting WIF1 and DKK1. *European Review for Medical and Pharmacological Sciences., 21*(21), 4844–4852.
69. Xie, Y.-H., Chen, Y.-X., & Fang, J.-Y. (2020). Comprehensive review of targeted therapy for colorectal cancer. *Signal Transduction and Targeted Therapy., 5*(1), 1–30.
70. Wu, Q.-B., Sheng, X., Zhang, N., Yang, M.-W., & Wang, F. (2018). Role of microRNAs in the resistance of colorectal cancer to chemoradiotherapy. *Molecular and Clinical Oncology., 8*(4), 523–527.
71. Jiang, S., Miao, D., Wang, M., Lv, J., Wang, Y., & Tong, J. (2019). MiR-30-5p suppresses cell chemoresistance and stemness in colorectal cancer through USP 22/Wnt/β-catenin signaling axis. *Journal of Cellular and Molecular Medicine., 23*(1), 630–640.
72. Yu, X., Shi, W., Zhang, Y., Wang, X., Sun, S., Song, Z., et al. (2017). CXCL12/CXCR4 axis induced miR-125b promotes invasion and confers 5-fluorouracil resistance through enhancing autophagy in colorectal cancer. *Scientific Reports, 7*(1), 1–13.
73. Liu, X., Xie, T., Mao, X., Xue, L., Chu, X., & Chen, L. (2016). MicroRNA-149 increases the sensitivity of colorectal cancer cells to 5-fluorouracil by targeting forkhead box transcription factor FOXM1. *Cellular Physiology and Biochemistry., 39*(2), 617–629.
74. Wan, L.-Y., Deng, J., Xiang, X.-J., Zhang, L., Yu, F., Chen, J., et al. (2015). miR-320 enhances the sensitivity of human colon cancer cells to chemoradiotherapy in vitro by targeting FOXM1. *Biochemical and Biophysical Research Communications., 457*(2), 125–132.
75. Lu, Y., Zhao, X., Liu, Q., Li, C., Graves-Deal, R., Cao, Z., et al. (2017). lncRNA MIR100HG-derived miR-100 and miR-125b mediate cetuximab resistance via Wnt/β-catenin signaling. *Nature Medicine, 23*(11), 1331.
76. Thomas, H. (2017). Mir-100 and miR-125b induce cetuximab resistance in CRC. *Nature Reviews Gastroenterology & Hepatology., 14*(12), 691.

77. Xie, T., Huang, M., Wang, Y., Wang, L., Chen, C., & Chu, X. (2016). MicroRNAs as regulators, biomarkers and therapeutic targets in the drug resistance of colorectal cancer. *Cellular Physiology and Biochemistry., 40*(1–2), 62–76.
78. Li, T., Lai, Q., Wang, S., Cai, J., Xiao, Z., Deng, D., et al. (2016). MicroRNA-224 sustains Wnt/β-catenin signaling and promotes aggressive phenotype of colorectal cancer. *Journal of Experimental & Clinical Cancer Research., 35*(1), 1–11.
79. Gregorieff, A., & Clevers, H. (2005). Wnt signaling in the intestinal epithelium: From endoderm to cancer. *Genes & Development., 19*(8), 877–890.
80. Sansom, O. J., Meniel, V. S., Muncan, V., Phesse, T. J., Wilkins, J. A., Reed, K. R., et al. (2007). Myc deletion rescues Apc deficiency in the small intestine. *Nature, 446*(7136), 676–679.
81. Wasylishen, A. R., & Penn, L. Z. (2010). Myc: The beauty and the beast. *Genes & Cancer., 1*(6), 532–541.
82. Sun, W., Li, J., Zhou, L., Han, J., Liu, R., Zhang, H., et al. (2020). The c-Myc/miR-27b-3p/ATG10 regulatory axis regulates chemoresistance in colorectal cancer. *Theranostics., 10*(5), 1981.
83. Kwak, B., Kim, D. U., Kim TO, Kim, H.-S., & Kim, S.-W. (2018). MicroRNA-552 links Wnt signaling to p53 tumor suppressor in colorectal cancer. *International Journal of Oncology., 53*(4), 1800–1808.
84. Tao, J., Zhao, X., & Tao, J. (2014). c-MYC–miRNA circuitry: A central regulator of aggressive B-cell malignancies. *Cell Cycle, 13*(2), 191–198.
85. Kim, H.-Y., Kim, Y.-M., & Hong, S. (2019). Astaxanthin suppresses the metastasis of colon cancer by inhibiting the MYC-mediated downregulation of microRNA-29a-3p and microRNA-200a. *Scientific Reports, 9*(1), 1–10.
86. Yu, F.-B., Sheng, J., Yu, J.-M., Liu, J.-H., Qin, X.-X., & Mou, B. (2020). MiR-19a-3p regulates the Forkhead box F2-mediated Wnt/β-catenin signaling pathway and affects the biological functions of colorectal cancer cells. *World Journal of Gastroenterology., 26*(6), 627.
87. Nagel, R., le Sage, C., Diosdado, B., van der Waal, M., Vrielink, J. A. O., Bolijn, A., et al. (2008). Regulation of the adenomatous polyposis coli gene by the miR-135 family in colorectal cancer. *Cancer Research., 68*(14), 5795–5802.
88. Wan, Y. C., Li, T., Han, Y. D., Zhang, H. Y., Lin, H., & Zhang, B. (2016). MicroRNA-155 enhances the activation of Wnt/β-catenin signaling in colorectal carcinoma by suppressing HMG-box transcription factor 1. *Molecular Medicine Reports., 13*(3), 2221–2228.
89. Guo, X., Zhang, Y., Li, J., & Liao, M. (2016). miR-15b facilitates the progression of colorectal cancer via targeting β-catenin and Axin2. *International Journal of Clinical and Experimental Pathology, 9*(9), 8990–8996.
90. He, Z., Dang, J., Song, A., Cui, X., Ma, Z., & Zhang, Y. (2020). The involvement of miR-150/β-catenin axis in colorectal cancer progression. *Biomedicine & Pharmacotherapy., 121*, 109495.
91. Guo, Y.-H., Wang, L.-Q., Li, B., Xu, H., Yang, J.-H., Zheng, L.-S., et al. (2016). Wnt/β--catenin pathway transactivates microRNA-150 that promotes EMT of colorectal cancer cells by suppressing CREB signaling. *Oncotarget, 7*(27), 42513.
92. Kapinas, K., Kessler, C., Ricks, T., Gronowicz, G., & Delany, A. M. (2010). miR-29 modulates Wnt signaling in human osteoblasts through a positive feedback loop. *Journal of Biological Chemistry., 285*(33), 25221–25231.
93. Lai, P.-S., Chang, W.-M., Chen, Y.-Y., Lin, Y.-F., Liao, H.-F., & Chen, C.-Y. (2021). Circulating microRNA-762 upregulates colorectal cancer might through Wnt-1/b-catenin signaling. *Cancer Biomarkers, 32*(2), 111–122.
94. Zhou, A., Diao, L., Xu, H., Xiao, Z., Li, J., Zhou, H., et al. (2012). β-Catenin/LEF1 transactivates the microRNA-371-373 cluster that modulates the Wnt/β-catenin-signaling pathway. *Oncogene, 31*(24), 2968–2978.

95. Bu, P., Wang, L., Chen, K.-Y., Srinivasan, T., Murthy, P. K. L., Tung, K.-L., et al. (2016). A miR-34a-Numb feedforward loop triggered by inflammation regulates asymmetric stem cell division in intestine and colon cancer. *Cell Stem Cell, 18*(2), 189–202.
96. Zhang, H., Hao, Y., Yang, J., Zhou, Y., Li, J., Yin, S., et al. (2011). Genome-wide functional screening of miR-23b as a pleiotropic modulator suppressing cancer metastasis. *Nature Communications, 2*(1), 1–11.
97. Zhang, N., Li, X., Wu, C., Dong, Y., Cai, M., Mok, M., et al. (2013). microRNA-7 is a novel inhibitor of YY1 contributing to colorectal tumorigenesis. *Oncogene, 32*(42), 5078–5088.
98. Zhang, Z., Kim, K., Li, X., Moreno, M., Sharp, T., Goodheart, M. J., et al. (2014). MicroRNA-26b represses colon cancer cell proliferation by inhibiting lymphoid enhancer factor 1 expression. *Molecular Cancer Therapeutics., 13*(7), 1942–1951.
99. Tang, Q., Zou, Z., Zou, C., Zhang, Q., Huang, R., Guan, X., et al. (2015). MicroRNA-93 suppress colorectal cancer development via Wnt/β-catenin pathway downregulating. *Tumor Biology, 36*(3), 1701–1710.
100. Wang, H., Cao, F., Li, X., Miao, H., Jifu, E., Xing, J., et al. (2015). miR-320b suppresses cell proliferation by targeting c-Myc in human colorectal cancer cells. *BMC Cancer, 15*(1), 1–9.
101. Subramanian, M., Rao, S. R., Thacker, P., Chatterjee, S., & Karunagaran, D. (2014). MiR-29b downregulates canonical Wnt signaling by suppressing coactivators of β-catenin in human colorectal cancer cells. *Journal of Cellular Biochemistry, 115*(11), 1974–1984.
102. Chang, S., Sun, G., Zhang, D., Li, Q., & Qian, H. (2020). MiR-3622a-3p acts as a tumor suppressor in colorectal cancer by reducing stemness features and EMT through targeting spalt-like transcription factor 4. *Cell Death & Disease., 11*(7), 1–19.
103. Cao, J., Yan, X. R., Liu, T., Han, X. B., Yu, J. J., Liu, S. H., et al. (2017). MicroRNA-552 promotes tumor cell proliferation and migration by directly targeting DACH1 via the Wnt/β--catenin signaling pathway in colorectal cancer. *Oncology Letters., 14*(3), 3795–3802.
104. Gu, C., Cai, J., Xu, Z., Zhou, S., Ye, L., Yan, Q., et al. (2019). MiR-532-3p suppresses colorectal cancer progression by disrupting the ETS1/TGM2 axis-mediated Wnt/β-catenin signaling. *Cell Death & Disease., 10*(10), 1–14.
105. Ba, S., Xuan, Y., Long, Z.-W., Chen, H.-Y., & Zheng, S.-S. (2017). MicroRNA-27a promotes the proliferation and invasiveness of colon cancer cells by targeting SFRP1 through the Wnt/β-catenin signaling pathway. *Cellular Physiology and Biochemistry., 42*(5), 1920–1933.
106. Zhou, H., Lin, C., Zhang, Y., Zhang, X., Zhang, C., Zhang, P., et al. (2017). miR-506 enhances the sensitivity of human colorectal cancer cells to oxaliplatin by suppressing MDR 1/P-gp expression. *Cell proliferation., 50*(3), e12341.

Small Activating RNA Therapy for Angiogenesis

Pia Laitinen, Tiia A. Turunen, and Mikko P. Turunen

Abstract RNA activation has been known about for 15 years. In essence, it means activating the endogenous specific gene of cells by small RNA, which targets the regulatory area of the gene in the chromatin, the promoter. There are various ways by which gene expression activation can occur. In the end, epigenetic marks are modified in the promoter, for example, silencing modifications such as trimethylation of histone3 lysine 27 are reduced, and activating modifications such as histone3 lysine 4 are increased. Increasing target gene activity by epigenetic modifications using small RNAs offers unique possibilities for next-generation gene therapy. There are several methods available to inhibit gene expression/activity, such as RNA interference and antibodies, but for increasing expression of a therapeutic gene, only delivery of a full cDNA has been an option so far. Recently, mRNA technology has also become available for increasing gene expression. However, small RNAs have an advantage over longer RNAs for some purposes, such as cardiovascular gene therapy, as they are much more stable and induce fewer immunological responses. Vascular endothelial growth factor A, a key regulator of angiogenesis, has been a popular target in RNA activation studies. Effective epigenetic upregulation of VEGF-A offers novel possibilities for treating ischemic diseases, such as myocardial infarction and peripheral artery disease.

P. Laitinen
A.I. Virtanen Institute, University of Eastern Finland, Kuopio, Finland
e-mail: pia.laitinen@uef.fi

T. A. Turunen · M. P. Turunen (✉)
A.I. Virtanen Institute, University of Eastern Finland, Kuopio, Finland

RNatives Oy, Kuopio, Finland
e-mail: tiia.a.turunen@uef.fi; mikko.turunen@uef.fi

Y. H. Yun, K. E. Yoder (eds.), *Biotechnologies for Gene Therapy*,
https://doi.org/10.1007/978-3-030-93333-3_3

1 Introduction

RNA activation has been known about for 15 years. In essence, it means activating the endogenous specific gene of cells by a small RNA, which targets the regulatory area of the gene in the chromatin, the promoter. There are various ways by which gene expression activation can occur. In the end, epigenetic marks are modified in the promoter, for example, silencing modifications such as trimethylation of histone3 lysine 27 are lowered and activating modifications such as histone3 lysine 4 are increased. Increasing target gene activity by epigenetic modifications using small RNAs offers unique possibilities for next-generation gene therapy. There are several methods available for inhibiting gene expression/activity, such as RNA interference and antibodies, but for increasing expression of a therapeutic gene, only delivery of a full cDNA has been an option so far. Recently, mRNA technology has also become available for increasing gene expression. However, small RNAs have an advantage over longer RNAs for some purposes, such as cardiovascular gene therapy, as they are much more stable and induce fewer immunological responses. Vascular endothelial growth factor A, a key regulator of angiogenesis, has been a popular target in RNA activation studies. Effective epigenetic upregulation of VEGF-A offers novel possibilities for treating ischemic diseases, such as myocardial infarction and peripheral artery disease.

2 Small Noncoding RNAs

Small noncoding RNAs (sncRNAs) are a group of different sized RNAs, excluding only over 200 nucleotides long ncRNAs (lncRNAs). As the group includes many different small RNAs, their functions in the cells are also diverse. Small noncoding RNAs are linked to many cellular functions such as angiogenesis [1, 2], epithelial-to-mesenchymal transition (EMT) [3], migration [4], and proliferation [5–7]. To emphasize their critical function in the cell, their misfunction has been linked to many diseases, for example, various cancers [1, 6, 7] such as glioma [8].

2.1 Small Activating RNA Classes

Small ncRNAs consist of many groups of ncRNAs. In this review, we focus on the small ncRNA classes that have been shown to activate gene expression (small activating noncoding RNAs, saRNAs) (Fig. 1). Yet, there are many groups of ncRNAs with undiscovered functions in the cell; thus, new studies for this area are still needed.

MicroRNAs (miRNAs) are perhaps the most frequently studied ncRNA group. As the name infers, miRNAs are micro-sized RNAs that in their mature form are 19–25 nucleotides in length. miRNAs are transcribed from the genome from gene

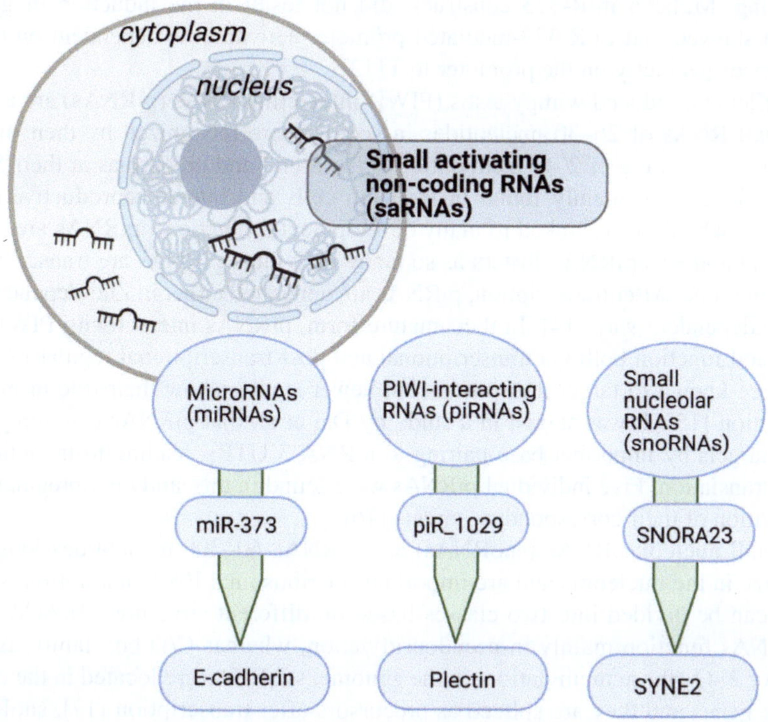

Fig. 1 Endogenous small RNAs have been shown to activate gene expression. Besides artificially designed small RNAs, cellular RNAs have been indicated in gene activation. These examples include miRNAs (miR-373 activating E-cadherin), piRNAs (piR_1029 activating plectin), and snoRNAs (SNORA23 activating SYNE2). Figure created with BioRender.com

introns, exons, or intergenic regions. Primary miRNA transcript is sliced into precursor miRNA by the Drosha enzyme and then moved from nucleus to cytoplasm. In cytoplasm, the Dicer enzyme cleaves loop formation from an miRNA precursor, leaving only double-stranded miRNA that continues its travel. dsRNA structure is further unwound in single-stranded miRNAs that then bind to proteins to form the RNA-induced silencing complex (RISC) and execute the miRNA function. Canonical studies have shown that one miRNA strand is incorporated into the RISC, having functions in the cell, whereas the other strand (the so-called passenger strand) is cleaved [9]. However, newer studies show that both arms of the same miRNA are found as the mature form within the cells, also in the nucleus [10], thus indicating their nuclear function as well. MicroRNAs are considered to function via the RISC to mainly inhibit gene function. Yet, mature miRNAs and active RISC proteins are found in the nuclear fractions of the cells [11] leading to suspicion of the nuclear function of miRNAs. Although the exact description of the mechanism remains unclear, there are some studies showing activating miRNAs in the cell. For example, Place et al. showed that miR-373 induces expression of E-cadherin and cold-shock domain-containing protein C2 (CSDC2) expression by promoter targeting. Mutated miR-373 constructs did not result in the induction of genes, which showed that miR-373-mediated promoter activation is dependent on target sequence specificity on the promoter loci [12].

P-Element induced wimpy testis (PIWI)-interacting RNAs (piRNAs) are a class of small RNAs of 26–30 nucleotides in size that are recognized by their unique structure consisting of 2′-O-methylation at the 3-end and uracil bias at their 5-end [13]. piRNAs are mainly found in the stem cells and human reproductive cells, which is why they are linked to many developmental functions. piRNAs are found in the genome as piRNA clusters as so many individual piRNAs are transcribed at the same time. After transcription, piRNAs are generated either in Zuc-dependent or Zuc-independent way [14]. In their mature form, piRNAs interact with PIWI proteins and function both via transcriptional and post-transcriptional regulators. piRNAs are known to target transposons but newer studies show their role in mRNA regulation [15]. It was shown in a study by Dai et al. that piRNAs can upregulate their targets by imperfect base-pairing to mRNA 3'UTRs, leading to induction of their translation. Five individual piRNAs were found in this study to upregulate the translation of their corresponding targets [16].

Small nucleolar RNAs (snoRNAs) are ncRNAs 60–300 nucleotides long that localize in the nucleolus and are important for ribosomal RNA maturation. snoRNAs can be divided into two classes based on different structures. H/ACA class snoRNAs function mainly in pseudouridylation, whereas C/D box family is crucial for 2′-O-ribose methylation. In the genome, snoRNAs are located in the intron of the genes and they are spliced as precursors after transcription [17]. snoRNAs are often found in the introns of their target genes, suggesting their important role in target gene regulation. snoRNAs can also be transcribed together with other small ncRNAs such as miRNAs or piRNAs [18], which indicates that the regulation network has more than one type of small ncRNA. As snoRNAs are important, especially in ribosomal RNA function, they are thought to be very conserved housekeeping regulators. That is why it is no surprise that they play a role in tumor formation, as Xia et al. showed, for example [7]. In that study, SNORD44 was found to be downregulated in four glioma cell lines and its upregulation was shown to decrease cell proliferation both in vitro and in vivo. In another study, Cui et al. showed that snoRNA can activate gene expression. In that study, snoRNA SNORA23 was shown to be increased in human pancreatic ductal adenocarcinoma (PDAC) cells and its expression was linked to tumor invasion and patient survival. Knockdown of SNORA23 using antisense oligos (ASOs) decreased the expression of SYNE2. However, SYNE2 expression was not affected by double-knockdown of AGO2 and SNORA23, which implies that SNORA23 potentially functions via ribosome biogenesis instead of via the miRNA-like AGO-dependent pathway [19].

2.2 Small Activating ncRNA Functions in the Cells

Small noncoding RNAs have been known to exist and function for many decades. Small ncRNAs are traditionally thought to only function in the cells as repressors of gene expression. One of the most well-known mechanisms for post-translational gene silencing (PTGS) is miRNA-regulated mRNA inhibition. This RNA interference (RNAi) functions via mature miRNAs connecting to specific proteins and together forming functional RISC. Then, by perfect or imperfect base-pairing, miRNA binds to the 3'UTR of mRNA and either degrades mRNA or inhibits mRNA translation respectively. Most papers focus on these inhibitory effects of the small ncRNAs. Still, many studies today show small ncRNAs as playing a role in activating gene expression.

There are many potential mechanisms for saRNA function. First papers showing the induction of gene expression by small RNAs were published in 2006 and 2007 and utilized exogenous, artificial saRNAs. Li et al. showed that synthetic promoter targeted double-stranded RNA (dsRNA) was able to transcriptionally activate E-cadherin, p21, and VEGF genes [20]. Also, Janowski et al. showed that dsRNA targeted the promoter region of progesterone receptor (PR) gene and by epigenetic changes it induced the transcription of PR gene, which lead to increased mRNA and protein levels [21]. Interestingly, in contrast to the study by Janowski et al., Li et al. also showed long-lasting induction of gene expression by dsRNA, which indicates possible different regulation mechanisms for different small RNAs.

In the study by Huang et al., miR-744 caused induction of cyclin B1 (Ccnb1) mRNA levels. Induction of Ccnb1 was achieved by miR-744-mediated epigenetic changes on transcription start site (TSS) of Ccnb1 [5]. He et al. showed that snoRNA-derived piRNA (pi-sno75) increased its target mRNA tumor necrosis factor (TNF)-related apoptosis-inducing ligand (TRAIL) expression both at the mRNA and at the protein levels. Interestingly, pi-sno75 was shown to target the TRAIL promoter region and induce its transcription by epigenetic modifications [18].

In many studies, miRNAs are suspected to convey their function via long ncRNAs. Long ncRNAs would function as a target for miRNA binding, and gene function at the transcriptional level is regulated by affecting the lncRNA. For instance, the PR promoter was further studied by Schwartz et al., and it was shown that dsRNA can target the PR promoter antisense transcript and thereby induce or decrease promoter activity in a sequence-specific manner. It was also shown that the binding of dsRNA to antisense transcripts recruit AGO protein to form a complex that acts as a scaffold on the PR promoter. This complex further lures other transcription factors to the site to enable a full regulatory system of promoter activation [22]. Huang et al. showed in their paper that antisense lncRNA NCK1-AS1 can be a target of miR-138-2-3p, preventing the miRNA from silencing its target mRNA TRIM24. In this case, lncRNA acts as a competing endogenous RNA (ceRNA) target for miRNA, which leads to increased target mRNA function (translation) as miRNA is not silencing it [8].

ceRNA hypothesis was first described by Salmena et al. when they showed the competing mechanism between mRNA and lncRNA as an sncRNA target [23]. Later on, other groups showed similar results [24]. It has been suggested that miRNA might act either in the cytoplasm (targeting mRNA) or in the nucleus (targeting nuclear transcripts), depending on where its main targets are located. Also, Quan et al. showed that lncRNA and mRNA compete for miRNA targets. They found that lncRNA AK131850 acts as sponge for miR-93-5p, inducing VEGF-A expression, as miRNA is not inhibiting VEGF-A mRNA [25]. In addition to other mechanisms described here, it cannot be ruled out that saRNA would directly bind the chromatin, as triplex formation has been shown to occur [26]. Meng et al. studied the saRNA duplex regulating the PR promoter. Interestingly, it was shown that saRNA induced PR expression by binding to sense transcripts on PR promotor or straight to promoter DNA (Fig. 2) [27].

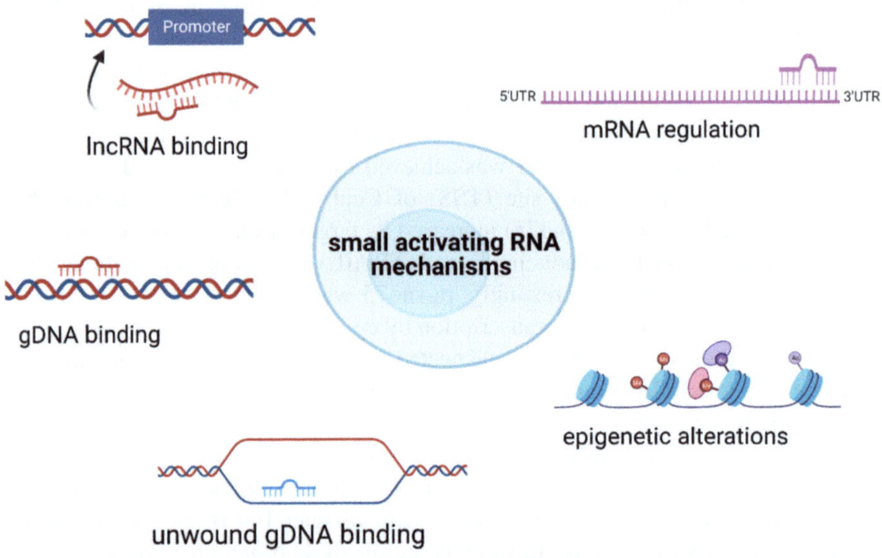

Fig. 2 Different potential mechanisms of saRNA function. saRNAs are most often considered to bind promoter-associated long noncoding RNA at the promoter of their target genes and thus regulate gene expression at transcriptional level. However, direct targeting of the chromatin is also a possible mechanism of action (triplex forming, or binding to unwound gDNA). Promoter targeting is thought to induce epigenetic alterations on the chromatin that induce gene transcription. Activating effects on mRNA expression by saRNAs can also come from the action of ceRNAs, longer RNA transcripts that act as a sponge and inhibit miRNA to bind to its mRNA target, or by binding mRNA directly and inducing translation. Figure created with BioRender.com

3 Small Activating RNAs in the Regulation of Angiogenesis

Angiogenesis, i.e., vessel reformation is an important function for cell development and growth. Angiogenesis starts in a lack-of-oxygen (hypoxia) environment, which makes cells produce growth factors, especially VEGF-A. Arrival of growth factors to a hypoxic environment leads to existing blood vessels to sprout, which leads to the formation of new blood vessels.

Many studies show sncRNA interaction on angiogenesis-related genes (Fig. 3). Exosome-delivered miR-21-3p was able to induce angiogenesis and wound healing in the study by Hu et al. In that study, exosomes from human umbilical cord blood plasma (UCB-Exos) were extracted and injected to skin wounds in mice. This lead to downregulation of PTEN and SPRY1 mRNA levels, which increased fibroblast proliferation and migration [28]. Also, Yang et al. showed that the miR-497~195 cluster is active in a specific bone vessel endothelium and by targeting F-box and WD-40 domain protein (Fbxw7) and by Prolyl 4-hydroxylase possessing a transmembrane domain (P4HTM), it takes part in Notch and Hif1a stability. This indicates the ability of the miRNA cluster to regulate angiogenesis. Interestingly, upregulation of the miR-497~195 cluster induced the Hif1a protein level via stabilized mRNA levels, not through increased transcription of mRNA [2].

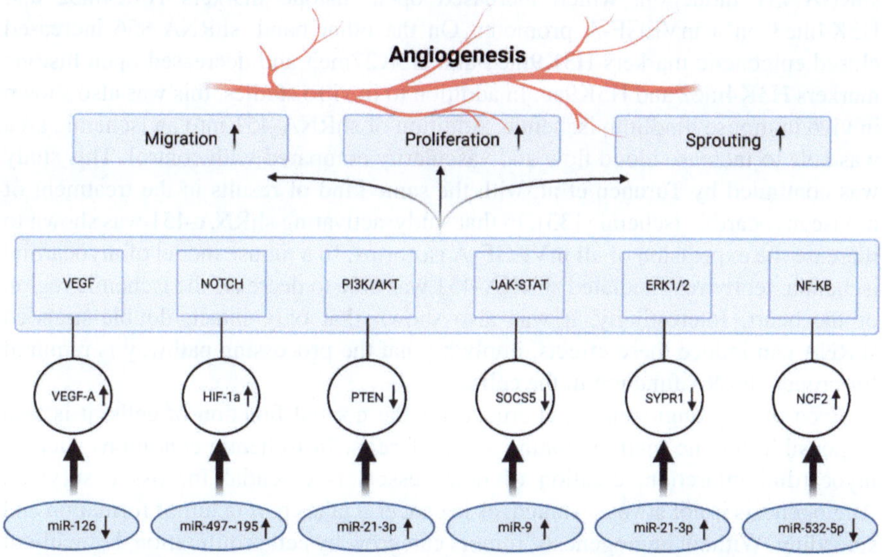

Fig. 3 Angiogenesis is regulated by noncoding RNAs. The angiogenesis pathway is carefully regulated and many miRNAs participate in the maintenance of healthy vasculature. On the other hand, angiogenesis also plays a part in pathological conditions such as cancer, where many miRNAs have been linked to induced angiogenesis and metastasis in the tumor. Figure created with BioRender.com

The VEGF family is a large family of growth factors that include VEGF-A, VEGF-B, VEGF-C, VEGF-D, VEGF-E (viral VEGF), VEGF-F (snake venom VEGF), and placenta growth factor (PIGF) [29]. Of these, VEGF-A is the most common and crucial for angiogenesis. VEGF-A gene contains eight exons and seven introns [30]. By alternative splicing, different isoforms of VEGF-A are formed. All the isoforms have five first exons transcribed, and differences are formed by combination and splicing of other exons. There are nine different VEGF-A isoforms of which VEGF165 is the most common form. VEGF proteins act by binding to receptors VEGFR-1 (also known as Flt-1), VEGFR-2 (KDR), VEGFR-3 (Flt-4), Neuropilin-1 (NP-1), Neuropilin-2 (NP-2), and soluble VEGF receptors (sVEGFR). VEGFR1 and VEGFR2 can be found on the surface of many cell types, but they are mainly expressed in endothelial cells, whereas VEGFR3 is characteristic of endothelial lymphatic cells [29]. VEGF-A is a dimeric glycoprotein containing two subunits, and requires binding to heparin for its functions. VEGF-A is specific to endothelial cells and can induce angiogenesis. It can also be secreted from the cells, thus communicating with the surrounding tissue [31].

Angiogenesis was the targeted pathway for the first in vivo study for saRNA-mediated gene activation, which was published by Turunen et al. [32]. In that study, mouse VEGF-A (mVEGF-A) promoter was targeted by artificial small hairpin RNAs (shRNAs), which lead promoter activity to increase or to decrease, depending on the targeted sequence. Activity was achieved by lentivirus-mediated shRNA-451 induction, which increased open histone markers H3K4me2 and H3K4me3 on a mVEGF-A promoter. On the other hand, shRNA-856 increased closed epigenetic markers H3K9me2 and H3K27me3 and decreased open histone markers H3K4me2 and H3K9ac. In addition to in vitro studies, this was also shown in vivo in mouse hindlimb ischemia. Addition of shRNA-451 into an ischemic area was able to increase blood flow and vascularity compared with control. This study was continued by Turunen et al. with the same kind of results in the treatment of mouse myocardial ischemia [33]. In that study, activating shRNA-451 was shown to increase the expression of all mVEGF-A isoforms. In a mouse model of myocardial ischemia, lentivirus-mediated shRNA-451 was able to decrease the ischemic region of the heart. Interestingly, it was also shown that only intact, double-stranded shRNA can induce these effects, implying that the processing pathway is required for proper shRNA function in the cells.

Even though angiogenesis is crucial for the normal function of cells, it is also responsible for the malfunctioning stage of cells. In ischemic conditions such as myocardial infarction, creation of new vessels is essential for tissue survival. Angiogenesis is not always wanted, as in cancer it takes part in tumor formation and spreading. Without angiogenesis, tumors can grow by cell proliferation, but without new vessel formation tumors remain in a benign state, which indicates a good prognosis for the patient. When angiogenesis takes place, the tumor is spreading and evolving to a malignant form, also called cancer.

As previously stated, besides maintaining normal activities within the body, many sncRNAs are also linked to the malfunctioning of cells (Fig. 3). Zhuang et al. showed that miR-9 is secreted by tumors and it promotes endothelial cell migration and angiogenesis. This regulation functions via the JAK-STAT pathway [34]. Zhu et al. showed that endothelial-specific miR-126 could target PIK3R2 and VEGF-A and its expression was decreased in breast cancer tissue. This led authors to suspect that miR-126 could take part in tumor formation by regulation of the VEGF/PI3K/AKT signal pathway and thus participate in the angiogenic features, proliferation, and migration of tumor [35]. Also, Zhang et al. showed that miR-532-5p regulates the NF-κB pathway by NCF2 in a gastric cancer model, where it was linked to angiogenesis and metastasis. Downregulation of miR-532-5p leads to upregulation of NCF2, whereas upregulation of miR-532-5p decreased NCF2 levels. Interestingly, miR-532-5p functioned through lncRNA LINC01410 where the expression changes of lncRNA changed the expression of miRNA [36].

4 Nucleic Acid-based Therapy

Nucleic acids have been developed as therapeutics in various forms. Lately, mRNA vaccines used in the battle against the COVID-19 pandemic have made RNA medicines known all around the world. The Moderna and Pfizer vaccines are based on mRNA encoding SARS-CoV2 spike protein, encapsulated in lipid nanoparticles [37]. This approach is new in the vaccine world and creates the potential for more easily and rapidly adaptable new vaccines in the future as well. Indeed, companies such as Moderna have been developing more vaccines for different viruses, such as cytomegalovirus (CMV) [38]. Along with vaccines, Moderna's pipelines include other medicines based on mRNAs to treat diseases such as cancer and myocardial infarction [39]. Besides the COVID-19 vaccine, none of these is yet on the market, but some have progressed to phase II clinical trials.

Antisense oligonucleotides (ASOs) are currently in the widest use as they come to clinical development, as multiple drugs using the chemistry have been approved to enter the market by regulatory officials. ASOs are short single-stranded DNA molecules where the bases can be chemically modified, with different aims, such as improving stability and half-life in the tissue. ASOs target specific mRNA in the cell, binding it and inducing inhibition of translation to protein or directing splicing of the pre-mRNA. The first ASO drug that was approved by the Food and Drug Administration (FDA) in 1998 was fomivirsen for the treatment of CMV retinitis [40]. It contained phosphorothioate chemistry for resistance to nucleases. Mipomersen, approved by the FDA in 2013 for the treatment of homozygous familial hypercholesterolemia, has additional modifications to increase stability: deoxyribose modifications in the middle of the molecule and 2′-O-methoxyethyl-modified ribose at the both ends [41]. Eteplirsen is a morpholino phosphorodiamidate ASO, affecting exon-skipping in DMD gene and was designed to treat Duchenne muscular dystrophy [42]. Inotersen, a 2′-O-methyoxyethyl-modified antisense oligonucleotide, was developed to treat hereditary transthyretin amyloidosis and was approved in 2018 by the European Medicines Agency (EMA) and the FDA [43]. Nusinersen is an ASO with 2′-O-2-methoxyethyl and phosphorothioate chemistry and is used to treat spinal muscular atrophy [40]. It affects alternative splicing of the SMN2 gene. Nusinersen was approved by the FDA in 2016 and by the EMA in 2017. Volanesorsen, an ASO by Akcea Therapeutics and Ionis Pharmaceuticals, was accepted by the EMA in 2019, for treatment of familial chylomicronemia syndrome [44].

In addition to DNA-based ASOs, RNA can be developed for therapy as siRNA (RNAi) or miRNA. The first medicine based on RNAi or siRNA that was approved by the FDA was patisiran from Alnylam Pharmaceuticals to treat peripheral nerve disease in 2018 [43]. The siRNA drug inclisiran (Novartis) was approved for treatment of low-density lipoprotein cholesterol in Europe in 2020, in the hope of alleviating the increasing burden of cardiovascular diseases [45]. Locked nucleic acid (LNA)-based antimir-drug MRG-110 from miRAgen Therapeutics (now Viridian Therapeutics) inhibits miR-92 function for treatment of cardiovascular disease and showed promising results in phase 1 trials, where it increased angiogenesis [46]. Pegaptanib (Pfizer) is a 28-bp RNA-aptamer that is used for treatment of age-related macular degeneration (AMD). It specifically binds and blocks the VEGF165 isoform, which is critical for abnormal growth and permeability of vessels, occurring in AMD [47].

Many types of RNA-based drugs are being developed and are already on the market, but most of them are based on inhibiting their targets in the cell. Only one saRNA is undergoing clinical trials to date. Small RNA targeting CEBPA is developed by MiNa Therapeutics and is activating CEBPA expression for the treatment of hepatocellular carcinoma [48]. CEBPA is a master regulator of myeloid lineage and activating its expression positively affects the microenvironment in the tumor. saRNA-induced activation of CEBPA can be combined with tyrosine kinase inhibitor to enhance the effect of the drugs. Data from a phase I clinical trial with saRNA MTL-CEBPA were promising and pave the way for more saRNA medicines.

5 Future Vision

Classical function of genetic information was thought for a long time to mainly include genetic information as a form of double-stranded DNA, from which genes are expressed to produce messenger RNAs, which are then translated to proteins. The initial findings from Fire and Mello, to whom the Nobel prize was awarded for the discovery of RNAi, opened a new era of research where the function of noncoding RNAs was realized. For more than two decades the biotech industry has been developing novel therapeutics based on noncoding RNAs, mostly based on classical RNAi. However, as the data have been accumulating at an ever increasing pace from RNA-sequencing studies, because the technology has advanced and such studies have become more readily available owing to decreasing costs, it has become evident that noncoding RNAs have many other functions, as has been previously thought. The actions of noncoding RNAs in the regulation of chromatin function have been starting to unravel in more detail. However, a huge amount of data from sequencing studies also represents a major challenge for the field. How to make

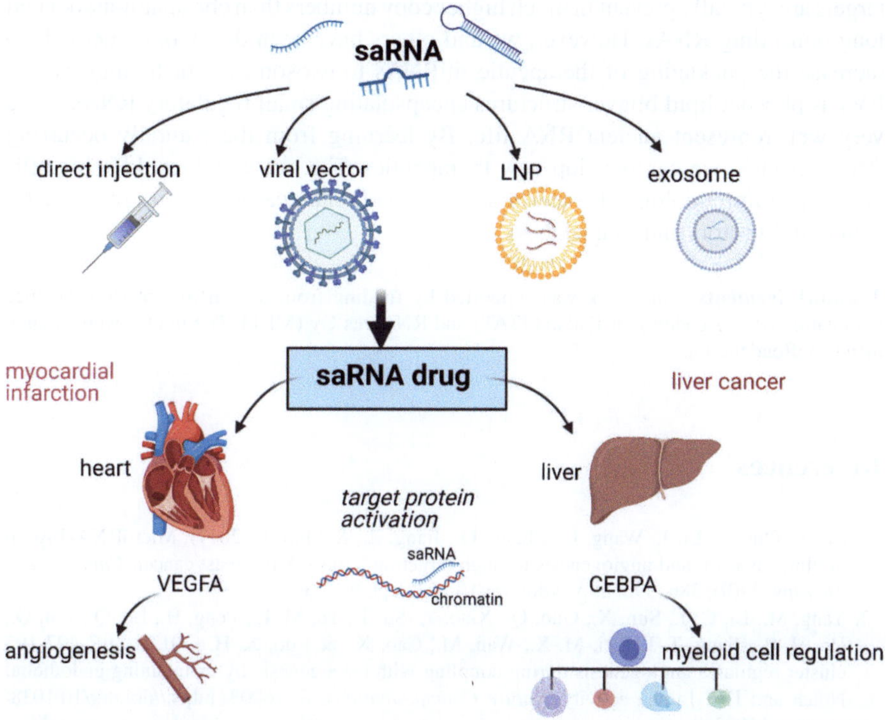

Fig. 4 saRNA medicine. saRNAs can be used as small hairpins, single-stranded small RNAs, or duplex RNAs. For efficient delivery, saRNAs are often loaded into carriers such as lipid nanoparticles (LNP), exosomes, or expressed from viral vectors, but direct injection to the tissue can also be used. saRNA drug is delivered to target tissue, for example, to the heart (treatment of myocardial infarction) or liver (treatment of liver cancer). saRNA binds its targets in the nucleus of target organ cells and regulates, e.g., VEGFA, thus inducing angiogenesis, or CEBPA, inducing myeloid cell differentiation. The first saRNA drug undergoing clinical trials is MTL-CEBPA by MiNa Therapeutics for treatment of liver cancer. Figure created with BioRender.com

sense of the complex relations between various RNAs spatially, in a living 3D growing cell? Maybe future technology that could combine RNA sequencing with RNA imaging continuously in a living cell could reveal interesting new aspects of RNA-mediated chromatin regulation. In addition, it should be taken into account that cells are not isolated units in a multicellular organism but communicate with each other. One field of research that has been actively growing in the last decade is research on extracellular vesicles (EVs), such as exosomes. It is known that cells transfer material, including noncoding RNAs, among themselves using EVs. Therefore, it is not surprising that developing therapeutics based on exosomes have recently gained much attention. However, it has been noted that naturally produced exosomes contain typically only small amounts of miRNAs. This has been surprising for some research in the field. Our hypothesis is that many of these miRNAs could have functions in the nucleus of the target cells and therefore much lower amounts are needed for chromatin regulatory effects compared with classical RNAi, where mRNA targets are typically present in much higher copy numbers than chromatin-associated long noncoding RNAs. However, we and others have been developing methods to increase the packaging of therapeutic miRNAs to exosomes. Our thought is that EVs as phospholipid bilayer structures encapsulating small regulatory RNAs could very well represent ancient RNA life. By learning from the naturally occurring RNA vesicles, we can develop new therapeutics (Fig. 4), which could solve efficiency and immunological issues associated with classical gene therapy, typically using viral vectors and long cDNAs.

Acknowledgements This work was supported by funding from the Instrumentarium Science Foundation (PL), Academy of Finland (TAT) and RNatives Oy (MPT). The figures were created using BioRender.com.

References

1. Lu, Y., Qin, T., Li, J., Wang, L., Zhang, Q., Jiang, Z., & Mao, J. (2017). MicroRNA-140-5p inhibits invasion and angiogenesis through targeting VEGF-A in breast cancer. *Cancer Gene Therapy, 24*(9), 386–392. https://doi.org/10.1038/cgt.2017.30
2. Yang, M., Li, C.-J., Sun, X., Guo, Q., Xiao, Y., Su, T., Tu, M.-L., Peng, H., Lu, Q., Liu, Q., He, H.-B., Jiang, T.-J., Lei, M.-X., Wan, M., Cao, X., & Luo, X.-H. (2017). MiR-497-195 cluster regulates angiogenesis during coupling with osteogenesis by maintaining endothelial Notch and HIF-1alpha activity. *Nature Communications, 8*, 16003. https://doi.org/10.1038/ncomms16003
3. Gregory, P. A., Bert, A. G., Paterson, E. L., Barry, S. C., Tsykin, A., Farshid, G., Vadas, M. A., Khew-Goodall, Y., & Goodall, G. J. (2008). The miR-200 family and miR-205 regulate epithelial to mesenchymal transition by targeting ZEB1 and SIP1. *Nature Cell Biology, 10*(5), 593–601. https://doi.org/10.1038/ncb1722
4. Zhou, Q., Fan, D., Huang, K., Chen, X., Chen, Y., & Mai, Q. (2018). Activation of KLF4 expression by small activating RNA promotes migration and invasion in colorectal epithelial cells. *Cell Biology International, 42*(4), 495–503. https://doi.org/10.1002/cbin.10926
5. Huang, V., Place, R. F., Portnoy, V., Wang, J., Qi, Z., Jia, Z., Yu, A., Shuman, M., Yu, J., & Li, L. C. (2012). Upregulation of Cyclin B1 by miRNA and its implications in cancer. *Nucleic Acids Research, 40*(4), 1695–1707. https://doi.org/10.1093/nar/gkr934
6. Pauli, C., Liu, Y., Rohde, C., Cui, C., Fijalkowska, D., Gerloff, D., Walter, C., Krijgsveld, J., Dugas, M., Edemir, B., Pabst, C., & M, L. P., & Zhou, F. (2020). Site-specific methylation of 18S ribosomal RNA by SNORD42A is required for acute myeloid leukemia cell proliferation. *Blood, 135*(23), 2059–2070. https://doi.org/10.1182/blood.2019004121
7. Xia, X. R., Li, W. C., Yu, Z. T., Li, J., Peng, C. Y., Jin, L., & Yuan, G. L. (2020). Effects of small nucleolar RNA SNORD44 on the proliferation, apoptosis and invasion of glioma cells. *Histochemistry and Cell Biology, 153*(4), 257–269. https://doi.org/10.1007/s00418-020-01848-y
8. Huang, L., Li, X., Ye, H., Liu, Y., Liang, X., Yang, C., Hua, L., Yan, Z., & Zhang, X. (2020). Long non-coding RNA NCK1-AS1 promotes the tumorigenesis of glioma through sponging microRNA-138-2-3p and activating the TRIM24/Wnt/β-catenin axis. *Journal of Experimental and Clinical Cancer Research, 39*(63), 1–15. https://doi.org/10.1186/s13046-020-01567-1
9. Liu, H., Lei, C., He, Q., Pan, Z., Xiao, D., & Tao, Y. (2018). Nuclear functions of mammalian MicroRNAs in gene regulation, immunity and cancer. *Molecular Cancer, 17*, 1–14. https://doi.org/10.1186/s12943-018-0765-5

10. Turunen, T. A., Roberts, T. C., Laitinen, P., Väänänen, M.-A., Korhonen, P., Malm, T., Ylä-Herttuala, S., & Turunen, M. P. (2019). Changes in nuclear and cytoplasmic microRNA distribution in response to hypoxic stress. *Scientific Reports, 9*(1), 10332. https://doi.org/10.1038/s41598-019-46841-1
11. Gagnon, K. T., Li, L., Chu, Y., Janowski, B. A., & Corey, D. R. (2014). RNAi factors are present and active in human cell nuclei. *Cell Reports, 6*, 211–221. https://doi.org/10.1016/j.celrep.2013.12.013
12. Place, R. F., Li, L.-C., Pookot, D., Noonan, E. J., & Dahiya, R. (2008). MicroRNA-373 induces expression of genes with complementary promoter sequences. *Proceedings of the National Academy of Sciences of the United States of America, 105*(5), 1608–1613. https://doi.org/10.1073/pnas.0707594105
13. Ohara, T., Sakaguchi, Y., Suzuki, T., Ueda, H., Miyauchi, K., & Suzuki, T. (2007). The 3′ termini of mouse Piwi-interacting RNAs are 2′-O-methylated. *Nature Structural & Molecular Biology, 14*(4), 349–350. https://doi.org/10.1038/nsmb1220
14. Li, M., Yang, Y., Wang, Z., Zong, T., Fu, X., Aung, L. H. H., Wang, K., Wang, J.-X., & Yu, T. (2021). Piwi-interacting RNAs (piRNAs) as potential biomarkers and therapeutic targets for cardiovascular diseases. *Angiogenesis, 24*(1), 19–34. https://doi.org/10.1007/s10456-020-09750-w
15. Rajan, K. S., Velmurugan, G., Gopal, P., Ramprasath, T., Venkatesh Babu, D., Krithika, S., Christy Jenifer, Y., Freddy, A., William Jn, G., Kalpana, K., & Ramasamy, S. (2016). Abundant and altered expression of PIWI-interacting RNAs during cardiac hypertrophy. *Heart, Lung and Circulation, 25*, 1013–1020. https://doi.org/10.1016/j.hlc.2016.02.015
16. Dai, P., Wang, X., Gou, L. T., Li, Z. T., Wen, Z., Chen, Z. G., Hua, M. M., Zhong, A., Wang, L., Su, H., Wan, H., Qian, K., Liao, L., Li, J., Tian, B., Li, D., Fu, X. D., Shi, H. J., Zhou, Y., & Liu, M. F. (2019). A translation-activating function of MIWI/piRNA during mouse spermiogenesis. *Cell, 179*(7), 1566–1581. https://doi.org/10.1016/j.cell.2019.11.022
17. Makarova, J. A., & Kramerov, D. A. (2005). Mammalian U87 small nucleolar RNA and its host gene. *Molecular Biology, 39*(4), 564–571. http://www.ncbi.nlm.nih.gov/dbEST
18. He, X., Chen, X., Zhang, X., Duan, X., Pan, T., Hu, Q., Zhang, Y., Zhong, F., Liu, J., Zhang, H., Luo, J., Wu, K., Peng, G., Luo, H., Zhang, L., Li, X., & Zhang, H. (2015). An Lnc RNA (GAS5)/SnoRNA-derived piRNA induces activation of TRAIL gene by site-specifically recruiting MLL/COMPASS-like complexes. *Nucleic Acids Research, 43*(7), 3712–3725. https://doi.org/10.1093/nar/gkv214
19. Cui, L., Nakano, K., Obchoei, S., Setoguchi, K., Matsumoto, M., Yamamoto, T., Obika, S., Shimada, K., & Hiraoka, N. (2017). Small nucleolar noncoding RNA SNORA23, up-regulated in human pancreatic ductal adenocarcinoma, regulates expression of spectrin repeat-containing nuclear envelope 2 to promote growth and metastasis of xenograft tumors in mice. *Gastroenterology, 153*, 292–306. https://doi.org/10.1053/j.gastro.2017.03.050
20. Li, L. C., Okino, S. T., Zhao, H., Pookot, D., Place, R. F., Urakami, S., Enokida, H., & Dahiya, R. (2006). Small dsRNAs induce transcriptional activation in human cells. *Proceedings of the National Academy of Sciences of the United States of America, 103*(46), 17337–17342. https://doi.org/10.1073/pnas.0607015103
21. Janowski, B. A., Younger, S. T., Hardy, D. B., Ram, R., Huffman, K. E., & Corey, D. R. (2007). Activating gene expression in mammalian cells with promoter-targeted duplex RNAs. *Nature Chemical Biology, 3*(3), 166–173. https://doi.org/10.1038/nchembio860
22. Schwartz, J. C., Younger, S. T., Nguyen, N.-B., Hardy, D. B., Monia, B. P., Corey, D. R., & Janowski, B. A. (2008). Antisense transcripts are targets for activating small RNAs. *Nature Structural & Molecular Biology, 15*(8), 842–848. https://doi.org/10.1038/nsmb.1444
23. Salmena, L., Poliseno, L., Tay, Y., Kats, L., & Pandolfi, P. P. (2011). A ceRNA hypothesis: the Rosetta stone of a hidden RNA language? *Cell, 146*(3), 353–358. https://doi.org/10.1016/j.cell.2011.07.014
24. Roberts, T. C. (2014). The MicroRNA biology of the mammalian nucleus. *Molecular Therapy - Nucleic Acids, 3*, e188. https://doi.org/10.1038/mtna.2014.40

25. Quan, H., Liang, M., Li, N., Dou, C., Liu, C., Bai, Y., Luo, W., Li, J., Kang, F., Cao, Z., Yang, X., Jiang, H., & Dong, S. (2018). LncRNA-AK131850 sponges MiR-93-5p in newborn and mature osteoclasts to enhance the secretion of vascular endothelial growth factor a promoting vasculogenesis of endothelial progenitor cells. *Cellular Physiology and Biochemistry, 46*(1), 401–417. https://doi.org/10.1159/000488474
26. Paugh, S. W., Coss, D. R., Bao, J., Laudermilk, L. T., Grace, C. R., Ferreira, A. M., Waddell, M. B., Ridout, G., Naeve, D., Leuze, M., LoCascio, P. F., Panetta, J. C., Wilkinson, M. R., Pui, C.-H., Naeve, C. W., Uberbacher, E. C., Bonten, E. J., & Evans, W. E. (2016). MicroRNAs form triplexes with double stranded DNA at sequence-specific binding sites; a eukaryotic mechanism via which microRNAs could directly alter gene expression. *PLoS Computational Biology, 12*(2), e1004744. https://doi.org/10.1371/journal.pcbi.1004744
27. Meng, X., Jiang, Q., Chang, N., Wang, X., Liu, C., Xiong, J., Cao, H., & Liang, Z. (2016). Small activating RNA binds to the genomic target site in a seed-region-dependent manner. *Nucleic Acids Research, 44*(5), 2274–2282. https://doi.org/10.1093/nar/gkw076
28. Hu, Y., Rao, S. S., Wang, Z. X., Cao, J., Tan, Y. J., Luo, J., Li, H. M., Zhang, W. S., Chen, C. Y., & Xie, H. (2018). Exosomes from human umbilical cord blood accelerate cutaneous wound healing through miR-21-3p-mediated promotion of angiogenesis and fibroblast function. *Theranostics, 8*(1), 169–184. https://doi.org/10.7150/thno.21234
29. Yamazaki, Y., & Morita, T. (2006). Molecular and functional diversity of vascular endothelial growth factors. *Molecular Diversity, 10*(4), 515–527. https://doi.org/10.1007/s11030-006-9027-3
30. Tischer, E., Mitchell, R., Hartman, T., Silva, M., Gospodarowicz, D., Fiddes, J., & Abraham, J. (1991). The human gene for vascular endothelial growth factor. *The Journal of Biological Chemistry, 266*(18), 11947–11954.
31. Leung, D. W., Cachianes, G., Kuang, W.-J., Goeddel, D. V., & Ferrara, N. (1989). Vascular endothelial growth factor is a secreted angiogenic mitogen. *Science, 4935*(246), 1306–1309.
32. Turunen, M. P., Lehtola, T., Heinonen, S. E., Assefa, G. S., Korpisalo, P., Girnary, R., Glass, C. K., Väisänen, S., & Ylä-Herttuala, S. (2009). Efficient regulation of VEGF expression by promoter-targeted lentiviral shRNAs based on epigenetic mechanism a novel example of epigenetherapy. *Circulation Research, 105*(6), 604–609. https://doi.org/10.1161/CIRCRESAHA.109.200774
33. Turunen, M. P., Husso, T., Musthafa, H., Laidinen, S., Dragneva, G., Laham-Karam, N., Honkanen, S., Paakinaho, A., Laakkonen, J. P., Gao, E., Vihinen-Ranta, M., Liimatainen, T., & Ylä-Herttuala, S. (2014). Epigenetic upregulation of endogenous VEGF-A reduces myocardial infarct size in mice. *PLoS One, 9*(2), e89979. https://doi.org/10.1371/journal.pone.0089979
34. Zhuang, G., Wu, X., Jiang, Z., Kasman, I., Yao, J., Guan, Y., Oeh, J., Modrusan, Z., Bais, C., Sampath, D., & Ferrara, N. (2012). Tumour-secreted miR-9 promotes endothelial cell migration and angiogenesis by activating the JAK-STAT pathway. *The EMBO Journal, 31*(17), 3513–3523. https://doi.org/10.1038/emboj.2012.183
35. Zhu, N., Zhang, D., Xie, H., Zhou, Z., Chen, H., Hu, T., Bai, Y., Shen, Y., Yuan, W., Jing, Q., & Qin, Y. (2011). Endothelial-specific intron-derived miR-126 is down-regulated in human breast cancer and targets both VEGFA and PIK3R2. *Molecular and Cellular Biochemistry, 351*, 157–164. https://doi.org/10.1007/s11010-011-0723-7
36. Zhang, J. X., Chen, Z. H., Chen, D. L., Tian, X. P., Wang, C. Y., Zhou, Z. W., Gao, Y., Xu, Y., Chen, C., Zheng, Z. S., Weng, H. W., Ye, S., Kuang, M., Xie, D., & Peng, S. (2018). LINC01410-miR-532-NCF2-NF-kB feedback loop promotes gastric cancer angiogenesis and metastasis. *Oncogene, 37*(20), 2660–2675. https://doi.org/10.1038/s41388-018-0162-y
37. Meo, S. A., Bukhari, I. A., Akram, J., Meo, A. S., & Klonoff, D. C. (2021). COVID-19 vaccines: comparison of biological, pharmacological characteristics and adverse effects of Pfizer/BioNTech and Moderna Vaccines. *European Review for Medical and Pharmacological Sciences, 25*(3), 1663–1669. https://doi.org/10.26355/eurrev_202102_24877
38. John, S., Yuzhakov, O., Woods, A., Deterling, J., Hassett, K., Shaw, C. A., & Ciaramella, G. (2018). Multi-antigenic human cytomegalovirus mRNA vaccines that elicit potent

humoral and cell-mediated immunity. *Vaccine, 36*(12), 1689–1699. https://doi.org/10.1016/j.vaccine.2018.01.029

39. Carlsson, L., Clarke, J. C., Yen, C., Gregoire, F., Albery, T., Billger, M., Egnell, A. C., Gan, L. M., Jennbacken, K., Johansson, E., Linhardt, G., Martinsson, S., Sadiq, M. W., Witman, N., Wang, Q. D., Chen, C. H., Wang, Y. P., Lin, S., Ticho, B., & Fritsche-Danielson, R. (2018). Biocompatible, purified VEGF-A mRNA improves cardiac function after intracardiac injection 1 week post-myocardial infarction in swine. *Molecular Therapy – Methods and Clinical Development, 9*, 330–346. https://doi.org/10.1016/j.omtm.2018.04.003
40. Stein, C. A., & Castanotto, D. (2017). FDA-approved oligonucleotide therapies in 2017. *Molecular Therapy, 25*(5), 1069–1075. https://doi.org/10.1016/j.ymthe.2017.03.023
41. Parham, J. S., & Goldberg, A. C. (2019). Mipomersen and its use in familial hypercholesterolemia. *Expert Opinion on Pharmacotherapy, 20*(2), 127–131. https://doi.org/10.1080/14656566.2018.1550071
42. Iftikhar, M., Frey, J., Shoha, M. J., Malek, S., & Mouase, S. A. (2021). Current and emerging therapies for Duchenne muscular dystrophy and spinal muscular atrophy. *Pharmacology & Therapeutics*, 107719. https://doi.org/10.1016/j.pharmthera.2020.107719
43. Benbrahim, M., Norman, K., Sanchorawala, V., Siddiqi, O. K., & Hughes, D. (2021). A review of novel agents and clinical considerations in patients with ATTR cardiac amyloidosis. *Journal of Cardiovascular Pharmacology, 77*(5), 544–548. https://doi.org/10.1097/FJC.0000000000001004
44. Esan, O., & Wierzbicki, A. S. (2020). Volanesorsen in the treatment of familial chylomicronemia syndrome or hypertriglyceridaemia: design, development and place in therapy. *Drug Design, Development and Therapy, 14*, 2623–2636. https://doi.org/10.2147/DDDT.S224771
45. Ruscica, M., Ferri, N., Santos, R. D., Sirtori, C. R., & Corsini, A. (2021). Lipid lowering drugs : present status and future developments. *Current Atherosclerosis Reports, 23*(17), 1–13. https://doi.org/10.1007/s11883-021-00918-3
46. Abplanalp, W. T., Fischer, A., John, D., Zeiher, A. M., Gosgnach, W., Darville, H., Montgomery, R., Pestano, L., Allée, G., Paty, I., Fougerousse, F., & Dimmeler, S. (2020). Efficiency and target derepression of anti-miR-92a: Results of a first in human study. *Nucleic Acid Therapeutics, 30*(6), 335–345. https://doi.org/10.1089/nat.2020.0871
47. Vinores, S. A. (2006). Pegaptanib in the treatment of wet, age-related macular degeneration. *International Journal of Nanomedicine, 1*(3), 263–268.
48. Sarker, D., Plummer, R., Meyer, T., Sodergren, M. H., Basu, B., Chee, C. E., Huang, K. W., Palmer, D. H., Ma, Y. T., Evans, T. R. J., Spalding, D. R. C., Pai, M., Sharma, R., Pinato, D. J., Spicer, J., Hunter, S., Kwatra, V., Nicholls, J. P., Collin, D., & Habib, N. (2020). MTL-CEBPA, a small activating RNA therapeutic upregulating C/EBP-α, in patients with advanced liver cancer: a first-in-human, multicenter, open-label, phase I trial. *Clinical Cancer Research, 26*(15), 3936–3946. https://doi.org/10.1158/1078-0432.CCR-20-0414

Muscular Dystrophy Therapy Using Viral Vector-based CRISPR/Cas

Eunyoung Choi and Taeyoung Koo

Abstract Duchenne muscular dystrophy (DMD) is a fatal X-linked genetic disorder caused by mutations in the *DMD* gene, which encodes the dystrophin protein. A lack of dystrophin disrupts the skeletal musculature, resulting in severe muscle degeneration. Currently, the clustered regularly interspaced short palindromic repeats (CRISPR)/CRISPR-associated (Cas) system offers an evolved and precise tool of programmed genomic modification and is being widely studied as a therapeutic tool for treating various genetic diseases. Recently, CRISPR-mediated DMD therapy has been intensively studied as a means of correcting or bypassing disease-causing mutations, resulting in the permanent repair of mutated *DMD* gene and rescues of dystrophin expression. However, delivery methods remain a major barrier for CRISPR-mediated genome editing. Various viral vectors have been utilized as vehicles for sequences encoding CRISPR/Cas components. Efforts have been made to optimize the viral vector systems for efficient delivery of these components to treat DMD. Herein, we review diverse aspects of several viral vectors combined with CRISPR/Cas systems for DMD therapy and discuss their therapeutic potential and the challenges ahead.

E. Choi
Department of Life and Nanopharmaceutical Sciences, Graduate School, Kyung Hee University, Seoul, Republic of South Korea

T. Koo (✉)
Department of Life and Nanopharmaceutical Sciences, Graduate School, Kyung Hee University, Seoul, Republic of South Korea

Department of Biomedical and Pharmaceutical Sciences, Graduate School, Kyung Hee University, Seoul, Republic of South Korea

Department of Pharmaceutical Science, College of Pharmacy, Kyung Hee University, Seoul, Republic of South Korea
e-mail: taeyoungkoo@khu.ac.kr

Y. H. Yun, K. E. Yoder (eds.), *Biotechnologies for Gene Therapy*,
https://doi.org/10.1007/978-3-030-93333-3_4

1 Introduction

Duchenne muscular dystrophy (DMD), the most common form of muscular dystrophy, is an X-linked recessive disorder of muscle caused by mutations in the *DMD* gene that affects approximately 1 in 3,500 newborn males worldwide. DMD patients exhibit severe, progressive muscle weakening due to complete abolishment of dystrophin, the protein encoded by *DMD*, and they eventually die from cardiac or respiratory failure at ages ranging from their early twenties to thirties. A loss of dystrophin disrupts the skeletal musculature and increases membrane permeability owing to abnormal ion channel function, resulting in increased calcium influx and myofiber necrosis. Dystrophin deficiency also leads to disruption of the dystrophin-associated protein (DAP) complex, as do mutations in DAP complex genes in other forms of congenital and autosomal muscular dystrophy. Different gene therapies for DMD are aimed at compensating for dystrophin loss-of-function. To prevent muscle degeneration, around 30% of normal dystrophin levels are likely to be required [1].

Becker muscular dystrophy (BMD), an allelic form of muscular dystrophy, is also caused by mutations in the *DMD* gene. BMD muscle pathology is relatively benign compared with that of DMD; BMD patients exhibit mild symptoms with slow disease progression or can even be virtually asymptomatic because of the continued expression of truncated but partially functional dystrophin proteins in the affected muscles. A large gene deletion that removes the central rod domain of dystrophin causes mild BMD symptoms because the DMD open reading frame (ORF) is maintained upstream and downstream of the mutation boundaries, resulting in partial expression of an internally deleted dystrophin protein. Thus, different mutations affecting various exon/intron regions of *DMD* can give rise to either DMD or BMD, depending upon the nature of the deletion boundaries. Dystrophin levels that are about 20% of the wild type results in a milder BMD phenotype [2]. Thus, alleviating severe symptoms for DMD patients by expressing semi-functional dystrophin that mimics BMD-like phenotype could be an effective therapeutic strategy.

Various pharmacological therapeutic approaches have focused on converting the DMD phenotype to a BMD-like phenotype by restoring the disrupted *DMD* reading frame. With such a mechanism, eteplirsen (Exondys 51; Sarepta Therapeutics), an antisense phosphorodiamidate morpholino oligomer (PMO) chemistry, was first approved by the Food and Drug Administration (FDA) for the treatment of DMD in 2016 [3]. It induces the expression of truncated but semi-functional dystrophin by exon 51 skipping in the *DMD* gene, resulting in BMD-like phenotype in eteplirsen-treated DMD patients. Additionally, two PMO drugs, golodirsen (Vyondys 53™; Sarepta Therapeutics) [4] and viltolarsen (Viltepso; NS Pharma) are prescription medications [5]. They induce exon 53 skipping in the *DMD* gene, restoring the expression of semi-functional dystrophin. These approaches have reduced disease symptoms, but none has yet eliminated the disease-causing mutation and allowed for long-term dystrophin expression. To address this goal, the adeno-associated virus (AAV) vector 2.5 was used in a phase I clinical trial to transfer a mini-dystrophin cDNA to biceps muscles of DMD patients with two doses (2.0×10^{10}

vector genomes (vg)/kg or 1.0×10^{11} vg/kg) [6]. However, this trial was suspended owing to T cell immunity against dystrophin epitopes in two patients following treatments.

Gene editing is a powerful method in which sequence-specific programmable nucleases are used to edit the target genomes in organisms ranging from animals to plants. Possible edits include gene knockout, base editing, or the addition of specific DNA fragments. These endonucleases have been developed from zinc-finger nucleases [7], engineered homing endonuclease [8], transcription activator-like effector nucleases [9], and RNA-guided nucleases in the clustered regularly interspaced short palindromic repeats (CRISPR)/CRISPR-associated (Cas) system [10]. Owing to its simplicity, broad target range, and high editing efficiency, the CRISPR/Cas system has become the most widely used platform for genome editing.

Editing of the CRISPR-mediated genome has been harnessed for permanent gene correction in a variety of genetic diseases. With advances in CRISPR-mediated therapeutic approaches, several strategies for treating DMD have exhibited great potential for compensating for the loss of dystrophin. Delivery of CRISPR/Cas components into diseased muscle tissues led to successful modification of the genome and the expression of semi-functional dystrophin protein, resulting in conversion from DMD- to BMD-like phenotype. For efficient CRISPR delivery, a range of viral and nonviral gene transfer vectors encoding the Cas nuclease together with guide RNA have been tested in systems ranging from patient cells to animal models of DMD. Even with the aid of sophisticated nonviral delivery strategies, such as electrotransfer, microinjection, or chemical delivery, CRISPR delivery methods still have major hurdles to overcome to reach the gene-transfer efficiencies required for therapeutic effectiveness. Therefore, a range of viral vectors have been evaluated as potentially attractive delivery systems for the transfer CRISPR/Cas to target organs for DMD therapy (Table 1). Safe and effective DMD gene therapy protocols based on the use of viral vectors are currently being developed. Viral vector-mediated rescue of dystrophin deficiency, which had been induced by precise CRISPR-mediated editing of the *DMD* gene in model organisms in conjunction with associated physiological improvement, has been evaluated. Here, we discuss diverse aspects of viral vector-mediated CRISPR systems in the field of DMD therapy (Table 2).

2 CRISPR/Cas: From Immune Systems to Genome Editing Tool

The CRISPR system is a prokaryotic adaptive immune system that provides resistance against invading foreign phages and plasmids [10]. When foreign genetic materials invade the cells, the DNA fragments are incorporated as spacers in a CRISPR array, providing sequence-specific memory. When the same foreign material reinvades, the CRISPR array is transcribed into a pre-CRISPR RNA (crRNA),

Table 1 Comparison of viral vectors used for clustered regularly interspaced short palindromic repeat delivery

Virus types	Genetic materials	Vector genome forms	Capacity (kb)	Advantages	Disadvantages
AAV	Single-stranded DNA	Mainly episomal	~4.7	• Long-term transgene expression • Minimal immunogenicity and toxicity • High transfection efficiency in vivo • Non-integrating	• Restriction in packaging capacity • Pre-existing immunity against natural serotypes
AdV	Double-stranded DNA	Episomal	~36	• Large packaging capacity • Non-integrating • Transient transgene expression	• Strong immunogenicity • Difficult large scaled production
LV	Single-stranded RNA	Integrated, non-integrated	~8	• High transduction efficiency in vitro • Suitable for generating stable cell lines	• Off-target effects (insertional mutagenesis and random integration) • Low delivery efficiency in vivo

Abbreviations: *AAV* adeno-associated viral vector, *AdV* adenoviral vector, *LV* lentiviral vector

which is then processed to become a mature crRNA that corresponds to the foreign DNA. In some systems, a second trans-activating crRNA (tracrRNA) is involved in crRNA maturation. Cas nucleases are guided to specific sequences associated with a protospacer adjacent motif (PAM) by the generated crRNAs, after which the Cas protein cleaves the targeted DNA. The two guide RNA (crRNA and tracrRNA) can be linked into a single form to generate a programmable, single guide RNA (sgRNA) that shows similar efficiency and is now widely used for genome engineering because of its simplicity. CRISPR/Cas-mediated genome editing involves two endogenous cellular DNA repair mechanisms: nonhomologous end joining (NHEJ) and homology directed repair (HDR). After DSBs are produced by the Cas protein, NHEJ induces indels that can result in reframing of the open reading frame (ORF) or generating of premature stop codons. In contrast, the HDR pathway inserts the donor template containing desired sequences into the target site after indel formation to induce precise genome editing or gene knock-in. However, the HDR mechanism is less efficient than NHEJ. Moreover, HDR is restricted to the S and G2 phases of the cell cycle, when sister chromatids are available to accept the template DNA; therefore, G1-arrested cells (post-mitotic cells), such as mature myofibers and cardiomyocytes, are not corrected efficiently by HDR-mediated gene editing.

The CRISPR systems are divided into two classes. Class I includes types I, III, and IV and requires several Cas proteins, such as Cas3, Cas10, and Csf1, to make up a functional complex. Class 2 includes types II, V, and VI and uses a single Cas

Table 2 Summary of viral vector-mediated clustered regularly interspaced short palindromic repeat systems used for treating Duchenne muscular dystrophy (DMD)

Origin	Types of vector	Tropism or generation	Strategies	Nucleases	DMD mutation	Therapeutic target gene region(s)	Model(s)	References
AAV	All-in-one AAV	AAV9	Exon reframing	CjCas9	*Dmd* frameshift mutation in exon 23	*Dmd* exon 23	*Dmd* knockout mice	[27]
		AAVrh.74	Exon deletion	SaCas9	*Dmd* nonsense mutation in exon 23	*Dmd* introns 20-23	*mdx* mice	[25]
	Dual AAVs	AAV9	Exon deletion	SaCas9	*Dmd* nonsense mutation in exon 23	*Dmd* introns 22-23	*mdx; Ai9* mice	[28]
		AAV9	Exon skipping	SpCas9	*Dmd* exon 44 deletion	*Dmd* splice site of exon 45	*Dmd* ΔEx44 mice	[19]
		AAV9	Exon skipping	SpCas9	*DMD* exon 50 deletion	*DMD* splice site of exon 51	Canine model of DMD	[21]
		AAV9	Exon deletion	SpCas9	*Dmd* nonsense mutation in exon 23	*Dmd* introns 22-23	*mdx* mice	[29]
		AAV8	Exon deletion	SaCas9	*Dmd* nonsense mutation in exon 23	*Dmd* introns 22-23	*mdx* mice	[24]
	Trans-splicing AAVs	AAV9	Base editing	ABE7.10	*Dmd* nonsense mutation in exon 20	*Dmd* exon 20	*Dmd* knockout mice	[30]
	Intein-split AAVs	AAV9	Exon deletion	SpCas9	*DMD* exon 52 deletion	*DMD* introns 50-51	*DMD* ΔEx52 pig	[32]

(continued)

Table 2 (continued)

Origin	Types of vector	Tropism or generation	Strategies	Nucleases	DMD mutation	Therapeutic target gene region(s)	Model(s)	References
AdV	AdV	First-generation (E1and E3 deletion)	Exon deletion	SpCas9	*Dmd* nonsense mutation in exon 23	*Dmd* introns 20-23	*mdx* mice	[36]
		Second-generation (E1 and E2A deletion)	Exon deletion	SpCas9	*DMD* exons 45–52 deletions, exons 48–50 deletions	*DMD* introns 52-53, introns 43-54	Human DMD myoblasts	[37]
			Exon skipping	SpCas9	*DMD* exons 45–52 deletions, exons 48–50 deletions	*DMD* splice site of exon 51, exon 53	Human DMD myoblasts	[38]
	HCAdV	Third-generation	Exon deletion	SpCas9	Wild type	*DMD* introns 50-51	Human DMD myoblasts	[40]
				eSpCas9	*DMD* exons 45–52 deletions, exons 48–50 deletions	*DMD* introns 43-54	Human DMD myoblasts	[42]
LV	LV	Third-generation	Exon deletion	SpCas9	*Dmd* nonsense mutation in exon 23	*Dmd* introns 22-23	Mouse *Dmd* iPSCs	[44]
				SaCas9	*DMD* exons 49–50 deletions, exons 50–52 deletions, exons 51–53 deletions, exons 51–56 deletions	*DMD* exons 47–58	Human DMD myoblasts	[45]
				SpCas9	*DMD* exon 2 duplications	*DMD* intron 2 to duplicated intron 2	Human DMD myoblasts	[46]
				SpCas9	*DMD* exons 18–30 duplications	*DMD* intron 27 to duplicated intron 27	Human DMD myoblasts	[47]
			Integration of transgene		*DMD* exons 45–52 deletions, exons 48–50 deletions	*DMD* exon 51, exon 53	Human DMD myoblasts	[38]

Abbreviations: *AAV* adeno-associated viral vector, *AdV* adenoviral vector, *LV* lentiviral vector, *ABE* adenine base editor, *hiPSC* human induced pluripotent stem cell protein with multiple domains, such as Cas9, Cas12a, and C2C2. Cas protein derived from *Streptococcus pyogenes* (SpCas9) is the most widely used and heavily engineered Cas protein, which belongs to class 2, type II. SpCas9 recognizes 'NGG' or 'NAG' PAMs located at the 3′ end of a protospacer. After targeting recognition, SpCas9 generates double-strand breaks (DSBs) at a site 3 bp upstream of the PAM. Small Cas orthologs, such as *Staphylococcus aureus* Cas9 (SaCas9), *Campylobacter jejuni* Cas9 (CjCas9), and CRISPR from *Prevotella* and *Francisella 1* (Cpf1) respectively, recognize the 'NNGRRT', 'NNNVRYAC', and 'TTTV' PAMs. These endonucleases are smaller than SpCas9; thus, sequences encoding them can be efficiently packaged into small viral vectors with comparable editing efficiency with SpCas9 but with higher specificity [11–13].

Recently, various types of base editors, which are composed of a nickase or dead form of Cas protein fused to a deaminase enzyme, have been engineered to edit specific nucleotides without generating DSBs. The first representative example is cytosine base editors (CBEs), which consist of rat APOBEC1 deaminase fused to nickase Cas9 (nCas9) containing a D10A mutation [14]. After nCas9 nicks the non-edited strand, APOBEC1 converts cytidines (C) into uracils (U), which are then converted to thymines (T). To increase editing efficiency, cellular uracil DNA-glycosylase inhibitor, which inhibits cellular base-excision repair, is also included in the fusion protein [14]. Adenine base editors (ABEs), which convert adenines (A) into guanines (G), consist of a heterodimer of the *Escherichia coli* (*E. coli*) tRNA adenosine deaminase (TadA and TadA*) fused with nCas9(D10A) [15]. C-to-G base editor consists of nCas9(D10A), an *E. coli*-derived uracil DNA N-glycosylase and a rat APOBEC1 cytidine deaminase variant (R33A) [16].

These base editors can convert specific bases for precise editing at the target site yet still have limitations, such as the inability to correct deletions, insertions, or some classes of point mutations. More recently, prime editors were developed to allow more diverse editing. Prime editor consists of nCas9(H840A) and an engineered Moloney murine leukemia virus reverse transcriptase (RT) domain [17]. nCas9 guided by pegRNA including desired edits generates a nick in the DNA strand and then RT processes the reverse transcription of the RT template for precise editing at the target locus.

3 Strategies for CRISPR-mediated DMD Mutation Correction

Exon skipping. Approximately 83% of DMD cases are caused by genomic frame-shift deletions that juxtapose out-of-phase exons in the *DMD* gene. Abolishing conserved RNA splice sites to inhibit the splicing of specific exons, called exon skipping, can juxtapose in-phase exons to restore ORFs [18]. When CRISPR-induced NHEJ occurs at an exon junction, the conserved RNA splice donor site or splice acceptor site flanking the out-of-frame exon is abolished. Splicing to the next available exon leads to exclusion of the target exon. As a result, a semi-functional,

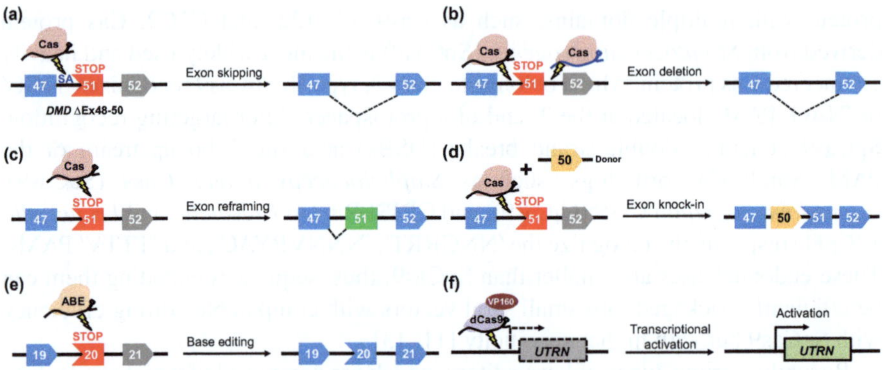

Fig. 1 Mechanisms of clustered regularly interspaced short palindromic repeat (CRISPR)-mediated Duchenne muscular dystrophy (DMD) correction. (**a-d**) A representative example of a frame-shifting mutation in the *DMD* gene (generated by deletion of exons 48–50 from the wild type sequence) and therapeutic strategies to correct the mutation. (**a**) Exon skipping induced by disruption of a splice acceptor (SA) site. (**b**) Exon deletion using two gRNAs targeting intronic regions flanking the targeted exon. (**c**) Exon reframing by nonhomologous end joining in the mutated exon. (**d**) Precise correction by homology directed repair-mediated exon knock-in. (**e**) Base editing by an adenine base editor to correct a nonsense mutation in the *Dmd* exon 20. (**f**) CRISPRa-mediated upregulation of utrophin expression

short dystrophin protein is expressed, converting the severe DMD phenotype into a milder BMD phenotype. With this strategy (Fig. 1a), efficient exon skipping with therapeutic outcomes has been demonstrated in human induced pluripotent stem cells (iPSCs) and myoblasts from DMD patients [19] as well as from mouse [20] and canine models of DMD [21].

Exon deletion. Exon deletion approaches can be used to restore disrupted ORFs. Approximately 65–72% of DMD patients have deletions of one or more exons that disturb the *DMD* gene ORF. Additional deletion of one or more exons using CRISPR systems can restore disrupted ORF. Moreover, deletion of a mutation hotspot spanning exons 45–55 could treat approximately 60% of DMD patients. For precise exon excision, a pair of sgRNAs targeting regions flanking exon(s) of interest can be designed. After CRISPR-mediated DSBs occur at those sites, the internal region is lost, and newly adjacent in-frame exons are spliced together (Fig. 1b). As a result, truncated but partially functional dystrophin protein can be expressed. Exon duplication mutations, which occur in ~5% of DMD patients, can also be corrected by the same strategy with an advantage of only requiring a single sgRNA targeting the intron region adjacent to the duplicated exon.

Exon reframing. Most DMD patients have deleterious mutations that interrupt *DMD* ORF. CRISPR-induced NHEJ has been used to introduce indels into a mutated exon to reframe the ORF (Fig. 1c). Of note, frameshift mutations can be reframed correctly with a one-in-three probability when small indels are generated through NHEJ upstream of the premature stop codons.

Exon knock-in. NHEJ-mediated exon reframing is limited because it cannot be applied to mutations located within exons encoding essential dystrophin domains.

In contrast, CRISPR-mediated HDR can accurately edit a mutated *DMD* gene to generate a wild type sequence, resulting in the expression of full-length dystrophin protein. HDR requires simultaneous delivery of Cas9, sgRNAs targeted to regions flanking the mutated region, and a single-stranded oligodeoxynucleotide or a donor template plasmid containing the desired sequences. Donor DNA is incorporated at the induced DSBs (Fig. 1d). This strategy offers precise repair, yet it has obvious disadvantages, such as it is not applicable to G1-arrested cells, repair is limited to the length of the donor template, integration of an unwanted DNA fragment is possible, and the frequency of HDR is lower than that of NHEJ. Prime editing via reverse transcription, the latest technology, has substantial potential to overcome these limitations.

Base editing. Base editing is a promising strategy for DMD therapy because it involves precise editing of a single base in the target site without generation of DSBs. Without the requirement for NHEJ-mediated repair, precise base editing of a single base in the targeted genome can be used to correct point mutations in the *DMD* gene, which account for approximately 27% of DMD cases (Fig. 1e) or disrupt splice sites for targeted exon skipping and ORF reframing. Base editing is also more efficient and generates lower frequencies of off-target effects than HDR.

Transcriptional modulation. Compensating for the lack of dystrophin function by modulating the expression level of other proteins with similar functions could be a novel strategy for DMD treatment. Using a CRISPR-mediated epigenetic editing approach to control protein expression has some advantages: it could be applied to all DMD patients regardless of which mutations are present and does not generate DNA DSBs, which can lead to unwanted indels, unlike techniques involving functional Cas9 nuclease. This strategy has been used to upregulate gene expression such as utrophin (Fig. 1f) and laminin or downregulate klotho and DUX4 by using catalytically inactivated or dead Cas9 with a transcriptional activator or repressor, called CRISPR activator [22] or CRISPR interference respectively [23].

4 Adeno-Associated Viral Vector-based Delivery of CRISPR/Cas

AAV vectors. AAV vectors (AAVs) are based on natural adeno-associated virus, which is a single-stranded, nonpathogenic, DNA parvovirus with an ability to efficiently transduce in both dividing and nondividing cells (such as myofibers and cardiomyocytes). These vectors are among the most efficient gene delivery vehicles and are extensively used for various gene therapies. Owing to long-term transgene expression without integrating into the host genome (they primarily exist episomally in circular head-to-tail concatemer), the expression of CRISPR components in AAV vectors-treated muscle has been shown to persist in vivo for several years, resulting in accumulated gene-edited transcripts with long-term therapeutic effects [24, 25]. The target cell preference depends on the AAV serotypes. In particular, AAV serotypes 8 and 9 exhibit high tropism for cardiac and skeletal muscles [26].

Therefore, AAV vectors with high muscle tropism have been utilized as vehicles to transfer CRISPR components to correct or bypass DMD mutations (Fig. 2). With respect to vector selection, the excellency of long-term expression of AAV-delivered transgenes is offset by the limited packaging capacity (<4.7 kb). To overcome the small cargo capacity of AAVs, various approaches to transferring CRISPR components are in development.

All-in-one AAV-CRISPR. Although SpCas9 (encoded by 4.10 kb sequence) is the most wildly used Cas nucleases, sequences encoding both SpCas9 and a sgRNA together are too large to be packaged in a single AAV. As an alternative to SpCas9, smaller Cas9 orthologs such as SaCas9 (encoded by 3.19 kb) or CjCas9 (encoded by 2.95 kb) can be carried in an all-in-one AAV. The gene encoding CjCas9 has been packaged into a single AAV serotype 9 together with sgRNA targeting *Dmd* exon 23 (Fig. 2a). Intramuscular injection of AAV9-CRISPR (AAV9-CjCas9-sgRNA) into the tibialis anterior (TA) muscles (5×10^{11} vg/muscle) of *Dmd* knockout mice that carry a frameshift mutation in exon 23-induced indels with a frequency of up to 8% and improved dystrophin levels to 39% of the wild type levels with no observable off-target effects [27]. Given that AAV exhibits long-term transgene expression, in vivo therapeutic effects induced by CRISPR have been shown up to 19 months post-injection [25]. In this study, delivery of two AAV-CRISPRs (AAVrh.74-SaCas9-sgRNA targeting intron 20 + AAVrh.74-SaCas9-sgRNA targeting intron 23) into neonatal *mdx* mice (a model of DMD harboring a nonsense mutation in *Dmd* exon 23) induced precise excision of exons 21–23 with a frequency of 76.3%, leading to restoration of dystrophin levels to 11.1% of wild type level and improved cardiac function at 19 months of age [25]. Treated *mdx* mice did not show any detectable signs of toxicity or immunity against AAV-CRISPR for time periods up to 19 months indicating that a significant host response against this system was avoided by administration in early life [25]. These studies demonstrate the long-term therapeutic effects of CRISPR by using an AAV delivery system to treat DMD.

Dual AAV-CRISPR. To overcome the small cargo capacity of AAVs, Cas9 and sgRNA can be delivered using dual AAVs. This system has been utilized for exon deletion or skipping. One of these studies reported that use of dual AAVs resulted in higher efficiency in genome editing than a single AAV platform [28]. Intramuscular delivery of SaCas9 coupled with paired gRNAs targeting regions flanking *Dmd* exon 23 by two AAV-CRISPR (AAV9-SaCas9 + AAV9-*Dmd*23 gRNAs targeting *Dmd* intron 22 and intron 23; 7.5×10^{11} vg each) resulted in excision of the targeted exon with a frequency of 39% in TA muscles of *mdx* mice harboring a nonsense mutation in *Dmd* exon 23, whereas treatment of a single vector rendered lower efficiency than dual AAVs [28]. Furthermore, systemic injection of *mdx* mice with dual AAV-CRISPR induced exon 23-deleted dystrophin transcripts at a frequency of up to ~18% in multiple skeletal muscles and cardiac muscles, demonstrating the potential value of dual AAVs for CRISPR delivery *in vivo* for DMD therapy [28].

With the exon skipping strategy, disrupted ORF can be corrected, as demonstrated previously. As an example of the study, two AAVs were generated to target the exon 45 splice acceptor site in DMD mice harboring a *Dmd* exon 44 deletion (Fig. 2b). Delivery of AAV-CRISPR (AAV9-SpCas9 and AAV9-sgRNA, 5×10^{10} vg/

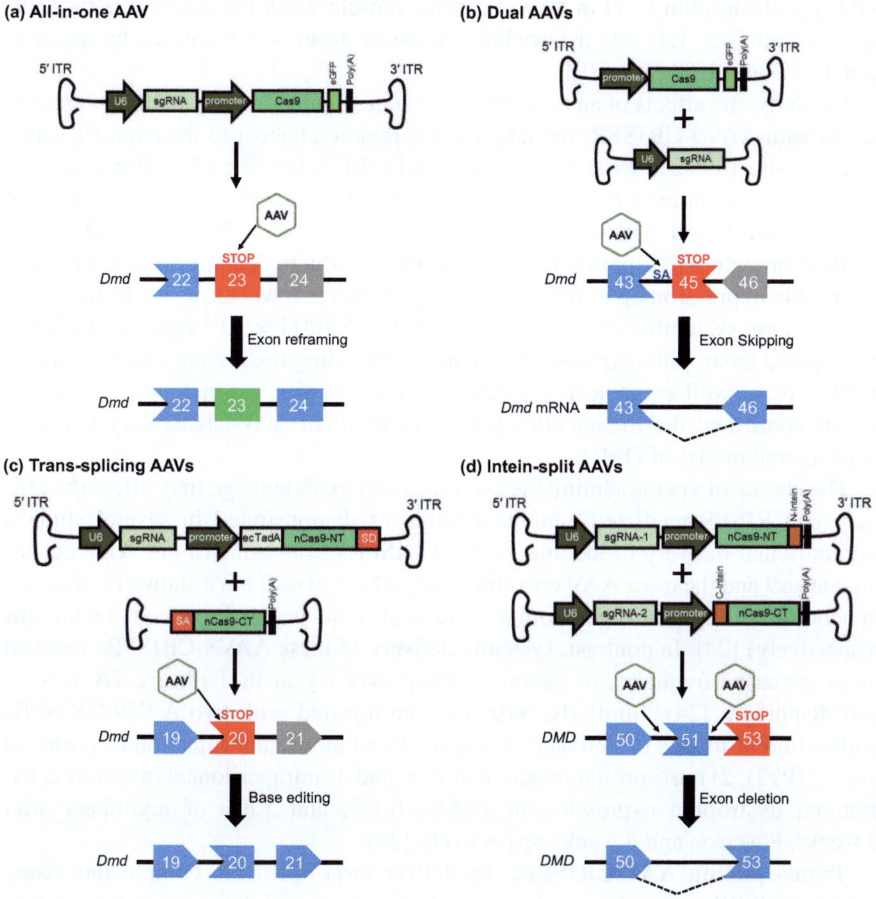

Fig. 2 Summary of AAV-mediated clustered regularly interspaced short palindromic repeat (CRISPR) systems used for treating Duchenne muscular dystrophy (DMD). Four types of recombinant adeno-associated virus (AAV) vectors encoding CRISPR components; (**a**) all-in-one AAV encoding Cas9 and sgRNA targeting the *Dmd* exon 23 for reframing of the open reading frame. (**b**) dual AAVs consisting of one AAV encoding Cas9 and the other AAV encoding a sgRNA targeting the splice acceptor (SA) for *Dmd* exon 45 skipping. (**c**) trans-splicing AAVs consisting of one AAV encoding engineered *E. coli* TadA (ecTadA), conjugated to the first half of nCas9 with splicing donor (SD) and sgRNA targeting *Dmd* exon 20 containing a nonsense mutation and the other AAV encoding the second half of nCas9 with SA for adenine base editing. (**d**) intein-split AAVs consisting of one AAV encoding the first half of Cas9 and sgRNA targeting intron 50 and the other AAV encoding the second half of Cas9 and sgRNA targeting intron 51 for exon 51 deletion. ITRs, inverted terminal repeats

muscle) resulted in exon 45 skipping with a frequency of 9.8%, leading to ORF reframing and restoration of dystrophin levels to 74% of wild type levels in treated TA muscles [19]. In addition, to find the optimal dose of gene editing components, 1:1, 1:5, and 1:10 ratios of AAV9-Cas9:AAV9-sgRNA were tested with the dose of AAV-Cas9 fixed at 5×10^{13} vg/kg [19]. The dystrophin level reached 82% of the wild type dystrophin level in treated cardiac muscle when the AAV9-Cas9:AAV9-sgRNA ratio was 1:1; this dystrophin expression level was increased by an additional 12% at a 1:10 ratio [19].

Similarly, the effects of an exon 50 deletion in a canine model of DMD were corrected with AAV9-CRISPR that targeted the region adjacent to the exon 51 splice acceptor site to cause exon 51 skipping and ORF reframing [21]. Intramuscular injection of the cranial tibialis muscles of DMD dogs with two AAV9-CRISPR (AAV9-SpCas9 and AAV-sgRNA targeting exon 51; 1.2×10^{13} vg/muscle each) resulted in exon 51 skipping in 73% of cDNA products, leading to restoration of dystrophin expression up to 60% of wild type levels at 6 weeks post-injection [21]. Furthermore, systemic delivery of the AAV9-CRISPR (1×10^{14} vg/kg each) led to widespread dystrophin expression in muscles including the cranial tibialis, triceps, and biceps, as well as the heart in treated dogs at 8 weeks post-injection [21]. These results confirmed the therapeutic efficacy of the dual AAV-CRISPR system in a large animal model of DMD.

The choice of vector administration routes and treatment age may affect the efficacy of CRISPR-mediated genome editing, as demonstrated in several studies. Intramuscular delivery of the dual AAV8-CRISPRs, consisting of one AAV encoding SaCas9 and the other AAV encoding two gRNAs in *mdx* mice showed a decrease in genome editing over time (from 8% to 2% of frequency at 8 weeks and 6 months respectively) [24]. In contrast, systemic delivery of these AAV8-CRISPRs resulted in an increased frequency of genome editing over 1 year in the heart, TA muscle, and diaphragm [24]. Similarly, *mdx* mice co-injected with two AAV9-CRISPRs with different modes of delivery, including 1) intramuscular injection at postnatal day 12 (P12), 2) retro-orbital injection at P18, and 3) intraperitoneal injection at P1, restored dystrophin expression in 25.5%, 6.1%, and 1.8% of myofibers after 6 weeks, 8 weeks, and 8 weeks respectively [29].

Trans-splicing AAV-CRISPR. To deliver large CRISPR components using AAVs, CRISPR molecules can be split into two parts and delivered independently in trans-splicing AAV systems. Each trans-splicing AAV carries one of the partial sequences with appropriate splice signal sequence in an AAV. In one example of *DMD* base editing using trans-splicing AAVs, ABEs have been used to correct a nonsense mutation in exon 20 of the *Dmd* gene by converting a stop codon (***T***AG) into a Gln (***C***AG) [30]. Because ABE is relatively large, a delivery method involving a trans-splicing AAV9 encoding ABE has been developed (Fig. 2c). The 6.1-kb sequence encoding one version of ABE (ABE7.10, a TadA-TadA*-nSpCas9 fusion), was split into two parts to overcome the packaging limits of AAV. A donor vector that carries the 5′ fragment of the ABE gene conjugated to a splicing donor together with the sgRNA. An acceptor vector carries the 3′ fragment of the ABE gene fused to a splicing acceptor. After co-infection of single cells in the TA muscles of a DMD mouse with trans-splicing AAV9 encoding ABE, one full length ABE mRNA would be produced following head-to-tail intermolecular recombination between the two independent viral genomes and subsequent splicing across the inverted terminal repeat (ITR) junction. The base editing efficiency (A>G) in the treated mouse was 3.3%, leading to an increase in the dystrophin expression level to 17% of wild type dystrophin with co-localization of neuronal nitric oxide synthase at the sarcolemma. This study demonstrates that the trans-splicing AAV system successfully carried the ABE into muscles for *Dmd* editing in the genome, demonstrating the therapeutical potential of base editing in a DMD animal model [30]. Additionally, HNHx-ABE is a more compact version of ABE in which the SpCas9-HNH domain is replaced with TadA and can be an option in conjunction with a single AAV vector system for treating DMD.

Intein-mediated split AAV-CRISPR. Intein-mediated split AAV system is another potential vehicle for the delivery of large CRISPR components. An intein is a segment of protein that has the ability to self-excise from a protein without leaving residual amino acids in the final product, similar to an intron being excised from an RNA [31]. Split-inteins (N-intein and C-intein polypeptides at the ends of two proteins) catalyze protein trans-splicing to generate one longer protein. This ability is harnessed in intein-mediated trans-splicing AAV, which has been used to deliver large CRISPR components for in vivo gene editing of a pig model of DMD [32]. Intramuscular delivery of AAV9 carrying an intein-split Cas9 and a pair of gRNAs targeting sequences flanking exon 51 (2×10^{13} vg/kg each) induces expression of truncated dystrophin, leading to improvement of muscle function (Fig. 2d).

5 Adenoviral Vector-based Delivery of CRIPSR/Cas

Adenoviral vectors. Adenoviral vectors (AdVs) are alternative delivery vehicles that have been extensively used for therapeutic applications. Owing to the large gene capacity of the AdVs, it can deliver CRISPR components in a single vector system. Cell entry by AdV generally involves binding to a cell receptor, such as the coxsackievirus and adenovirus receptors (CAR) or the ubiquitous cell receptor CD45, which is recognized by adenoviral fiber motifs [33]. There are more than 50 different human serotypes with six groups (A to F) that have been identified [34]. Among them, serotype 2 and 5 belonging to group C are the most widely used as gene delivery vectors because of their high nuclear transfer efficiency, low pathogenicity, and broad tissue tropism [34] including dividing, nondividing, and quiescent primary cells, which broaden the possibilities for CRISPR-based DMD therapy (Fig. 3).

First-generation AdVs with a deletion of the E1 viral gene can transfer cDNA cassettes of up to 5.1 kb in length, whereas E1/E3 deleted AdV can package up to 8.2 kb, which has enough capacity to carry the Cas9 gene and a gRNA expression cassette in a single vector. However, these vectors are highly immunogenic because of the viral capsid and viral gene expressions, which may cause the loss of the transgene expression through immune responses. To attenuate the chronic immunogenicity, second-generation AdVs have been created by deletion of the E1 and E2 or E4 viral genes, rendering an increased packaging capacity of up to 14 kb. Furthermore, helper-dependent or "gutless" AdVs have been developed as third-generation with high-capacity AdVs (HC-AdVs) and devoid of all viral genes. HC-AdVs contain

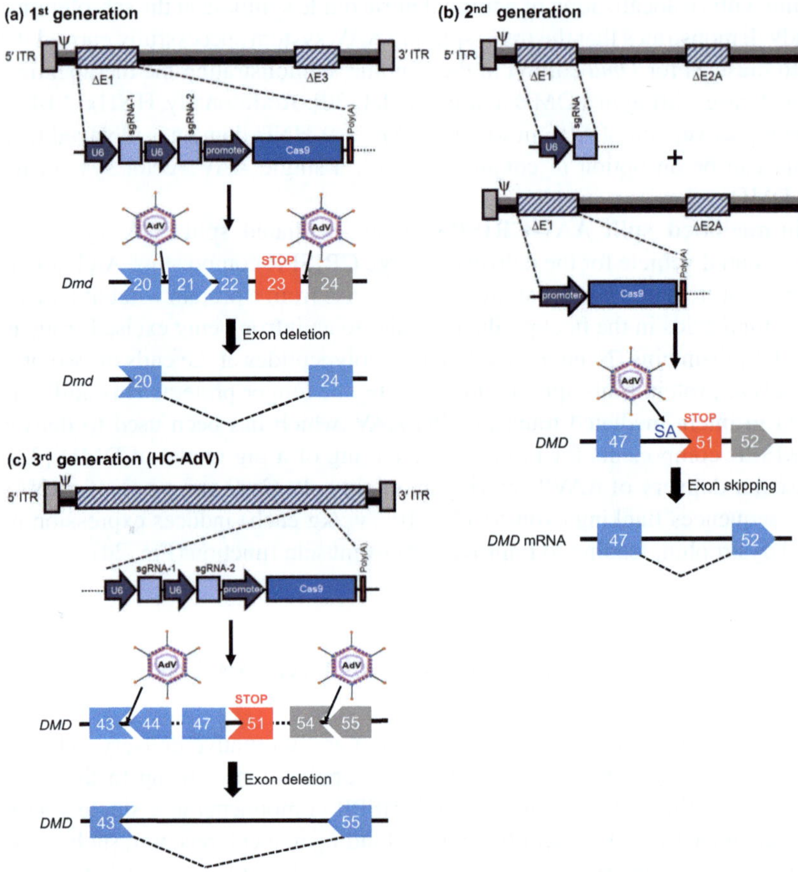

Fig. 3 Three types of recombinant adenoviral vectors (AdVs) encoding CRISPR components; (**a**) first-generation adenoviral vectors (AdVs) encode Cas9 and two sgRNAs targeting intronic regions flanking *Dmd* exons 21–23. (**b**) second-generation AdVs consisting of one AdV encoding Cas9 and the other AdV encoding sgRNA targeting the splice acceptor (SA) for *DMD* exon 51 skipping, (**c**) third-generation AdVs encode Cas9 and two sgRNAs targeting the sites flanking *DMD* exons 44–54 for exon deletion. Ψ, packaging signal; ITRs, inverted terminal repeats.

only the ITR and packaging signal, increasing the packaging capacity up to 36 kb, enabling large CRISPR components to be delivered in a single vector. Another important advantage of AdVs as a delivery vector for CRISPR-based gene editing is the non-integration of the viral genes into the host genome, which can avoid unwanted and sustained editing at off-target sites. In addition, these vectors do not elicit chronic immune responses owing to the lack of viral genes; however, acute immune responses elicited by the viral capsid are a major issue [35].

AdV-CRISPR. To rescue the lack of dystrophin protein in *mdx* mice, AdVs encoding SpCas9 and sgRNA have been utilized. With the aim of targeted exon(s) deletion, a first-generation AdV serotype 5 containing E1 and E3 deletions was modified to encode either SpCas9 or two sgRNAs targeting intronic regions flanking *Dmd* exons 21 to 23 (Fig. 3a). The AdVs were co-injected into the gastrocnemius muscles of neonatal *mdx* mice, and treated muscles exhibited exons 21–23 deletions (23 kb) at 3 weeks post-injection, which led to restored dystrophin expression in 50% of muscle fibers [36]. Unlike first-generation AdVs, second-generation AdVs can infect the human muscle progenitor cells and nonmuscle cells with myogenic capacity expressing CD46 but not CAR, with low levels of "leaky" viral gene expression [33]. With these advantages for efficient transduction into DMD muscle cells, the "all-in-one" second-generation AdV particles with E1 and E2A deletion were constructed to encode SpCas9 and sgRNAs targeting intronic regions to excise mutated *DMD* exon 53 or exons 44–54 [37]. Upon transduction into patient-derived myoblasts (DMDΔ45–52 and DMDΔ48–50), these AdV constructs induced excision of the targeted exon(s) with a frequency of up to 13% and 18% at 7 days post-transduction respectively, at multiplicity of infection (MOIs) of 50 IU/cell and 75 IU/cell, leading to robust dystrophin expression in differentiated myoblasts [37]. This study indicates that AdV is a powerful platform for delivering the CRISPR system to patient-derived DMD cells.

DMD splice sites have also been targeted with AdV-CRISPR to modulate splicing [38]. The AdV-Cas9 transduction at MOIs ranging from 25 to 200 IU/cell in a DMDΔ48–50 cell line that stably expressed sgRNAs targeting splicing acceptor sites adjacent to exons 51 introduced indels at the exon 51 splice acceptor site, leading to exon 51 skipping (Fig. 3b). As a result, exons 47 and 52 were joined. Similarly, DMDΔ45–52 cells that stably expressed sgRNAs targeting splicing acceptor sites adjacent to exon 53 were modified such that exon 53 was skipped, resulting in juxtaposition of exons 44 and 54 in the mRNA and expression of truncated dystrophin protein.

High-capacity adenoviral vectors (HC-AdVs). Because the packaging capacity of HC-AdV is up to 36 kb, it is possible to deliver not only CRISPR but also additional genes. Despite these advantages, CRISPR delivery with HC-AdVs has not been widely applied owing to the complicated handling required for large constructs and the time-intensive procedure necessary for vector production. To overcome these limitations, intermediate shuttle plasmids for fast cloning of sgRNA and subsequent insertion of all CRISPR components into the HC-AdV genome were generated in a bacterial artificial chromosome [39] or the established pAdFTC plasmid [41]. Using this system, HC-AdV-CRISPR targeting intronic regions flanking *DMD* exon 51 was generated to treat DMD. This construct led to successful deletion of exon 51 at frequencies of 5.9% (at an MOI of 1) to 93.3% (at an MOI of 10) in primary human skeletal myoblasts [40]. At high doses, HC-AdV-CRISPR showed comparable genome editing efficiency but less cytotoxicity than second-generation AdVs [42]. HC-AdV-CRISPR targeting intronic regions flanking exons 43 to 54 also induced deletion of the *DMD* mutation hotspot, leading to correction of ORFs with a frequency of up to 42% in DMDΔ48–50 myoblasts (Fig. 3c) [42]. These studies showed that HC-AdV is an effective tool for delivering CRISPR components for gene editing in immortalized primary myoblasts with low cytotoxicity.

6 Lentiviral Vector-based Delivery of CRISPR/Cas

Lentiviral vectors. Lentiviruses are spherical, single-stranded RNA viruses belonging to the *Retroviridae* family that reverse transcribe RNA to DNA. Lentiviral vectors (LVs) have been generated by deleting the pre-viral genome of a parent lentivirus, most often human immunodeficiency virus type 1 (HIV-1). Thus, the first-generation LVs were replication incompetent but maintained the ability to integrate into the host genome. LVs can accommodate up to 8 kb of exogenous DNA, a capacity that is larger than that of AAVs, making it possible to deliver CRISPR components in a single LV (Fig. 4). Similar to AAV, LVs lead to long-term expression of transgene in both dividing and nondividing cells. Thus far, LVs have been harnessed primarily for ex vivo delivery of the CRISPR components and for creating libraries of gene knockouts by delivering a pool of sgRNAs to Cas9-expressing cell lines. However, the ability to integrate the viral genes into the host genome raises concerns for potential genotoxicity, immunogenicity, and accumulated off-target effects by prolonged expression of Cas nucleases. To address these issues, integration-defective LVs have been utilized to deliver CRISPR components, exhibiting efficiency equal to that of integration-competent LVs in vivo, despite the reduced gene editing efficiency in vitro. Second-generation LVs have been engineered by deleting accessory genes in packaging vector and utilizing a split-plasmid system for improved safety [43]. The third-generation system further improves on the safety of the second-generation LVs by split-packaging two plasmids, one encoding Rev and one encoding Gag and Pol.

LV-CRISPR. LV-CRISPR has therapeutic potential in that it modifies *DMD* genes in patient-derived myoblasts for treating DMD. In particular, third-generation LV-mediated CRISPR delivery to treat DMD has been demonstrated in several in vitro studies [38, 44–47]. To take advantage of the ability of LVs to integrate transgenes into the host genome, LV-CRISPR was utilized to generate cell lines that stably expressed the sgRNA targeting *DMD* exon 51 or 53 in patient-derived DMDΔ48–50 and DMDΔ45–52 myoblasts respectively (Fig. 4a) [38]. Treatment of an AdV-Cas9 in these cell lines resulted in indel formation in exon 51 or 53 with frequencies of 8% to 53% and expression of truncated but semi-functional dystrophin protein. Similarly, the delivery of LVs carrying SaCas9 and two sgRNAs targeting exons 47 and 58 in DMD patient-derived myoblasts corrected the reading frame, in which 73% of hybrid exon contained the correct junction sequences [45]. LV-CRISPR has also been used to treat DMD by removing duplication mutations. DMD patient-derived myoblasts containing a large duplication of exons 18–30 were transduced, with LV-CRISPR targeting the *DMD* intron 27 and duplicated intron 27 (Fig. 4b). It resulted in the removal of a duplication of DMD exons 18–30 in myotubes, leading to an increase in the dystrophin expression level to 4.4% of wild type dystrophin [47].

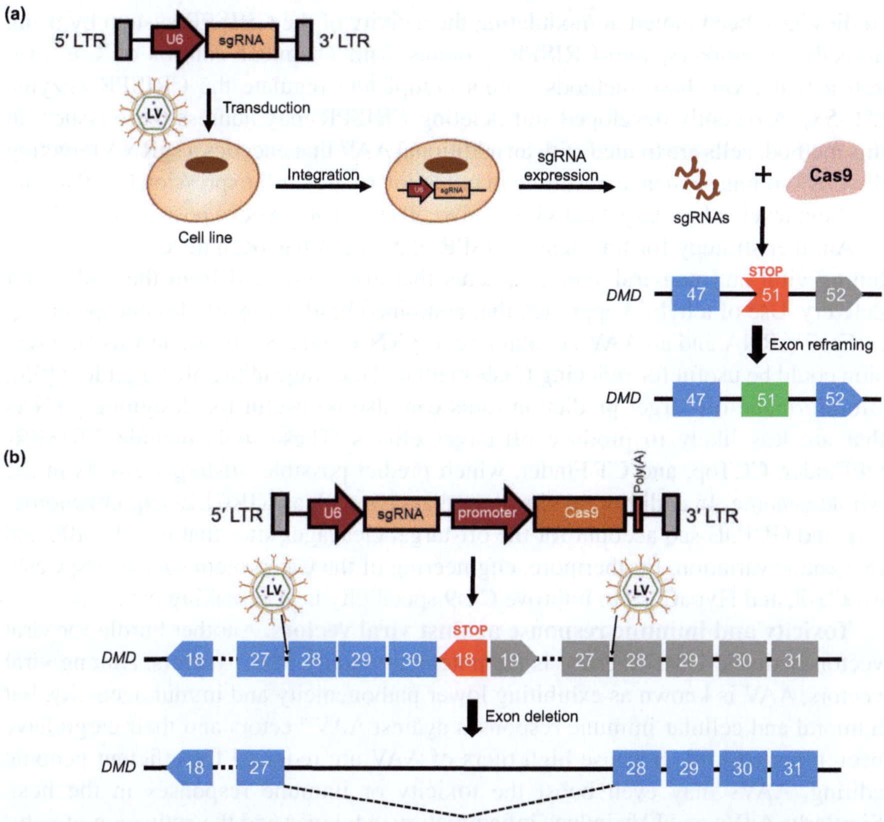

Fig. 4 Recombinant lentiviral vectors (LVs) encoding CRISPR components; (**a**) Generation of a cell line expressing a sgRNA targeting the *DMD* exon 51 by stable integration of a sgRNA-encoding LVs. (**b**) LVs encoding Cas9 and a sgRNA targeting the intronic regions of duplicated exons for exons 18–30 deletion. LTR, long terminal repeats

7 Challenges

Off-target effects. Viral vector-mediated delivery of the CRISPR system has great therapeutic potential, given its high delivery efficiency and ability to result in long-term stable expression of CRISPR components; however, the potential for generating CRISPR-mediated off-target mutations in the genome must be intensively investigated. CRISPR-mediated off-target effects are a well-known issue for in vitro and in vivo genome editing [48]. Despite no accumulative off-target effects being described in the AAV-CRISPR treated mice over 14 months [49], long-term Cas9 expression raises safety concerns. Off-target editing at nontargeted loci may cause genomic instability and interfere with normal gene functionality, raising the risk of targeted cells being tumors [50]. Therefore, efforts have been made to develop methods for the transient or conditional expression of CRISPR components. Several studies have been aimed at modulating the activity of the CRISPR system by using inducible promoters, anti-CRISPR proteins, and chemical control of Cas protein activity, yet these methods cannot completely regulate the CRISPR enzyme [51–53]. A recently developed self-deleting CRISPR may address these issues. In this method, cells are treated with an additional AAV that encodes a sgRNA targeting the Cas9 coding sequence, resulting in reduction of the Cas9 expression by 79%, and has high levels of on-target activity in vivo observed at 4 weeks post-injection [54].

Another strategy for transient CRISPR expression has been developed by combining viral and nonviral vector systems that are eliminated from the body after delivery. Use of a hybrid approach that combined lipid nanoparticles encapsulating SpCas9 mRNA and an AAV8 containing a sgRNA to induce transient Cas9 expression could be useful for reducing Cas9-mediated cleavage at the off-target loci [55]. Bioinformatic off-target prediction tools can also be useful for designing gRNAs that are less likely to produce off-target effects. These tools include CRISPR-OFFinder, CCTop, and CT-Finder, which predict possible off-target effects in the whole genome. In addition, in vitro-based assays such as CIRCLE-seq, Digenome-seq, and GUIDE-seq account for the off-target cleavages sites that may be affected by genetic variation. Furthermore, engineering of the Cas protein such as eSpCas9, evoCas9, and HypaCas9 to improve Cas9 specificity is also making progress.

Toxicity and immune response against viral vectors. Another hurdle for viral vector-mediated DMD therapy is the potential toxicity of the vectors. Among viral vectors, AAV is known as exhibiting lower pathogenicity and immunogenicity, but humoral and cellular immune responses against AAV vectors and their cargo have been reported [56]. Because high titers of AAV are required for efficient genome editing, AAVs may even boost the toxicity or immune responses in the host. Similarly, AdVs and LVs induce inflammatory cytokines and the activation of natural killer cells and dendritic cells with associated antibody and T cell responses [35]. This phenomenon will likely result in stronger restrictions against using a viral vector-mediated CRISPR system in clinical trials. Therefore, safe dosages of viral vectors must be determined as a prerequisite to clinical trials that examine CRISPR-mediated therapies.

Immune response to CRISPR components. Immune responses against Cas protein or CRISPR RNAs represent a great barrier to the therapeutic success of CRISPR-based approaches. CRISPR components can elicit an immune response in vitro owing to their origin from external bacteria [57]. If a patient had been previously exposed to Cas proteins and re-exposed during CRISPR-based therapy, specific memory T cells against the Cas antigen may be activated, leading to the elimination of Cas endonuclease-containing cells. This process could trigger a serious immune response, greatly reduce the effectiveness of treatment, and lead to organ impairment. It has been reported that anti-Cas9 antibodies exist in 58–78% of healthy people, and T cells specific to this antigen exist in 67–78% of healthy people [58]. These results present the possibility that numerous DMD patients may exhibit severe immune responses against the Cas protein after treatment with CRISPR components, which could severely limit the application of CRISPR-based gene therapy. Several studies have been conducted with the aim of overcoming the immune response by decreasing the immunogenic properties of Cas protein through protein engineering [59], the use of Cas9-reactive regulatory T cells [60], or modification of CRISPR RNA by removing the 5'-triphosphate [57], but further studies are required.

8 Conclusions

Rapid progress is being made in the field of DMD genome editing because engineering of the CRISPR system and the development of viral vectors are diverse and continuously advancing. Although CRISPR systems offer therapeutic tools that are highly effective and precise for correcting the genome in vitro and in vivo, the long-term safety and efficacy of these strategies need to be fully evaluated in larger animal models prior to progression into clinical trials for viral vector-mediated CRISPR-based approaches to DMD gene therapy. In conclusion, the CRISPR system is indeed an innovative gene-editing tool with great potential for DMD therapy and will offer the most effective tool for gene therapy in the near future.

Acknowledgements This work was supported by grants from the National Research Foundation of Korea grant funded by the Korea government [2022R1A2C1013352 and 2020R1A4A1016142].

Conflicts of Interest The authors declare no conflicts of interest.

References

1. Neri, M., Torelli, S., Brown, S., Ugo, I., Sabatelli, P., Merlini, L., Spitali, P., Rimessi, P., Gualandi, F., Sewry, C., Ferlini, A., & Muntoni, F. (2007). Dystrophin levels as low as 30% are sufficient to avoid muscular dystrophy in the human. *Neuromuscular Disorders, 17*, 913–918. https://doi.org/10.1016/j.nmd.2007.07.005
2. Beggs, A. H., Hoffman, E. P., Snyder, J. R., Arahata, K., Specht, L., Shapiro, F., Angelini, C., Sugita, H., & Kunkel, L. M. (1991). Exploring the molecular basis for variability among patients with Becker muscular dystrophy: Dystrophin gene and protein studies. *American Journal of Human Genetics, 49*, 54–67.
3. Lim, K. R. Q., Maruyama, R., & Yokota, T. (2017). Eteplirsen in the treatment of Duchenne muscular dystrophy. *Drug Design, Development and Therapy, 11*, 533–545. https://doi.org/10.2147/DDDT.S97635
4. Heo, Y. A. (2020). Golodirsen: First approval. *Drugs, 80*, 329–333. https://doi.org/10.1007/s40265-020-01267-2
5. Roshmi, R. R., & Yokota, T. (2021). Pharmacological Profile of viltolarsen for the treatment of duchenne muscular dystrophy: A Japanese experience. *Clinical Pharmacology: Advances and Applications, 13*, 235–242. https://doi.org/10.2147/CPAA.S288842
6. Mendell, J. R., Campbell, K., Rodino-Klapac, L., Sahenk, Z., Shilling, C., Lewis, S., Bowles, D., Gray, S., Li, C., Galloway, G., Malik, V., Coley, B., Clark, K. R., Li, J., Xiao, X., Samulski, J., McPhee, S. W., Samulski, R. J., & Walker, C. M. (2010). Dystrophin immunity in Duchenne's muscular dystrophy. *The New England Journal of Medicine, 363*, 1429–1437. https://doi.org/10.1056/nejmoa1000228

7. Kim, Y. G., Cha, J., & Chandrasegaran, S. (1996). Hybrid restriction enzymes: Zinc finger fusions to Fok I cleavage domain. *Proceedings of the National Academy of Sciences of the United States of America, 93*, 1156–1160. https://doi.org/10.1073/pnas.93.3.1156
8. Smith, J., Grizot, S., Arnould, S., Duclert, A., Epinat, J. C., Chames, P., Prieto, J., Redondo, P., Blanco, F. J., Bravo, J., Montoya, G., Pâques, F., & Duchateau, P. (2006). A combinatorial approach to create artificial homing endonucleases cleaving chosen sequences. *Nucleic Acids Research, 34*, e149. https://doi.org/10.1093/nar/gkl720
9. Boch, J., Scholze, H., Schornack, S., Landgraf, A., Hahn, S., Kay, S., Lahaye, T., Nickstadt, A., & Bonas, U. (2009). Breaking the code of DNA binding specificity of TAL-type III effectors. *Science, 326*, 1509–1512. https://doi.org/10.1126/science.1178811
10. Jinek, M., Chylinski, K., Fonfara, I., Hauer, M., Doudna, J. A., & Charpentier, E. (2012). A programmable dual-RNA-guided DNA endonuclease in adaptive bacterial immunity. *Science, 337*, 816–821. https://doi.org/10.1126/science.1225829
11. Zetsche, B., Gootenberg, J. S., Abudayyeh, O. O., Slaymaker, I. M., Makarova, K. S., Essletzbichler, P., Volz, S. E., Joung, J., Van Der Oost, J., Regev, A., Koonin, E. V., & Zhang, F. (2015). Cpf1 is a single RNA-guided endonuclease of a class 2 CRISPR-Cas system. *Cell, 163*, 759–771. https://doi.org/10.1016/j.cell.2015.09.038
12. Ran, F. A., Cong, L., Yan, W. X., Scott, D. A., Gootenberg, J. S., Kriz, A. J., Zetsche, B., Shalem, O., Wu, X., Makarova, K. S., Koonin, E. V., Sharp, P. A., & Zhang, F. (2015). In vivo genome editing using Staphylococcus aureus Cas9. *Nature, 520*, 186–191. https://doi.org/10.1038/nature14299
13. Kim, E., Koo, T., Park, S. W., Kim, D., Kim, K., Cho, H. Y., Song, D. W., Lee, K. J., Jung, M. H., Kim, S., Kim, J. H., Kim, J. H., & Kim, J. S. (2017). In vivo genome editing with a small Cas9 orthologue derived from Campylobacter jejuni. *Nature Communications, 8*, 14500. https://doi.org/10.1038/ncomms14500
14. Komor, A. C., Kim, Y. B., Packer, M. S., Zuris, J. A., & Liu, D. R. (2016). Programmable editing of a target base in genomic DNA without double-stranded DNA cleavage. *Nature, 533*, 420–424. https://doi.org/10.1038/nature17946
15. Gaudelli, N. M., Komor, A. C., Rees, H. A., Packer, M. S., Badran, A. H., Bryson, D. I., & Liu, D. R. (2017). Programmable base editing of A•T to G•C in genomic DNA without DNA cleavage. *Nature, 551*, 464–471. https://doi.org/10.1038/nature24644
16. Kurt, I. C., Zhou, R., Iyer, S., Garcia, S. P., Miller, B. R., Langner, L. M., Grünewald, J., & Joung, J. K. (2021). CRISPR C-to-G base editors for inducing targeted DNA transversions in human cells. *Nature Biotechnology, 39*, 41–46. https://doi.org/10.1038/s41587-020-0609-x
17. Anzalone, A. V., Randolph, P. B., Davis, J. R., Sousa, A. A., Koblan, L. W., Levy, J. M., Chen, P. J., Wilson, C., Newby, G. A., Raguram, A., & Liu, D. R. (2019). Search-and-replace genome editing without double-strand breaks or donor DNA. *Nature, 576*, 149–157. https://doi.org/10.1038/s41586-019-1711-4
18. Kole, R., & Krieg, A. M. (2015). Exon skipping therapy for Duchenne muscular dystrophy. *Advanced Drug Delivery Reviews, 87*, 104–107. https://doi.org/10.1016/j.addr.2015.05.008
19. Min, Y. L., Li, H., Rodriguez-Caycedo, C., Mireault, A. A., Huang, J., Shelton, J. M., McAnally, J. R., Amoasii, L., Mammen, P. P. A., Bassel-Duby, R., & Olson, E. N. (2019). CRISPR-Cas9 corrects Duchenne muscular dystrophy exon 44 deletion mutations in mice and human cells. *Sci Adv, 5*, eaav4324. https://doi.org/10.1126/sciadv.aav4324
20. Amoasii, L., Long, C., Li, H., Mireault, A. A., Shelton, J. M., Sanchez-Ortiz, E., McAnally, J. R., Bhattacharyya, S., Schmidt, F., Grimm, D., Hauschka, S. D., Bassel-Duby, R., & Olson, E. N. (2017). Single-cut genome editing restores dystrophin expression in a new mouse model of muscular dystrophy. *Sci Transl Med, 9*, eaan8081. https://doi.org/10.1126/scitranslmed.aan8081
21. Amoasii, L., Hildyard, J. C. W., Li, H., Sanchez-Ortiz, E., Mireault, A., Caballero, D., Harron, R., Stathopoulou, T. R., Massey, C., Shelton, J. M., Bassel-Duby, R., Piercy, R. J., & Olson, E. N. (2018). Gene editing restores dystrophin expression in a canine model of Duchenne muscular dystrophy. *Science, 362*, 86–91. https://doi.org/10.1126/science.aau1549

22. Liao, H. K., Hatanaka, F., Araoka, T., Reddy, P., Wu, M. Z., Sui, Y., Yamauchi, T., Sakurai, M., O'Keefe, D. D., Núñez-Delicado, E., Guillen, P., Campistol, J. M., Wu, C. J., Lu, L. F., Esteban, C. R., & Izpisua Belmonte, J. C. (2017). In vivo target gene activation via CRISPR/Cas9-mediated trans-epigenetic modulation. *Cell, 171*, 1495–1507. https://doi.org/10.1016/j.cell.2017.10.025
23. Himeda, C. L., Jones, T. I., & Jones, P. L. (2016). CRISPR/dCas9-mediated transcriptional inhibition ameliorates the epigenetic dysregulation at D4Z4 and represses DUX4-fl in FSH muscular dystrophy. *Molecular Therapy, 24*, 527–535. https://doi.org/10.1038/mt.2015.200
24. Nelson, C. E., Wu, Y., Gemberling, M. P., Oliver, M. L., Waller, M. A., Bohning, J. D., Robinson-Hamm, J. N., Bulaklak, K., Castellanos Rivera, R. M., Collier, J. H., Asokan, A., & Gersbach, C. A. (2019). Long-term evaluation of AAV-CRISPR genome editing for Duchenne muscular dystrophy. *Nature Medicine, 25*, 427–432. https://doi.org/10.1038/s41591-019-0344-3
25. Xu, L., Lau, Y. S., Gao, Y., Li, H., & Han, R. (2019). Life-long AAV-mediated CRISPR genome editing in dystrophic heart Improves cardiomyopathy without causing serious lesions in mdx mice. *Molecular Therapy, 27*, 1407–1414. https://doi.org/10.1016/j.ymthe.2019.05.001
26. Zincarelli, C., Soltys, S., Rengo, G., & Rabinowitz, J. E. (2008). Analysis of AAV serotypes 1–9 mediated gene expression and tropism in mice after systemic injection. *Molecular Therapy, 16*, 1073–1080. https://doi.org/10.1038/mt.2008.76
27. Koo, T., Lu-Nguyen, N. B., Malerba, A., Kim, E., Kim, D., Cappellari, O., Cho, H. Y., Dickson, G., Popplewell, L., & Kim, J. S. (2018). Functional rescue of dystrophin deficiency in mice caused by frameshift mutations using Campylobacter jejuni Cas9. *Molecular Therapy, 26*, 1529–1538. https://doi.org/10.1016/j.ymthe.2018.03.018
28. Tabebordbar, M., Zhu, K., Cheng, J. K. W., Chew, W. L., Widrick, J. J., Yan, W. X., Maesner, C., Wu, E. Y., Xiao, R., Ran, F. A., Cong, L., Zhang, F., Vandenberghe, L. H., Church, G. M., & Wagers, A. J. (2016). In vivo gene editing in dystrophic mouse muscle and muscle stem cells. *Science, 351*, 407–411. https://doi.org/10.1126/science.aad5177
29. Long, C., Amoasii, L., Mireault, A. A., McAnally, J. R., Li, H., Sanchez-Ortiz, E., Bhattacharyya, S., Shelton, J. M., Bassel-Duby, R., & Olson, E. N. (2016). Postnatal genome editing partially restores dystrophin expression in a mouse model of muscular dystrophy. *Science, 351*, 400–403. https://doi.org/10.1126/science.aad5725
30. Ryu, S. M., Koo, T., Kim, K., Lim, K., Baek, G., Kim, S. T., Kim, H. S., Kim, D. E., Lee, H., Chung, E., & Kim, J. S. (2018). Adenine base editing in mouse embryos and an adult mouse model of Duchenne muscular dystrophy. *Nature Biotechnology, 36*, 536–539. https://doi.org/10.1038/nbt.4148
31. Mills, K. V., Johnson, M. A., & Perler, F. B. (2014). Protein splicing: How inteins escape from precursor proteins. *The Journal of Biological Chemistry, 289*, 14498–14505. https://doi.org/10.1074/jbc.R113.540310
32. Moretti, A., Fonteyne, L., Giesert, F., Hoppmann, P., Meier, A. B., Bozoglu, T., et al. (2020). Somatic gene editing ameliorates skeletal and cardiac muscle failure in pig and human models of Duchenne muscular dystrophy. *Nature Medicine, 26*, 207–214. https://doi.org/10.1038/s41591-019-0738-2
33. Maggio, I., Holkers, M., Liu, J., Janssen, J. M., Chen, X., & Gonçalves, M. A. F. V. (2014). Adenoviral vector delivery of RNA-guided CRISPR/Cas9 nuclease complexes induces targeted mutagenesis in a diverse array of human cells. *Scientific Reports, 4*, 5105. https://doi.org/10.1038/srep05105
34. Volpers, C., & Kochanek, S. (2004). Adenoviral vectors for gene transfer and therapy. *The Journal of Gene Medicine, 6*(S164), S171. https://doi.org/10.1002/jgm.496
35. Shirley, J. L., de Jong, Y. P., Terhorst, C., & Herzog, R. W. (2020). Immune responses to viral gene therapy vectors. *Molecular Therapy, 28*, 709–722. https://doi.org/10.1016/j.ymthe.2020.01.001
36. Xu, L., Park, K. H., Zhao, L., Xu, J., El Refaey, M., Gao, Y., Zhu, H., Ma, J., & Han, R. (2016). CRISPR-mediated genome editing restores dystrophin expression and function in mdx mice. *Molecular Therapy, 24*, 564–569. https://doi.org/10.1038/mt.2015.192

37. Maggio, I., Liu, J., Janssen, J. M., Chen, X., & Gonçalves, M. A. F. V. (2016). Adenoviral vectors encoding CRISPR/Cas9 multiplexes rescue dystrophin synthesis in unselected populations of DMD muscle cells. *Scientific Reports, 6*, 37051. https://doi.org/10.1038/srep37051
38. Maggio, I., Stefanucci, L., Janssen, J. M., Liu, J., Chen, X., Mouly, V., & Gonçalves, M. A. F. V. (2016). Selection-free gene repair after adenoviral vector transduction of designer nucleases: Rescue of dystrophin synthesis in DMD muscle cell populations. *Nucleic Acids Research, 44*, 1449–1470. https://doi.org/10.1093/nar/gkv1540
39. Mück-Häusl, M., Solanki, M., Zhang, W., Ruzsics, Z., & Ehrhardt, A. (2015). Ad 2.0: A novel recombineering platform for high-throughput generation of tailored adenoviruses. *Nucleic Acids Res, 43*, e50. https://doi.org/10.1093/nar/gkv031
40. Ehrke-Schulz, E., Schiwon, M., Leitner, T., Dávid, S., Bergmann, T., Liu, J., & Ehrhardt, A. (2017). CRISPR/Cas9 delivery with one single adenoviral vector devoid of all viral genes. *Scientific Reports, 7*, 17113. https://doi.org/10.1038/s41598-017-17180-w
41. Ehrhardt, A., & Kay, M. A. (2002). A new aadnoviral helper-dependent vector results in long-term therapeutic levels of human coagulation factor IX at low doses in vivo. *Blood, 99*, 3923–3930. https://doi.org/10.1182/blood.V99.11.3923
42. Brescia, M., Janssen, J. M., Liu, J., & Gonçalves, M. A. F. V. (2020). High-capacity adenoviral vectors permit robust and versatile testing of DMD gene repair tools and strategies in human cells. *Cell, 9*, 869. https://doi.org/10.3390/cells9040869
43. Sarkis, C., Philippe, S., Mallet, J., & Serguera, C. (2008). Non-integrating lentiviral vectors. *Current Gene Therapy, 8*, 430–437. https://doi.org/10.2174/156652308786848012
44. Jin, Y., Shen, Y., Su, X., Weintraub, N., & Tang, Y. (2019). CRISPR/Cas9 technology in restoring dystrophin expression in iPSC-derived muscle progenitors. *Journal of Visualized Experiments, 151*, e59432. https://doi.org/10.3791/59432
45. Duchêne, B. L., Cherif, K., Iyombe-Engembe, J. P., Guyon, A., Rousseau, J., Ouellet, D. L., Barbeau, X., Lague, P., & Tremblay, J. P. (2018). CRISPR-induced deletion with SaCas9 restores dystrophin expression in dystrophic models in vitro and in vivo. *Molecular Therapy, 26*, 2604–2616. https://doi.org/10.1016/j.ymthe.2018.08.010
46. Lattanzi, A., Moiani, A., Izmiryan, A., Martin, S., Mavilio, F., Bovolenta, M., Duguez, S., Mamchaoui, K., Mouly, V., Barbon, E., & Bernardi, F. (2017). Correction of the exon 2 duplication in DMD myoblasts by a single CRISPR/Cas9 system. *Molecular Therapy Nucleic Acids, 7*, 11–19. https://doi.org/10.1016/j.omtn.2017.02.004
47. Wojtal, D., Kemaladewi, D. U., Malam, Z., Abdullah, S., Wong, T. W. Y., Hyatt, E., Baghestani, Z., Pereira, S., Stavropoulos, J., Mouly, V., Mamchaoui, K., Muntoni, F., Voit, T., Gonorazky, H. D., Dowling, J. J., Wilson, M. D., Mendoza-Londono, R., Ivakine, E. A., & Cohn, R. D. (2016). Spell checking nature: Versatility of CRISPR/Cas9 for developing treatments for inherited disorders. *American Journal of Human Genetics, 98*, 90–101. https://doi.org/10.1016/j.ajhg.2015.11.012
48. Wienert, B., Wyman, S. K., Richardson, C. D., Yeh, C. D., Akcakaya, P., Porritt, M. J., Morlock, M., Vu, J. T., Kazane, K. R., Watry, H. L., Judge, L. M., Conklin, B. R., Maresca, M., & Corn, J. E. (2019). Unbiased detection of CRISPR off-targets in vivo using DISCOVER-Seq. *Science, 364*, 286–289. https://doi.org/10.1126/science.aav9023
49. Jo, D. H., Koo, T., Cho, C. S., Kim, J. H., Kim, J. S., & Kim, J. H. (2019). Long-term effects of in vivo genome editing in the mouse retina using Campylobacter jejuni Cas9 expressed via adeno-associated virus. *Molecular Therapy, 27*, 130–136. https://doi.org/10.1016/j.ymthe.2018.10.009
50. Rayner, E., Durin, M.-A., Thomas, R., Moralli, D., O'Cathail, S. M., Tomlinson, I., Green, C. M., & Lewis, A. (2019). CRISPR-Cas9 causes chromosomal instability and rearrangements in cancer cell lines, detectable by cytogenetic methods. *CRISPR Journal, 2*, 406–416. https://doi.org/10.1089/crispr.2019.0006
51. Senturk, S., Shirole, N. H., Nowak, D. G., Corbo, V., Pal, D., Vaughan, A., Tuveson, D. A., Trotman, L. C., Kinney, J. B., & Sordella, R. (2017). Rapid and tunable method to temporally control gene editing based on conditional Cas9 stabilization. *Nature Communications, 8*, 14370. https://doi.org/10.1038/ncomms14370

52. Dow, L. E., Fisher, J., O'Rourke, K. P., Muley, A., Kastenhuber, E. R., Livshits, G., Tschaharganeh, D. F., Socci, N. D., & Lowe, S. W. (2015). Inducible in vivo genome editing with CRISPR-Cas9. *Nature Biotechnology, 33*, 390–394. https://doi.org/10.1038/nbt.3155
53. Harrington, L. B., Doxzen, K. W., Ma, E., Liu, J. J., Knott, G. J., Edraki, A., Garcia, B., Amrani, N., Chen, J. S., Cofsky, J. C., Kranzusch, P. J., Sontheimer, E. J., Davidson, A. R., Maxwell, K. L., & Doudna, J. A. (2017). A broad-spectrum inhibitor of CRISPR-Cas9. *Cell, 170*, 1224–1233. https://doi.org/10.1016/j.cell.2017.07.037
54. Li, A., Lee, C. M., Hurley, A. E., Jarrett, K. E., De Giorgi, M., Lu, W., Balderrama, K. S., Doerfler, A. M., Deshmukh, H., Ray, A., Bao, G., & Lagor, W. R. (2019). A self-deleting AAV-CRISPR system for in vivo genome editing. *Molecular Therapy – Methods and Clinical Development, 12*, 111–122. https://doi.org/10.1016/j.omtm.2018.11.009
55. Yin, H., Song, C. Q., Dorkin, J. R., Zhu, L. J., Li, Y., Wu, Q., et al. (2016). Therapeutic genome editing by combined viral and non-viral delivery of CRISPR system components in vivo. *Nature Biotechnology, 34*, 328–333. https://doi.org/10.1038/nbt.3471
56. Martino, A. T., & Markusic, D. M. (2020). Immune response mechanisms against AAV vectors in animal models. *Molecular Therapy – Methods and Clinical Development, 17*, 198–208. https://doi.org/10.1016/j.omtm.2019.12.008
57. Kim, S., Koo, T., Jee, H. G., Cho, H. Y., Lee, G., Lim, D. G., Shin, H. S., & Kim, J. S. (2018). CRISPR RNAs trigger innate immune responses in human cells. *Genome Research, 28*, 367–373. https://doi.org/10.1101/gr.231936.117
58. Charlesworth, C. T., Deshpande, P. S., Dever, D. P., Camarena, J., Lemgart, V. T., Cromer, M. K., Vakulskas, C. A., Collingwood, M. A., Zhang, L., Bode, N. M., Behlke, M. A., Dejene, B., Cieniewicz, B., Romano, R., Lesch, B. J., Gomez-Ospina, N., Mantri, S., Pavel-Dinu, M., Weinberg, K. I., & Porteus, M. H. (2019). Identification of preexisting adaptive immunity to Cas9 proteins in humans. *Nature Medicine, 25*, 249–254. https://doi.org/10.1038/s41591-018-0326-x
59. Ferdosi, S. R., Ewaisha, R., Moghadam, F., Krishna, S., Park, J. G., Ebrahimkhani, M. R., Kiani, S., Anderson, K. S. (2019) Multifunctional CRISPR-Cas9 with engineered immunosilenced human T cell epitopes. Nature Communications 10(1), https://doi.org/10.1038/s41467-019-09693-x
60. Wagner, D. L., Amini, L., Wendering, D. J., Burkhardt, L. M., Akyüz, L., Reinke, P., Volk, H. D., & Schmueck-Henneresse, M. (2019). High prevalence of Streptococcus pyogenes Cas9-reactive T cells within the adult human population. *Nature Medicine, 25*, 242–248. https://doi.org/10.1038/s41591-018-0204-6

CRISPR-Cas-Based Gene Therapy to Target Viral Infections

Mouraya Hussein, Ben Berkhout, and Elena Herrera-Carrillo

Abstract In recent years, there is an increasing demand for the development of new antiviral strategies due to the prevalence of viral infections such as those caused by the human immunodeficiency (HIV) or the hepatitis B and C viruses (HBV and HCV) and the emergence of a variety of "new" viruses including SARS-CoV-2. The pharmaceutical industry and the scientific community work on the development of new antiviral drugs, including the repurposing of previously approved drugs. But alternative strategies such as the development of novel RNA-based therapeutics based on the clustered regularly interspaced short palindromic repeats (CRISPR)-Cas system are also welcome. Here we discuss different aspects of the use of CRISPR-Cas technology in the broad field of virology, ranging from applications in the diagnosis of viral infections to the development of novel antiviral treatments. After a brief introduction on the CRISPR-Cas technology, we will explore the advantages and limitations of this antiviral approach and discuss the future prospects for improvement of diagnostics and therapeutics and their potential for future clinical application.

1 Introduction

Recent technological advances have spurred the development of conventional antiviral strategies like antiviral drugs and vaccines. However, viral infections remain among the main causes of disease globally [1]. Vaccines are the most efficient tools in terms of prevention [2]. However, production of effective and safe vaccines in a short timeframe is a challenge, as demonstrated in the current severe acute respiratory syndrome coronavirus 2 (SARS-CoV-2) pandemic [3]. Antiviral drugs face

M. Hussein · B. Berkhout · E. Herrera-Carrillo (✉)
Laboratory of Experimental Virology, Department of Medical Microbiology, Amsterdam UMC, University of Amsterdam, Amsterdam, The Netherlands
e-mail: e.herreracarrillo@amsterdamumc.nl

Y. H. Yun, K. E. Yoder (eds.), *Biotechnologies for Gene Therapy*,
https://doi.org/10.1007/978-3-030-93333-3_5

other problems, including the possibility that drug-resistance develops due to the high mutation rate of viruses [4].

The recent outbreak of COVID-19 at the end of 2019 and the immense burden that it has posed on the global community turned the attention to the necessity of developing durable antiviral approaches. The CRISPR-Cas system is a novel, but very attractive player in this respect [5]. This RNA-controlled gene-editing mechanism was recently discovered as a sequence-specific defense system of prokaryotes. It is highly efficient, also in eukaryotic cells, and fairly easy to use. The system has been much improved by molecular engineering to create a powerful genome-editing tool. The CRISPR technology allows researchers to target any specific nucleotide sequence and to modify any gene or noncoding sequence. Here we discuss different aspects of the use of CRISPR-Cas in the field of virology, ranging from its use in viral diagnostics to its application as antiviral approach by directly targeting the viral genomes or indirectly targeting of the mRNA coding for host factors that are critical for virus replication.

2 Discovery and Description of the CRISPR-Cas Mechanism

The first identification of a CRISPR mechanism dates back to 1987 [6], but the CRISPR acronym was only coined in 2001 by Mojica and Jansen [7–9]. In 2005, the CRISPR sequences were proposed to be part of an adaptive immune system of prokaryotes with spacer sequences derived from invading genomic elements like transposable elements and bacteriophages [10–12]. The protective effect of the CRISPR system was divided into three stages [13–15]. (1) "Adaptation," which starts with the integration of short sequences of the invading pathogen into the genome of the prokaryotic host, (2) "Expression" when the machinery of the host expresses CRISPR RNAs (crRNAs) based on these foreign sequence inserts, and (3) "Interference" when the Cas endonuclease, guided by such a crRNA, induces sequence-specific cleavage of the invading genetic element [16].

In 2012, the group of Doudna and Charpentier proposed to use this mechanism as gene-editing tool, which revolutionized not only the molecular biology field, but also the gene therapy discipline [17, 18]. Both scientists were awarded with the Nobel Prize in Chemistry in 2020. They demonstrated that site-specific double-stranded breaks (DSBs) are induced by the Cas9 endonuclease, which is guided by complex tandem RNA molecules composed of a crRNA and a trans-activating crRNA (tracrRNA) (Fig. 1). They redesigned this structure as a single chimeric RNA molecule called guide RNA (gRNA) and succeeded to induce sequence-specific DNA breaks, providing a simple, efficient, and specific gene-editing tool. Since then, the CRISPR-Cas system has been optimized for many different applications. CRISPR has also been tailored for targeting of a wide range of human pathogens and subsequent implementation in a clinical setting. In this chapter, the development of the CRISPR-Cas system will be discussed specifically in the context of the design of novel antiviral strategies.

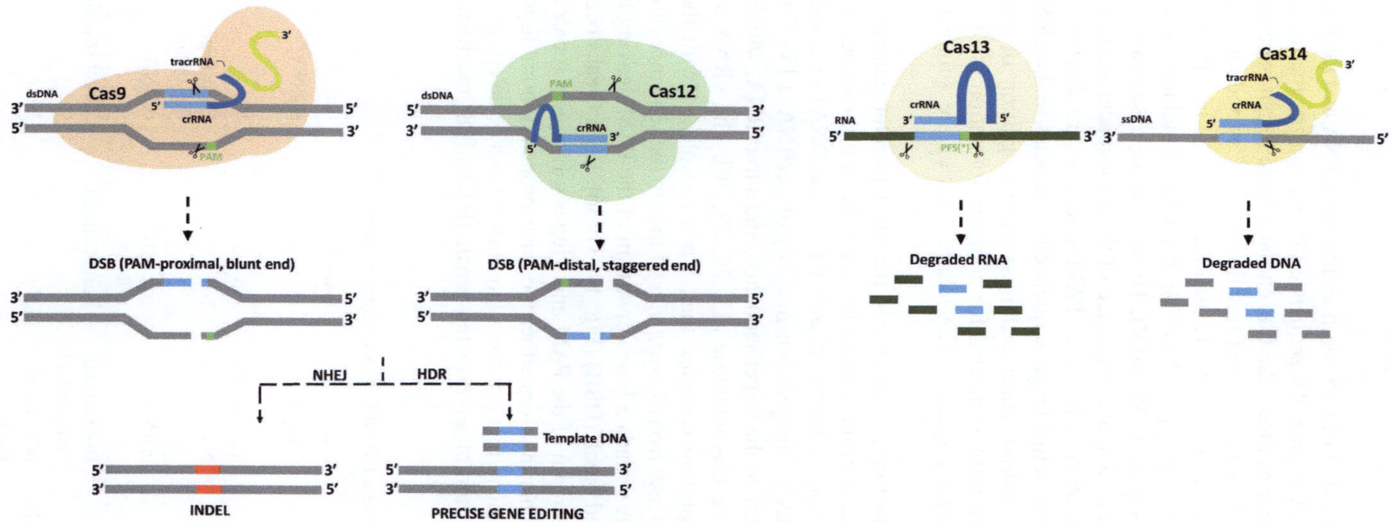

Fig. 1 Different CRISPR-Cas systems. Different CRISPR-Cas effector proteins (Cas9, Cas12, Cas13, and Cas14), their features, cleavage mechanism, and target genomes are represented schematically. The protospacer adjacent motif (PAM) required for Cas9 and Cas12 cleavage and the protospacer flanking site (PFS) required for some Cas13 orthologues are shown as green squares. The required crRNA and/or tracrRNA are depicted for each CRISPR-Cas system. Cellular repair mechanisms of double-strand breaks (DSB), induced upon cleavage by Cas9 and Cas12, are nonhomologous end joining (NHEJ) and homology-directed repair (HDR). HDR requires a donor DNA template and results in precise gene editing, while NHEJ produces insertions and deletions (INDELS). Cas13 recognition and cleavage of a target transcript leads to nonspecific degradation of nearby transcripts (collateral activity). Single-strand DNA cleavage by Cas14 can also exhibit collateral activity

3 CRISPR-Cas as a Gene-Editing Tool

The CRISPR-Cas system consists of two components: the gRNA and the Cas endonuclease [19–21]. The gRNA is an RNA sequence that recognizes the target DNA in a sequence-specific manner to direct the Cas endonuclease for site-specific editing [22, 23]. According to the structure and function of the Cas endonuclease, the CRISPR-Cas systems can be classified into two classes (class I, class II), which are further subdivided into six types (type I–VI) [24, 25]. Class I includes types I, III, and IV, and class II includes types II, V, and VI. These endonucleases have different characteristics such as protein size and sequence of the protospacer adjacent motif (PAM) region, among many other differences. PAM sequences are short sequences, 2–5 nt, located adjacent to the actual target sequence that anneals to the gRNA. The PAM is essential for Cas endonuclease action as the recognition of this short sequence by specific Cas proteins is needed for complex formation of the gRNA/Cas ensemble with the target sequence. Key characteristics are summarized in Table 1.

The type II CRISPR-Cas9 system is clearly the most popular genome-editing tool. The ability to facilitate genome editing by several Cas orthologues has been described [26–28]. Cas9 cleaves double-stranded DNA (dsDNA) in a sequence-specific manner led by a gRNA. The gRNA is made up of a crRNA, a 17–23 nucleotide sequence complementary to the target sequence, and a tracrRNA, which serves as a binding scaffold for the Cas nuclease [23, 26, 29, 30]. The gRNA and Cas protein form a ribonucleoprotein complex that, when complexed with the PAM sequence (e.g., NGG for spCas9 from *S. pyogenes*), binds to the sequence of interest and cleaves one of the DNA strands 3 bases upstream of the PAM, leaving a blunt-ended DNA double-stranded break (DSB) (Fig. 1) [31–36]. DNA breaks occur precisely three nucleotides upstream of the PAM and trigger subsequent repair by host DNA repair mechanisms. This involves either the error-prone nonhomologous end joining (NHEJ) [37] mechanism that produces small insertions or deletions (indels) at the cleavage site, or the homology-directed repair (HDR) [38] mechanism that

Table 1 Characteristics of different CRISPR-Cas effector proteins

	Cas9	Cas12	Cas14	Cas13
Target genome	dsDNA	dsDNA	ssDNA	ssRNA
PAM	Yes *(G-rich)*	Yes *(T-rich)*	No	No
PFS	No	No	No	Orthologue dependent
gRNA	tracrRNA + crRNA	crRNA	tracrRNA + crRNA	crRNA
Cleavage	Both strands (blund end)	Both strands (stagger end)	Target strand	Target strand
Repair	NHEJ and HDR	NHEJ and HDR	No	No

uses an homologous repair template to generate a precise genome modification at the cleavage site (Fig. 1). In toto, the Cas9-gRNA system presents a simple and sequence-specific tool for editing and disruption of any dsDNA target.

CRISPR-Cas9 systems have also been modified to accomplish gene activation in a sequence-specific manner. CRISPR activation uses an inactive or dead CRISPR-Cas9 protein (dCas9) that lacks endonuclease and thus editing activity, but that is able to bind to specific genes without the ability to edit the genome [39]. Gene activation by dCas9, also termed CRISPRa, was initially described in 2013 [40, 41] but in the following years, the system was optimized, thus expanding its popularity in the scientific community [42–44]. CRISPRa is composed of dCas9 fused to a transcriptional effector to modulate target gene expression. A gRNA complementary to the sequence of interest leads dCas9 to the genome locus of interest. The dCas9 is unable to make a cut, but the effector activates downstream gene expression. A variety of transcriptional activator domains have been coupled to CRISPRa for this purpose [42–45]. CRISPR interference or CRISPRi is another variation in which dCas9 is fused with a transcriptional effector that represses downstream gene expression instead of activating it. dCas9 binding alone can also interfere with transcription initiation when targeting a promoter, likely by blocking binding of RNA polymerases or transcription factors [46].

In addition, an atypical Cas9 from the bacterium *F. novicida* (FnCas9) was reported to target endogenously transcribed mRNA, which expands the RNA-editing toolbox [47]. FnCas9 was proven to target and inhibit human positively-stranded +ssRNA viruses within eukaryotic cells like hepatitis C virus [48].

Additional CRISPR-Cas systems have been discovered more recently. They were tested and optimized for genome-editing tasks, including the type V Cas12a and b systems that target and cleave dsDNA (Fig. 1) [49–54]. In the presence of a T-rich PAM sequence (e.g., NTTT for Cas12a), the Cas12 nuclease binds to the target sequence and cleaves both strands, creating a staggered-end dsDNA break with 4–5 nucleotides overhang distant from the PAM. The Cas12 endonuclease requires only a single crRNA for successful targeting, whereas the original Cas9 system requires both a crRNA and tracrRNA. Host DNA repair pathways ultimately create mutations at the cleavage site. Interestingly, the sequence signature of the edited sites can differ per endonuclease. Small indels (insertions or deletions) are most common for the Cas9 endonuclease, but we recently reported a distinct mutational pattern for Cas12a: for example, by creating typical deletions that are combined with a small insertion that we termed “delins” and the absence of pure sequence inserts (Fig. 1) [55].

More recently, an RNA-targeting type VI system (CRISPR-Cas13) with the crRNA guide was identified and engineered for RNA cleavage and knockdown in diverse environments like mammalian and plant cells (Fig. 1) [56–63]. Different subtypes have been recognized (a, b, c, d, X, and Y). For Cas13a-c, a so-called Protospacer Flanking Site (PFS) was reported important for cleavage [62, 64], although there is still conflicting evidence for requirement of a sequence motif adjacent to the target site. Cas13d and X do not have such a PFS-constraint, and Cas13Y has not been studied in further detail yet [65].

In addition, a type V CRISPR-Cas DNA-targeting enzyme (CRISPR-Cas14) was identified for cleavage of single-stranded DNA (ssDNA), without any restrictive sequence-requirement (PAM) (Fig. 1) [66]. Cas14 may have evolved in bacteria for the defense against ssDNA viruses or mobile genetic elements (MGEs) that propagate through ssDNA intermediates. In contrast to other Cas enzymes, Cas14 has not been found in bacterial genomes, but in the genome of a group of Archaea, which may suggest that it is a primitive version of the more complex Cas9 and Cas12 proteins. Cas14 associates with both the crRNA and tracrRNA molecules and cleaves next to the heteroduplex formed by the crRNA and target DNA. Moreover, target recognition by Cas14 triggers cleavage of ssDNA molecules [66]. Although ssDNA cleavage does not directly form a good addition to the genome-editing toolbox, Cas14 is a promising tool for the development of diagnostics. In the next section, we will describe in more detail the development of CRISPR-Cas based diagnostic kits.

4 CRISPR-Cas as a Diagnostic Tool

CRISPR-Cas systems can also be used for the detection of viral sequences and allow one to differentiate between viral strains. In order to detect specific sequences, sequence-specific Cas9-mediated DNA cleavage was coupled to a PCR step in methods like CARP (Cas9-gRNAs-associated reverse PCR) or ctPCR (CRISPR-typing PCR). These methods involve the PCR amplification of a particular target sequence with two physically separated Cas9 PAM sites. The PCR-generated dsDNA is targeted with two matching Cas9 gRNAs. Target identification is subsequently achieved by PCR amplification using reverse primers that only amplify a cleaved and ligated sequence [67], PCR amplification of the ligated adapters [67], or qPCR amplification [67]. These methods were used to detect the human papillomavirus (HPV) genome. In addition, NASBA-CRISPR cleavage (NASBACC) was developed by combining Cas9 with isothermal RNA amplification or nucleic acid sequence–based amplification (NASBA) and toehold sensors to quickly detect Zika virus (ZKV) genome sequences in plasma [68]. More recently, Cas9 was combined with lateral flow methodology to develop the CRISPR-Cas9-mediated lateral flow nucleic acid assay (CASLFA) to quickly identify the genomes of *Listeria monocytogenes* and African swine fever virus (ASFV) [69].

Further characterization of type V and type VI Cas proteins (Cas12, Cas14 and Cas13) identified a nonspecific nuclease activity termed collateral cleavage. For these Cas family members, cutting of the targeted nucleic acid can trigger the cleavage of irrelevant, but nearby single-strand DNA (ssDNA) or single-strand RNA (ssRNA) molecules [63, 66, 70–72]. This collateral cleavage activity can be exploited for nucleic acid detection purposes [66, 70, 71]. The collateral activity of Cas13 was used to develop the technology termed SHERLOCK (Specific High-Sensitivity Enzymatic Reporter unlocking) (Fig. 2a) [73, 74]. Once Cas13 binds the target sequence, it promiscuously cleaves nearby ssRNA molecules. In SHERLOCK a quenched fluorophore is coupled to the collateral ssRNA substrate

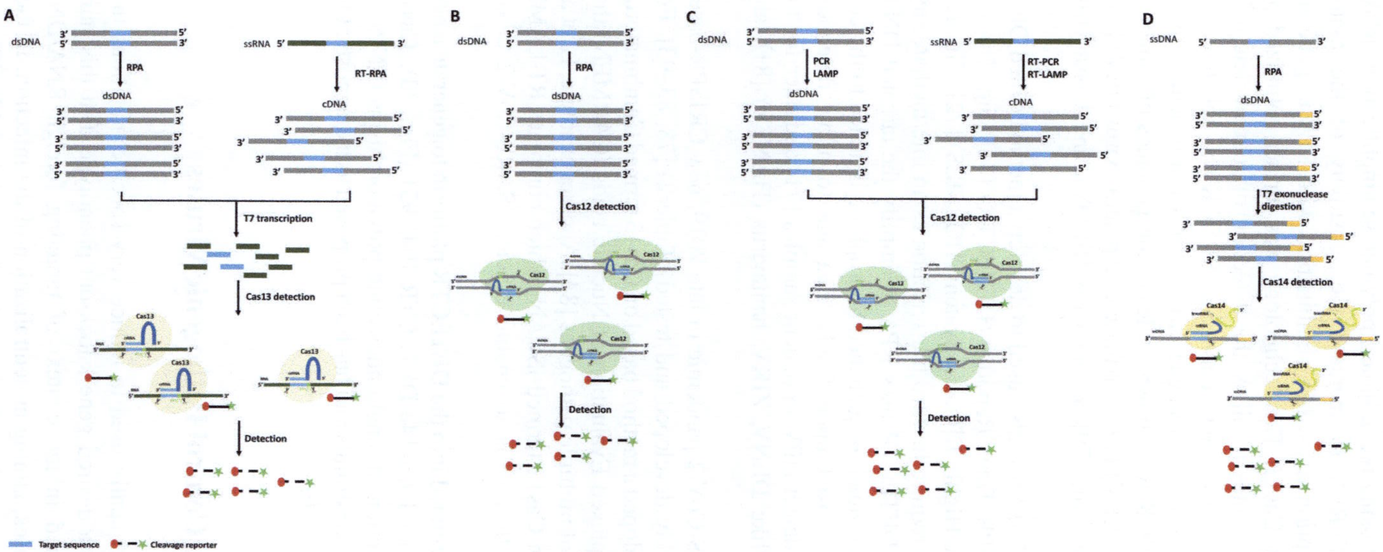

Fig. 2 CRISPR-Cas12 and CRISPR-Cas13 as sequence-specific detection methods. (**a**) The Cas13-based SHERLOCK detection assay. Input DNA or RNA is preamplified by recombinase polymerase amplification (RPA) and reverse transcription (RT), respectively. Amplified DNA is converted into RNA through T7 transcription and is detected by Cas13:crRNA complexes that activate and cleave fluorescent RNA sensors (red spheres and green stars). (**b**) The Cas12a-based DETECTR detection assay. The system follows the same principle as SHERLOCK, with the exception that the T7 transcription step is not required, enabling direct detection of the amplified target input material. (**c**) The Cas12b-based HOLMES detection assay. DNA target input material is amplified by polymerase chain reaction (PCR) and loop-mediated isothermal amplification (LAMP) or by RT-LAMP when the input material is RNA. Upon amplification, detection follows the same workflow as described for DETECTR. (**d**) The Cas14-based DETECTR detection assay. ssDNA input material is amplified by RPA using modified primers that will modify one end of the generated dsDNA products (yellow end). The strand with a modified end will be resistant to T7 exonuclease treatment that will degrade the unmodified DNA strand, leaving the modified strand as a substrate for Cas14-crRNA complex that is activated to cleave the fluorescent reporter

that is released to emit fluorescence once the substrate is cleaved. The target DNA or RNA can first be amplified using recombinase polymerase amplification (RPA) or reverse transcriptase (RT)-RPA to increase the sensitivity of the assay. T7-mediated in vitro transcription is used in combination with RPA to generate RNA that can be detected by Cas13. This technology was further optimized by combining 4 different Cas enzymes to allow the detection of four nucleic acid sequences in the same sample in a single reaction. SHERLOCK was able to detect the genomes of Zika virus (ZIKV) and Dengue virus (DENV) in patient samples [75]. More recently, SHERLOCK was shown to detect the genomes of influenza virus [76], Epstein-Barr virus (EBV) [77], Japanese encephalitis virus (JEV) [77], SARS-CoV-2 [78], Powassan virus [79], Ebola virus (EBOV), and Lassa virus (LASV) [80].

The collateral activity of Cas12 was also used to develop a method called DNA Endonuclease Targeted CRISPR Trans Reporter (DETECTR) [71] (Fig. 2b) and 1-Hour low-cost Multipurpose Highly Efficient System (HOLMES) [72] (Fig. 2c). These novel strategies do not require in vitro transcription as an intermediate step for the detection of dsDNA. DETECTR uses RPA to amplify the targeted DNA, whereas HOLMES uses PCR instead. A quenched fluorophore is added to the collateral substrate, which is released upon Cas12 cleavage to allow detection. DETECTR was first used to detect HPV in patient samples [71] and later for the detection of human viruses like DENV, ZIKV, hantavirus (HANTV) [81], and SARS-CoV-2 [82].

With the start of the SARS-CoV-2 pandemic in late 2019, new CRISPR-based detection methods were quickly developed and tested (Table 2) [78, 83–93]. For instance, Ackerman et al. developed a method based on Cas13 termed Combinatorial Arrayed Reactions for Multiplexed Evaluation of Nucleic acids (CARMEN) that allows multiplexed detection of multiple pathogens [84]. Another example, Li et al. developed a method based on Cas12 termed iSCAN, which combines RT-LAMP, CRISPR-Cas12, and lateral flow cells to enable efficient SARS-CoV-2 detection [83].

In addition, Cas14 was integrated into the DETECTR platform to generate a new ssDNA detection system termed Cas14a-DETECTR [94, 95] (Fig. 2d). Cas14 enables the high-fidelity detection of single nucleotide polymorphisms (SNPs) in DNA without PAM sequence constraints and has been proposed as a diagnostic tool to detect DNA virus infections [66].

5 CRISPR Editing To Control Pathogenic Viruses

CRISPR-Cas systems are frequently used to induce very specific mutations in a gene of interest, resulting in a desired gene knock-out phenotype. But this same basic property can be applied in the context of targeting foreign RNA/DNA sequences of pathogenic viruses, aiming at neutralization of the infection. In fact, this application reminds us of the physiological role of the original CRISPR-Cas

Table 2 CRISPR–Cas-based detection of SARS-CoV-2

Cas protein	Detection method	Amplification	Assay readout	Reference(s)
Cas12	DETECTR	RT-LAMP	Lateral flow	[82]
	CASdetec	RT-RPA	Fluorescence	[93]
	CRISPR-FDS-v1.0	RT-RPA	Fluorescence	[90]
	CRISPR-FDS-v2.0	RT-PCR		
	CRISPR/Cas12a-NER	RT-RPA	Fluorescence	[258]
	StopCovidv1	RT-LAMP	Fluorescence	[259]
	StopCovidv2		Lateral flow	
	iSCAN	RT-LAMP	Fluorescence	[83]
			Lateral flow	
	Electric-field-mediated SARS-CoV-2 detection	RT-LAMP	Fluorescence	[89]
	Contamination-free visual Cas12 assay	RT-LAMP	Fluorescence	[87]
Cas13	CARMEN	PCR	Fluorescence	[84]
	SHERLOCK	RT-RPA	Fluorescence	[78]
			Lateral flow	
	CRISPR-COVID	RT-RPA	Fluorescence	[86]
	SHINE	RT-RPA	Fluorescence	[260]
			Lateral flow	
	Direct, Cas13a detection	–	Fluorescence	[86]

defense systems in prokaryotes. In this chapter, we will discuss several strategies to develop Cas-mediated therapeutics against viral infections in humans. We will group the CRISPR-Cas therapeutics according to the type of viral genome that is targeted: DNA or RNA. We will explore the challenges of designing CRISPR-Cas therapeutics for the selective targeting of the viral genome, including the choice of a particular Cas system, the PAM constraints, and the selection of conserved viral target sequences as a key determinant to minimize or even avoid the emergence of resistant virus variants. We will deal with different representative viruses and discuss approaches directed at the virus itself versus indirect approaches that target critical cellular cofactors that support virus replication.

Although most research has focused on the direct targeting of viral genomes, targeting of host factors that play an important role in the viral replication cycle could be an attractive alternative strategy. A potential advantage is that the likelihood of viral resistance is minimized when a constant, non-evolving cellular cofactor is targeted. A possible disadvantage is toxicity due to expected or unexpected effects on important cellular pathways. We will explore different CRISPR-Cas-based approaches to target viral host factors, but will also discuss the reverse strategy of inducing cellular immunity by inducing the expression of antiviral restriction factors. Restriction factors are host cell proteins that interfere with a specific step of the viral replication cycle [96]. These host proteins are usually expressed at a low level, but their expression is induced by the interferon system that is activated upon virus infection of cells. We will also discuss off-target effects and the major bottlenecks for the clinical development of CRISPR-Cas therapeutics, including the major bottleneck of delivery of the therapeutics to the patient [97–99].

6 RNA Viruses

There are 219 species of viruses that are known to infect humans, of which 214 are RNA viruses. RNA viruses comprise pathogens such as hepatitis C and E virus (HCV and HEV, respectively), human immunodeficiency virus (HIV), influenza virus, coronaviruses, DENV, West Nile virus (WNV), EBOV, lymphocytic choriomeningitis virus (LCV), measles virus (MV), polio virus (PV), and many others. Classification of the RNA viruses is sometimes complicated by their high mutation rate and ability to recombine. RNA viruses are grouped based on the type of genome (negative- or positive-strand RNA, single-strand or double-stranded RNA), genome organization, and replication strategy. Related to, but distinct from the RNA viruses are the viroids and the RNA satellite viruses that will be also discussed in this section. For instance, hepatitis D virus (HDV) is a satellite of hepatitis B virus (HBV).

The Cas12, Cas13, and the atypical FnCas9 systems target RNA and could thus be used against RNA viruses, targeting either their RNA genome, the anti-genome RNA, the viral mRNA(s), or the cellular mRNAs that encode important host factors that support virus replication. In addition, retroviruses like HIV have an RNA genome, but replicate via a DNA intermediate, and could thus be targeted both by RNA- and DNA-targeting CRISPR-Cas systems. This will also be discussed in this section.

6.1 Hepatitis C Virus

The most common causes of viral hepatitis are a variety of diverse hepatotropic viruses: hepatitis A, B, C, D, and E virus (HAV, HBV, HCV, HDV, HEV). HAV and HEV cause transient or acute infections with a relatively rapid onset, while HBV HCV and HDV cause chronic infections that may eventually lead to cirrhosis and liver cancer. HBV has a double-stranded DNA genome that is replicated via an RNA intermediate by the process of reverse transcription; all other hepatitis viruses are RNA viruses. HCV has a plus-strand single-strand RNA genome, and its replication occurs exclusively within the cytoplasm [100]. The plus-strand RNA genome serves as the template for synthesis of the antisense RNA copy, which serves in turn as template for the production of many plus-strand viral RNA genomes that serve either as mRNA for translation of the viral polyprotein or as viral genome that is packaged in new virion particles. HCV, like HBV, is able to establish a chronic infection of the liver, thereby increasing the risk of cirrhosis that may eventually lead to hepatocellular carcinoma. Unlike for HBV, there is no effective vaccine to prevent HCV infection, but we nowadays have many potent antiviral drugs that target specific steps of the HCV replication cycle. The RNA-dependent RNA polymerase of HCV lacks proofreading activity, which creates genetic diversity that may trigger the selection of drug-resistant variants [101]. But as for HIV, a combinatorial drug approach can prevent this and result in a sterilizing cure, something that is unheard of for HIV and HBV.

6.1.1 Targeting Viral RNA

Although the majority of Cas9 orthologues target DNA, some systems were identified or engineered to target RNA. For instance, FnCas9 was optimized and employed to target HCV RNA genomes [48]. The first report on FnCas9 dates back to 2013, when this system was shown to cleave endogenous bacterial mRNA in the cytosol [47]. Besides the previously mentioned tracrRNA, FnCas9 requires a so-called small and unique CRISPR-Cas-associated RNA (scaRNA), which interacts with tracrRNA in the same manner as crRNA does [102]. Price et al. engineered an RNA-targeting guide RNA (rgRNA) made up of fused tracrRNA and customizable scaRNA, similar to the natural occurring structure, and demonstrated that FnCas9-mediated inhibition of HCV is PAM-independent [48]. The authors targeted highly conserved HCV sequences and demonstrated that FnCas9 can target both the negative- and positive-sense RNA strands to inhibit both viral translation and replication. This suggest that FnCas9 could be used to target diverse viruses, potentially including all RNA viruses, either with a plus- or minus-strand RNA genome.

In another more recent attempt to inhibit HCV replication, the CRISPR-Cas13a system was employed to target the HCV RNA genome [103]. Highly conserved regions of the important internal ribosomal entry site (IRES) motif were attacked with Cas13a, causing inhibition of HCV replication and translation in Huh-7.5 cells.

6.1.2 Targeting Host Factors

A number of different cellular cofactors required for HCV infection have been described in literature [104–106]. Cellular proteins that can be silenced by CRISPR-Cas attack include VAPA, PDPK1, RAF1, and EIF2AK2 kinases that bind to NS5A and EIF2S3 that binds to the 5′ nontranslated region of HCV [105–107]. Some restriction factors that inhibit HCV entry could be candidates for CRISPR-mediated activation, such as Ficolin 2 [108], interferon-induced transmembrane protein 1 (IFITM1) [109], and ezrin-moesin-radixin (EMR) protein [110]. One could also consider activation of TRIM14, NOS2, and IFITM3 that contribute to the suppression of HCV replication [111] or BST-2/tetherin that blocks the release of new viral particles [112, 113].

6.2 Human Immunodeficiency Virus

HIV is a positive-sense, enveloped RNA virus. The HIV genome in virion particles is made up of two identical ssRNA strands [114]. Upon entry into target T cells, the viral RNA genome is reverse transcribed into double-stranded DNA by the viral reverse transcriptase that requires both RNA copies to generate a single complete viral dsDNA (Fig. 3). The dsDNA is then imported into the cell nucleus and integrated into the cellular chromosomal DNA by the viral integrase. Once integrated, the virus may stay transcriptionally silent, thus creating a latent viral reservoir in these cells. Alternatively, the integrated viral DNA can be transcribed, producing new RNA genomes and viral proteins that assemble into new virion particles that are subsequently released from the cell to ignite a spreading infection. Despite the potency of current combinatorial antiviral drug cocktails to treat HIV infection, the

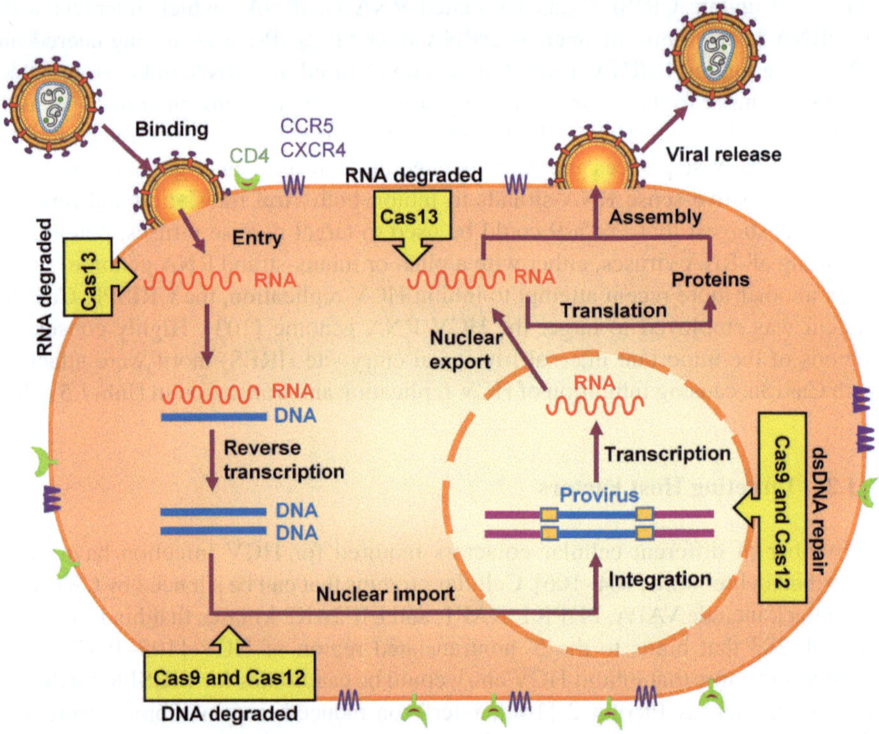

Fig. 3 Schematic representation of the HIV replication cycle. Upon binding, HIV enters the host cell and release the viral RNA genome into the cytoplasm of the target T cell. The positive-sense, single-strand RNA (+ssRNA) is reverse transcribed into double-stranded DNA (dsDNA). Viral dsDNA is transported into the nucleus and integrated into the host genome. HIV proviral DNA is then transcribed into viral messenger RNA (mRNA). Viral mRNA is translated into new viral proteins that assemble with the viral RNA to form new virion particles. New virions bud from the host cell and spread to infect new cells

patient is never cured [115]. The patient should therefore take these drugs lifelong, which may cause drug resistance and adverse effects in a minority of the patients. Therefore, the development of more durable therapeutic strategies remains of utmost importance. This includes a direct attack with CRISPR-Cas methods on the integrated viral DNA genome with the aim to inactivate it and to reduce and ideally inactivate the viral reservoir. Alternatively, one could also target the viral RNA genome or host cofactors or activate cellular restriction host factors. Much research has been done on this HIV system.

6.2.1 Targeting Viral RNA

CRISPR-Cas13a has been used to inhibit HIV infection by targeting HIV RNA, which greatly diminished viral gene expression in human cells [116]. When compared to antiviral strategies based on RNAi, CRISPR-Cas13a does not only inhibit RNA that is expressed from the integrated DNA, but also the "incoming" viral RNA of virion particles upon entry into cells [116, 117]

6.2.2 Targeting Viral DNA

HIV DNA in the viral reservoir cells can be targeted by nucleases like Cas9 and Cas12 [55, 118–126]. The viral long-terminal repeat (LTR) domain forms an attractive target as this motif flanks the viral genome at both the 5′ and 3′ ends. Thus, in principle, an LTR-targeting gRNA could result in the nearly complete removal of the viral DNA by a double cleavage event, excision of the intervening HIV sequences, and subsequent DNA ligation. Ebina et al. first demonstrated a significant loss of gene expression driven by the LTR promoter by LTR targeting with CRISPR-Cas9 in a cellular latency model with an integrated HIV viral genome. Sequence analysis confirmed that Cas9 attack resulted in mutation of the LTR via indel formation and to a lesser extent by excision of all viral sequences between the 5′ and 3′ LTR targets [120].

We and others demonstrated that HIV replication can be inhibited with a single antiviral gRNA, but also that the virus could escape by the acquisition of such indel mutations in the target site that prevent new rounds of gRNA binding and cleavage [121, 122, 127–129]. Interestingly, this viral escape mechanism is thus dependent on the cellular NHEJ system that creates these indels during DNA repair. A dual-gRNA attack demonstrated to be more effective in preventing viral escape [130, 131].

Several groups suggested that targeting of the HIV genome with two gRNAs caused deletion of the viral segment between the two target sites, which is called excision and obviously the ideal scenario toward a cure [118, 119, 123–125]. However, the DNA products were analyzed by PCR using primers immediately upstream and downstream of the two targets, which strongly favors amplification of the short excision product over alternative products. We and others demonstrated that dual-gRNA targeting can result in three products: mutation of both target sites occurs most frequently, excision can occur and in fact also reversion of the intervening viral segment occurs, although to a much lesser extent [120, 132–134]. We demonstrated that the introduction of indels at two otherwise very conserved viral targets can trigger virus inactivation, a process that we termed hypermutation to set it apart from excision. Thus, viral escape can be prevented by both excision and hypermutation. In fact, repeated Cas cleavage may occur if the edited target can still be recognized by the gRNA, leading to increased mutational inactivation of the viral genome and causing the gradual loss of replication-competent virus, which means a cure of the infected cells [55, 131]. We reported that DNA cleavage by the Cas12 endonuclease and subsequent DNA repair causes mutations with a distinct sequence profile compared to Cas9 [55]. For instance, Cas12 does not induce the pure DNA insertions that are frequently observed for Cas9. As the mutation type will affect the likelihood of HIV genome inactivation, this adds another variable to the selection criteria for chosing the best cure strategy.

More recently, CRISPR-Cas9 was demonstrated to cleave HIV DNA in transgenic mice. PCR analysis demonstrated the presence of both excised and non-excised HIV genomes in different cell types [123–125]. The non-excised PCR fragment could represent hypermutated or wild-type genomes, but that was not analyzed.

Given the potential of CRISPR-Cas therapy to control HIV infection, alternative systems such as Cas12a were also used to target HIV DNA [50, 135, 136]. With an optimized CRISPR-Cas12a system, long-term inhibition of HIV replication was recently demonstrated with only a single gRNA, whereas Cas9 required a dual gRNA attack [55]. This was explained by the different architecture of the endonucleases and their mechanism of action. Interestingly, repeated editing rounds are more likely for Cas12a as the PAM and crRNA-seed region of the target DNA do frequently remain unaltered upon DNA cleavage and repair due to the architecture of the endonuclease, thus increasing the chance of HIV inactivation.

In addition, the CRISPR-Cas technology can be used to reverse HIV latency. CRISPRa was used to target the U3 region of HIV LTR promoter to locate the best activation locus near the sites for binding of the NF-kB transcription factor [137–140]. Such a CRISPRa approach worked more efficiently than a treatment with latency-reversing small molecules such as prostratin and SAHA in HIV-latency models in T cells [137]. Other groups similarly reactivated the latent HIV reservoir by targeting other U3 elements like the Sp-1-binding sites and sequences in the R and U5 regions, which was also tested in combination with latency-reversing agents [139].

6.2.3 Targeting Host Factors

Targeting of host cell cofactors that support HIV replication has advantages and shortcomings. Inhibition of cofactors will be effective against all viral variants in an infected individual, and viral escape seems possible only by a major adaptation of HIV to an alternative cellular cofactor. An obvious danger of targeting cellular factors is that it may cause unwanted toxic effects on cell physiology. Important cellular cofactors to target would be the receptors at the cell surface needed for viral entry (CD4, CCR5, and CXCR4) [141]. But CD4-independent virus variants exist and CXCR4 is usually not used by primary virus isolates; thus CCR5 is the prime candidate, also because this protein does not seem to have a critical physiological function. A proof of concept was obtained by bone marrow transplantation of a long-term HIV-infected individual (the "Berlin" patient) with cells from a CCR5-minus donor. The recipient subsequently maintained an undetectable viral load in the absence of antiviral drugs [142, 143]. CRISPR-Cas9 has been widely used to disrupt the CCR5 gene, which resulted in acquired resistance to HIV in human T cell lines and an in vivo mouse model [21, 144–146]. For instance, Li et al. silenced CCR5 expression by CRISPR-Cas9 in human CD4+ T cells, which protected the cells from subsequent HIV infection [144]. Xu et al. targeted CCR5 in human CD34+ hematopoietic stem progenitor cells (HSPCs) and achieved long-term CCR5 disruption in vivo [146]. However, HIV could escape by switching to CXCR4 as alternative entry receptor, which seems dangerous as such virus variants have been linked to faster disease progression [147–149]. The CXCR4 coreceptor was efficiently disrupted by CRISPR-Cas9 attack in CD4+ T cells of human and rhesus macaques [119, 150, 151]. Reduced CXCR4 expression was scored in T cells that showed resistance to HIV infection. Targeting both CCR5 and CXCR4 coreceptors protected primary CD4+ T cells against infection by single- and dual-tropic HIV strains [152]. However, as CXCR4 plays a role in HSPC development, this may raise concerns when translating this therapeutic approach to the clinic [153, 154].

One could also consider activation of restriction factors as an antiviral alternative, and in fact, many anti-HIV restriction factors have been described, including APOBEC3G, TRIM5α, SAMHD1, and tetherin [155]. Bogerd et al. demonstrated efficient induction of APOBEC3G expression in human cells, which resulted in profound HIV infection [156]. Serinc5 was identified more recently as HIV restriction factor that could also be potentially targeted by CRISPRa [157, 158]. As some restriction factors target multiple human viruses, this strategy may be ideal to reach broad protection against several viral pathogens.

6.3 Dengue Virus

DENV is a single-strand positive-sense RNA virus that infects humans and results in a mild and self-limiting febrile disease. However, more severe disease forms such as dengue hemorrhagic fever and dengue shock syndrome occur less frequently. There is still no effective vaccine to prevent against DENV infection and with increased global travel, dengue is increasingly becoming a serious global health problem. Therefore, there is a need to develop novel therapeutics for dengue treatment.

6.3.1 Targeting Viral RNA

Engineered Cas13 proteins have been demonstrated to result in a knockdown of DENV RNA copies upon an attack on different regions of the dengue viral RNA genome. Li et al. demonstrated that CRISPR-Cas13a is capable of inhibiting dengue virus infection at the early stage of viral replication by cleaving the incoming plus-strand viral RNA genome [159]. Singsuksawat et al. delivered Cas13b-crRNA ribonucleoprotein (RNP) to an immortalized hepatocyte-like cell line (iMHC) and demonstrated efficient inhibition of dengue virus infection [160].

6.3.2 Targeting Host Factors

DENV restriction factors are a logical target for CRISPR-mediated activation of gene expression. A tripartite motif (TRIM) protein TRIM69 is induced by DENV infection and restricts DENV replication to contribute to the control of the infection in an immunocompetent mouse model [161]. TRIM69 binds to viral nonstructural protein 3 (NS3), which leads to NS3 ubiquitination and degradation. Thus, TRIM69 is a novel interferon-inducible host antiviral factor that could be activated by CRISPRa to cause cellular immunity by enhanced viral protein degradation.

6.4 Influenza A Virus

The genome of influenza viruses is segmented and composed of eight single-strand RNAs that differ in length and that have to be copackaged in a virion particle to constitute the complete viral genome. There are four types of influenza viruses: A, B, C, and D. Human influenza A and B viruses cause seasonal epidemics of disease known as the flu. Influenza A viruses (IAV) are the only influenza viruses known to cause flu pandemics. We will discuss here the possible CRISPR-Cas targets.

6.4.1 Targeting the Viral Genome

Several groups employed the Cas13 endonuclease enzymes to attack the IAV RNA genome. Freije et al. used Cas13b-crRNAs against highly conserved IAV sequences and measured a reduced level of viral particles produced in the cell culture [160]. Abbot et al. evaluated whether the alternative Cas13d system was effective in targeting IAV [162]. The crRNAs were designed against highly conserved IAV sequences and tested in cell culture, where a reduction of the viral RNA level and inhibition of virus replication was scored. Blanchard et al. tested the efficacy of anti-IAV Cas13a-crRNAs in infected mice [163]. They delivered Cas13a mRNA and the crRNAs to the respiratory tract using a nose nebulizer. The viral RNA in the lungs of these mice was quantified by qRT-PCR. A profound reduction of the viral RNA demonstrated robust knockdown of the viral RNA in vivo.

6.4.2 Targeting Host Factors

In the context of IAV, CRISPR-Cas9 has been used by several groups to target host factors. Two studies employed genome-wide gRNA libraries for gene disruption, which allows the researchers to identify important cellular cofactors that facilitate IAV infection [164, 165]. Such a genome-wide screen for host cofactors of virus replication is attractive, because custom-made gRNA-libraries are commercially available for human and mouse. Some host factors have been found to regulate the replication of IAV by interacting with the viral RNA-dependent RNA polymerase (RdRP), including the poly-ADP ribosyl polymerase (PARP1) [166]. Inactivation of such host factors by CRISPR-Cas could block IAV infection.

6.5 *Human Coronaviruses*

The human coronaviruses are positive-sense RNA viruses, which typically infect the upper or lower respiratory tract. The world is currently facing the serious pandemic of coronavirus disease 2019 (COVID-19) that is caused by the severe acute respiratory syndrome coronavirus 2 (SARS-CoV-2) of unknown animal origin. Although several effective and safe vaccines have been approved for human use, which allows one to temper the spreading infection, questions remain on the durability of the protection, both in terms of the induced immunity and the appearance of novel virus variants. Thus, there is an urgent need for the development of potent and specific antiviral strategies, including small molecule inhibitors and CRISPR-Cas approaches.

6.5.1 Targeting Viral RNA

The RNA-targeting Cas13d system can be used to target the SARS-CoV-2 RNA genome. Like other Cas13 proteins, Cas13d employs a crRNA with a customizable spacer sequence that can direct the Cas13d protein to specific RNA molecules for targeted degradation. Abbot et at. designed different crRNA sets targeting conserved viral genome sequences. SARS-CoV-2 reporter constructs were designed, and the inhibitory capacity of different crRNA pools of four crRNAs was tested [162]. The authors demonstrated that the crRNAs were able to cleave the RNA fragments in vitro, but their efficacy in spreading virus infections in cell culture remains to be tested. Blanchard et al. went a step further and tested these Cas13a-crRNAs in hamsters, in which the Cas13 mRNA and crRNA molecules were delivered into the lungs with a nebulizer [163]. Single versus combination crRNA attack was compared, showing a benefit of the combination with a slightly more pronounced reduction of the viral RNA level. Combination crRNA attack led to sustained inhibition of SARS-CoV-2 replication for 6 days and reduced the COVID-19 symptoms in this animal model. These initial findings suggest that Cas13 systems could be developed to mitigate SARS-CoV-2 infections.

6.5.2 Targeting Host Factors

Recently, some restriction factors have been identified that impair SARS-CoV-2 viral infection, such as the BST2 protein and the death-domain-associated protein 6 (DAXX) [167, 168]. Activation of such host factors by CRISPRa could be used as an innovative therapeutic approach in the current pandemic.

6.6 Lymphocytic Choriomeningitis Virus

LCMV is a negative strand RNA virus that has the capacity to induce meningitis in both human and murine hosts. Because LCMV is a noncytopathic virus, the damage induced in the central neural system is not caused by the virus, but rather by the subsequent immune response. Freije et al. performed a genome-wide LCMV screen and designed Cas13a-crRNAs, of which the majority was able to efficiently reduce the LCMV RNA levels in cell culture without an impact on the cell viability [169].

6.7 Vesicular Stomatitis Virus

VSV is a nonsegmented, negative-strand RNA virus that infects a variety of mammals including horses, cattle, swine, wild mammals, and humans, although human-to-human transmission does not readily occur. Freije et al. evaluated whether Cas13b could inhibit this neurotropic RNA virus [169]. The authors designed and tested crRNAs against conserved regions of the two main VSV serotypes, Indiana and New Jersey. Cells expressing Cas13b and VSV-specific crRNAs showed decreased levels of secreted VSV RNA. They demonstrated that Cas13b can efficiently and specifically cleave viral RNA. Delivery of multiple crRNAs was shown to enhance the antiviral effect of Cas13.

7 DNA Viruses

DNA viruses comprise serious pathogens such as herpesviruses, adenoviruses, papillomaviruses, and smallpox viruses, among many others. DNA viruses are divided into three major categories: double-strand DNA viruses (e.g. herpesviruses), single-strand DNA viruses (e.g. parvoviruses), and pararetroviruses (e.g. hepadnaviruses) that replicate their DNA genome through an RNA intermediate. In general, DNA viruses with a large genome (>10 kb) usually have double-stranded DNA, whereas DNA viruses with a small DNA genome usually have a circular DNA, either single or double stranded. Many research groups have developed CRISPR-Cas approaches to target the DNA genomes of several of these pathogens. In this section, we will present a few representative topics: CRISPR-Cas systems to target the "large" herpesviruses and the relatively "small" human papilloma virus (HPV) and hepatitis B virus (HBV).

7.1 *Herpesviruses*

Herpesviruses have a linear, double-stranded DNA genome that can vary in size from around 125 up to 290 kb [170]. Herpesviruses establish a latent infection in the host that can be interrupted by periodic or rare reactivation events during which active virus replication is triggered. They are divided into three subfamilies, the alpha subfamily with members such as herpes simplex virus 1 (HSV-1) and HSV-2, the beta subfamily that includes the human cytomegalovirus (HCMV), and the gamma subfamily with prominent members like the EBV and Kaposi's sarcoma–associated herpesvirus (KSHV) [171]. HSV-1 and HSV-2 infections are lifelong and can cause cold and genital sore, respectively. HCMV infection can cause congenital abnormalities in neonates and serious disease in immunocompromised individuals. EBV is associated with infectious mononucleosis and a broad range of carcinomas, including Burkitt's lymphoma or nasopharyngeal carcinoma.

7.1.1 Targeting Viral DNA

There are efficient antivirals to treat herpesvirus infection, but they are ineffective at eliminating the established viral reservoir. CRISPR-Cas9 systems were proposed as a novel strategy to target the DNA genome of these viruses [172–176]. Single gRNAs effectively impaired viral infection, but replication-competent virus emerged after prolonged culturing of the infected cells due to outgrowth of virus variants with mutations that allow the expression of functional viral proteins in most gRNA-expressing cells. As we first reported for HIV, these target mutations are introduced during repair of the CRISPR-cleaved target DNA [128]. Inhibition of viral replication was shown for four herpesviruses (HSV-1, HCMV, KSHV and EBV) and in some cases, even clearance of seemingly all viral genomes from the infected cultures was observed. For instance, Diemen et al. used CRISPR-Cas9 technology to suppress virus replication in both latent and lytic infection models [176]. The authors designed gRNAs to target EBV sequences that code for EBV miRNAs and tested the antiviral efficacy upon stable transduction of latently EBV-infected gastric carcinoma cells. They showed that the targeted miRNA genes were edited and their expression downregulated in this herpesvirus latent infection model. For the lytic infection model, the authors designed gRNAs targeting the viral nuclear antigen 1 (EBNA1) and several areas within the EBV origin of replication (OriP). They observed a complete loss of viral replication after a combinatorial treatment with multiple gRNAs. A similar approach was tested for HCMV and HSV. The effectiveness of CRISPR-Cas9 in treating latent herpesvirus infections has been also demonstrated in an in vivo mouse model [177]. Viral replication was reduced in mice receiving gRNAs targeting the HSV viral genome. The latent HSV viral genome was analyzed after the CRISPR-Cas treatment. Single DSB in the viral DNA was repaired, which often resulted in mutations that abrogate virus replication, while a double DSB attack may lead to further degradation and loss of HSV genomes [177].

7.1.2 Viral Genome Modification: Knock-out and Knock-in Strategies

Additionally, the CRISPR-Cas9 systems can be applied in basic herpesvirus research by engineering, targeting, activating, or repressing specific genes of interest [178–182]. CRISPR-Cas systems were used as a tool for the study of the interaction of the herpesvirus and the host cell. For instance, CRISPR-Cas9 technology was used to generate a knock-out of the viral DNA sensor IFI16 to investigate its role during herpesvirus infection, demonstrating that IFI16 acts as restriction factor of HSV-1 replication [179, 180]. Using a similar knock-out strategy, the analysis of KSHV-infected cells revealed a role for the viral ORF45 protein during lytic virus replication [182].

CRISPR-Cas technology can also be used to generate mutated DNA genomes of other herpesviruses [181, 183, 184]. Manipulation of large viral DNA genomes usually requires the use of bacterial artificial chromosomes (BACs). This complex technique allows maintenance of the complete infectious viral genome in the *Escherichia coli* bacterium and successful reconstitution of infectious virus in permissive eukaryotic cells after BAC transfection. CRISPR-mediated genome editing can also be used to introduce targeted mutations in the large viral genome by HDR using DNA templates that carry the relevant mutation. As CRISPR-Cas9 mutagenesis occurs directly in infected cells, this methodology avoids unwanted bottlenecks and selection pressures that may occur during propagation of the viral genome in bacteria [185]. For instance, CRISPR-Cas9-mediated mutagenesis enabled the generation of gastric-cancer-derived EBV strains [183]. However, the efficiency of the HDR repair route is rather low for the insertion of large foreign genes into such relatively large viral genomes. This problem was circumvented by the development of a nonhomologous insertion strategy that is based on the NHEJ DNA repair pathway

[186–189]. The use of a rationally designed nonhomologous insertion strategy allowed the rapid and efficient knock-in of foreign genes into the HSV genome and provided useful guidance for gene insertion strategies with other large DNA viral genomes [189]. This allows scientists to add novel functions to the viral gene repertoire and to study the relationship between viral genome variation and virus-associated disease. The different technical approaches used for large fragment knock-in were summarized by Erwood et al [190].

7.2 *Human Papilloma Virus*

Human papilloma virus (HPV) is a cancer-causing virus with a small circular dsDNA genome of 8 kb that may occasionally get integrated into the chromosome of the host cell. Integration is actually a prerequisite for the oncogenic process, such that each HPV-induced cancer will show integrated viral genomes. HPV infects cutaneous and mucosal epithelia, and different virus genotypes cause a variety of diseases characterized by uncontrolled cell growth, ranging from skin and genital warts to cervical cancer, but the latter can be prevented by vaccination strategies. Despite that, HPV-related cervical cancer remains the second most common cancer in women worldwide. There are more than 200 HPV genotypes [191], of which the high-risk HPV16 and HPV18 variants are associated with cervical and penile cancers [192], while HPV6 and HPV11 are linked to anogenital warts and laryngeal papillomatosis [193].

7.2.1 Targeting Viral DNA

Direct targeting of the HPV DNA genome by CRISPR-Cas could provide an alternative antiviral strategy. Several groups designed CRISPR-Cas9 gRNAs to target the ORF encoding the oncogenic E6 and E7 proteins that are important for malignant transformation of the host cell and maintenance of the malignant phenotype of cervical cancer [194]. These gRNAs were able to edit and inactivate those genes in HPV-transformed cells and reduced the tumor size in an experimental mouse model [195].

7.2.2 Activation of Viral Restriction Factors

There are several restriction factors expressed by the host cell that limit HPV replication, including as p53, p56, Sp100, APOBEC, miR-145, and gamma-interferon-inducible protein 16 (IFI16). One antiviral strategy is to activate such antiviral factors by CRISPRa to block HPV infection [196]. More recently, other potential targets for CRISPRa were described such as Sirtuin 1 (SIRT1) [197], microRNA-146a [198], origin recognition complex subunit 2 (ORC2) [199],

Bromodomain-containing protein (Brd4) [200], and CXCL12/CXCR4 [201]. For instance, Das et al. removed SIRT1 from cervical cancer cells, which resulted in enhanced HPV replication in those cells [197]. Depletion of ORC2 also stimulated viral replication [199].

7.3 *Hepatitis B Virus*

HBV has a double-stranded DNA genome that is replicated via an RNA intermediate by the process of reverse transcription. HBV persists in infected hepatocytes due to the presence of covalently closed circular DNA (cccDNA), which acts as the template for transcription of viral RNAs. But permanent integration of HBV DNA into the host genome may occasionally occur, conferring the risk of viral persistence and future reactivation, but also hepatocellular carcinoma [202]. Current antiviral therapies inhibit replication of HBV DNA in the cytoplasm of infected cells, but do not destroy the nuclear cccDNA [203]. A vaccine comprised of the HBV surface antigen is currently included in many national immunization programs. The vaccine prevents HBV infection, but has no therapeutic benefit for already infected persons. Consequently, developing novel therapeutics that are capable of eradicating HBV has become an international research priority.

7.3.1 Targeting Viral DNA

Several groups reported the use of CRISPR-Cas9 to edit and inactivate the HBV cccDNA [204–209]. First, Seeger et al. investigated the stability of cccDNA following CRISPR-Cas cleavage and studied whether activation of an antiviral innate immune response could boost the antiviral effect of CRISPR-Cas9 [206]. The authors designed several gRNAs targeting regions of the HBV genome that are important for transcription of pregenomic RNA. Infection of cells transduced with expression constructs for the Cas9 protein and gRNAs resulted in cleavage of the cccDNA, but also the integrated HBV DNA. It has been argued that a global reduction in HBV cccDNA may result in a high percentage of linear viral DNA that is not repaired, but rather destroyed [210]. Moreover, Kennedy et al. demonstrated inhibition of cccDNA accumulation and HBV DNA production in models of both chronic HBV infection and de novo acute infection [207]. At the same time, Ramanan et al. demonstrated that gRNAs targeting conserved regions of the HBV genome cause strong inhibition of virus replication both in vitro and in vivo [208]. In addition, Stone et al. used CRISPR-Cas9 to edit the HBV genome in liver-humanized FRG mice chronically infected with HBV [211].

Although these studies demonstrated the effectiveness of the CRISPR-Cas9 system in inhibiting HBV replication, supporting the potential of Cas-based therapeutics in treating chronic HBV infection, a remaining challenge is the high sequence heterogeneity among HBV genomes [212]. Ideally, one should design gRNAs that target multiple and ideally all HBV genotypes. One should also analyze the mismatch tolerance for HBV variants that will arise during chronic HBV infection. As the sequence-specificity of each Cas system is specified by the PAM sequence, the number of potential target sites is limited. Therefore, different CRISPR-Cas9 orthologues were investigated to broaden the reach of possible target sequences [209]. Kostyushev et al. proposed StCas9 from *Streptococcus thermophilus* as an ideal system for development of the HBV cure as it allowed one to effectively target and degrade three highly conservative regions in the HBV genome. CRISPR-StCas9 attack resulted in degradation of HBV cccDNA with very limited tolerance to nucleotide mismatches.

7.3.2 Targeting Host Factors

CRISPR-Cas-mediated targeting of HBV host factors could also prevent HBV replication. One could consider to target the bile acid pump sodium taurocholate cotransporting polypeptide (NTCP) as its inhibition prevents HBV replication [213, 214] or the heat stress cognate 70 (Hsc70) and Hsc90 proteins, which are essential cofactors for HBV replication [215, 216]. The Tyrosyl-DNA-phosphodiesterase 2 (TDP2) could also be targeted by CRISPR-Cas and its knock-down was shown to reduce the cccDNA levels [217, 218]. More recently, gamma2-Adaptin was found to mediate the attachment of HBV to liver cells [219, 220]. Altogether, these host factors could be targeted by the CRISPR-Cas system to inhibit HBV infection or replication.

HBV restriction factors could be a target for CRISPRa. SAMHD1, a cellular protein with dNTPase activity, could serve as CRISPRa target, which should lead to depletion of cellular dNTPs that are required for virus replication [221]. One could also consider activation of BST-2/tetherin, which is known to block the release of a wide range of enveloped viruses including HBV [222–224], the myxovirus resistance protein 2 (MX2) that inhibits HBV infection by reducing the cccDNA level [225], or the deoxyribonuclease 1 (DNASE1) enzyme that impairs HBV replication through degradation of the DNA genome within the capsid [226].

8 Challenges and Limitations of CRISPR-Cas Therapeutics

This chapter discusses the use of CRISPR-Cas technology in the context of virus infections, both for antiviral strategies and diagnostic purposes. The antiviral efficacy of different CRISPR-Cas systems was described for a variety of pathogenic viruses. The efficacy and specificity of the CRISPR-Cas systems were tested mainly in simple in vitro settings and in cell culture infections. Despite the promising early results, the application of CRISPR-Cas systems in vivo and translation of this technology towards clinical use remains a major endeavour [227]. Among the many challenges are potential off-target effects and the difficulty to selectively deliver the CRISPR-Cas components to the virus-infected tissue.

Although the targeting specificity of Cas9 is controlled by the gRNA sequence and the PAM adjacent to the target sequence in the viral genome, several studies have revealed that Cas9 can also bind to unintended DNA sites to induce cleavage, termed off-target effects [35]. Off-target effects could induce genetic mutations that could cause a loss of gene function. Therefore, it remains a key challenge to detect off-target sites in a highly sensitive way. A variety of methods to detect off-target effects have been developed, but experimental confirmation of a lack of such potentially adverse effects seems a prerequisite for further translation of this method into the clinic [228–231]. Moreover, scientists have altered gRNAs to minimize off-target mutagenesis [232], either by chemical modification, by modification of the GC-content [233] or the gRNA length [234]. Scientists have also developed high-fidelity Cas9 enzyme variants. For instance, a Cas9 variant from *Staphylococcus aureus* (SaCas9) recognizes an extended PAM sequence (5′-NNGRRT-3′ compared to 5′-NGG-3′ for SpCas9), which may potentially reduce off-target effects [235]. Engineered High-Fidelity SpCas9 and saCas9 variants have shown exceptional accuracy in human cells, exhibiting no detectable genome-wide off-target effects [236, 237]. Cas12 expanded the nuclease toolbox with an enzyme with low off-target editing rate [50, 53, 238]. Of note, many colleagues have argued that RNA editing by Cas13 forms a safe alternative as this system will likely never introduce changes in the host DNA genome [57].

Cas nucleases and gRNAs can be delivered as DNA in the form of plasmids or as ribonucleoprotein complexes (RNPs) or gRNA and Cas-encoding mRNA. However, the use of mRNA is limited due to its relative instability in biological fluids and in cells. Therefore, direct RNP delivery may be a preferred strategy for therapeutic applications, but was not studied yet for antiviral applications [239]. The delivery method is usually key for the specificity of the approach. Transfection and electroporation are the most common delivery methods used in vitro. Transfection exhibits minimal efficacy in primary cell types, often making electroporation the preferred delivery method. However, electroporation requires treatment of isolated cells (ex vivo) and can cause toxicity by permeabilization of cell membranes [240, 241]. To circumvent this problem, a number of nanoparticle materials were developed to create Cas-RNPs that can cross the cell membrane [242–244]. Although the in vivo applicability of these nanomaterials has been demonstrated in other fields, it was not applied yet in the antiviral field. On the other hand, delivery by means of viral vectors is efficient and can achieve a more stable expression of relevant genes compared to non-viral delivery systems, which makes it the preferred delivery method for some clinical applications [245].

Three viral vector systems have been used extensively: Adeno-associated viruses (AAV), Adenoviral vectors (AdV), and Lentiviral vectors (LV). AAV are a common viral vector for gene delivery as they provide a safe option due to the fact that AAV genomes remain episomal in the nucleus and are gradually diluted by cell division and not known to associate with human diseases, except a minor immune response [246]. However, AAV vectors have a limited packaging capacity (<4.7 kb) compared to AdVs and LVs (<8 kb), which remains one of the major bottlenecks for viral delivery methods. Packaging all CRISPR-Cas components in a single vector remains challenging due to the large size of the Cas protein (e.g., 4.2 kb for spCas9) [247]. Moreover, sometimes more than one gRNA is needed, for example, to avoid the selection of virus variants that escape from the therapy. LV vectors allow the packaging of both Cas9 and sgRNA into the same vector. Moreover, the production of LV vectors is less laborious than that of AAV and the transduction efficiency is usually high in a wide variety of cell types, including both dividing and nondividing cells [248, 249]. Although there is a theoretical concern that LV integration may lead to oncogene activation, this has only been observed with the first-generation retroviral vector and not for the LV vectors in a growing number of clinical trials. In fact, LV integrate within introns of active transcriptional units, thereby reducing the danger of insertional oncogenesis [250–253]. LV vectors are also amenable to pseudotyping, meaning that heterologous envelope proteins can be accommodated to mediate viral entry into a wide variety of cells. Another criterion to be considered is vector immunogenicity. AAV vectors have been shown to be less immunogenic than systems based on other viruses [246]. However, the immune responses elicited by LV or AdV do not hamper their use [254], and otherwise, a transient immunosuppressive treatment could be considered [255, 256].

More recently, nanoblades were engineered as protein-delivery system based on the friend murine leukemia virus (MLV). Nanoblades were effective in carrying RNPs into cell lines and primary cells in vitro and in vivo [257]. Nanoblades lack viral genomes, which means that they are non-infectious and non-replicative and they can be pseudotyped like viral vectors with different envelope glycoproteins. Indeed, nanoblades were successfully used for gene editing of primary hematopoietic cells. These results demonstrate that nanoblades form an attractive alternative delivery method for many applications, ranging from the generation of transgenic mice to non-invasive in vivo gene editing.

9 Conclusions

Translation of current CRISPR-Cas technologies into the clinic as a safe and effective antiviral treatment still poses many hurdles. However, the simplicity, specificity, and efficiency of this technology as demonstrated in simple in vitro settings hold much promise for future clinical applications. Scientists continue to search for better and safer CRISPR-Cas systems, and optimization of the delivery method is a topic that is pursued in many laboratories. The speed at which the CRISPR technology for genetic engineering became commonplace in many laboratories was perplexing. Let us hope that a similar avalanche of technical developments will effectively bring a variety of CRISPR therapies into the clinic.

CRISPR-Cas-based approaches have an enormous potential as novel antiviral therapy because of the sequence-specificity of the gene knock-out. In a sense, CRISPR-Cas pays off what antisense and RNA interference technologies once promised to deliver. The use of Cas nucleases enables disruption of any viral DNA and RNA molecule in a very specific manner. The method is also ideal for launching a quick attack against novel viral threats, for example, pathogenic viruses that are transferred from animals to humans in a zoonotic transmission event. CRISPR-Cas therapeutics are able to fully inactivate viral reservoirs, overcoming one of the hurdles of current antiviral drug treatments of chronic virus infections. The success of such a therapeutic approach relies on an impeccable gRNA design that must target highly conserved viral sequences and must combine several gRNAs to prevent viral escape. Optimal gRNA design combined with progressive improvement of the CRISPR-Cas safety and optimized delivery systems should come together to realize the enormous potential of CRISPR-Cas-based antiviral approaches. In addition, CRISPR-based viral detection systems have proven to be a rapid and sensitive manner to detect viral infections.

References

1. Andersen, P. I., Ianevski, A., Lysvand, H., Vitkauskiene, A., Oksenych, V., Bjoras, M., Telling, K., Lutsar, I., Dumpis, U., Irie, Y., et al. (2020). Discovery and development of safe-in-man broad-spectrum antiviral agents. *International Journal of Infectious Diseases, 93*, 268–276.
2. Graham, B. S. (2013). Advances in antiviral vaccine development. *Immunological Reviews, 255*, 230–242.
3. Calina, D., Docea, A. O., Petrakis, D., Egorov, A. M., Ishmukhametov, A. A., Gabibov, A. G., Shtilman, M. I., Kostoff, R., Carvalho, F., Vinceti, M., et al. (2020). Towards effective COVID19 vaccines: Updates, perspectives and challenges (Review). *International Journal of Molecular Medicine, 46*, 3–16.
4. De Clercq, E., & Li, G. (2016). Approved antiviral drugs over the past 50 years. *Clinical Microbiology Reviews, 29*, 695–747.
5. Jin, D. Y., & Berkhout, B. (2018). CRISPR-Cas antiviral strategies. *Virus Research, 244*, 285.
6. Ishino, Y., Shinagawa, H., Makino, K., Amemura, M., & Nakata, A. (1987). Nucleotide sequence of the iap gene, responsible for alkaline phosphatase isozyme conversion in Escherichia coli, and identification of the gene product. *Journal of Bacteriology, 169*, 5429–5433.
7. Jansen, R., van Embden, J. D. A., Gaastra, W., & Schouls, L. M. (2002). Identification of genes that are associated with DNA repeats in prokaryotes. *Molecular Microbiology, 43*, 1565–1575.
8. Ledford, H. (2016). The unsung heroes of CRISPR. *Nature, 535*, 342–344.
9. Lander, E. S. (2016). The heroes of CRISPR. *Cell, 164*, 18–28.
10. Mojica, F. J., Diez-Villasenor, C., Garcia-Martinez, J., & Soria, E. (2005). Intervening sequences of regularly spaced prokaryotic repeats derive from foreign genetic elements. *Journal of Molecular Evolution, 60*, 174–182.
11. Pourcel, C., Salvignol, G., & Vergnaud, G. (2005). CRISPR elements in Yersinia pestis acquire new repeats by preferential uptake of bacteriophage DNA, and provide additional tools for evolutionary studies. *Microbiology, 151*, 653–663.
12. Bolotin, A., Ouinquis, B., Sorokin, A., & Ehrlich, S. D. (2005). Clustered regularly interspaced short palindrome repeats (CRISPRs) have spacers of extrachromosomal origin. *Microbiology-SGM, 151*, 2551–2561.

13. van der Oost, J., Jore, M. M., Westra, E. R., Lundgren, M., & Brouns, S. J. (2009). CRISPR-based adaptive and heritable immunity in prokaryotes. *Trends in Biochemical Sciences, 34*, 401–407.
14. Patterson, A. G., Yevstigneyeva, M. S., & Fineran, P. C. (2017). Regulation of CRISPR-Cas adaptive immune systems. *Current Opinion in Microbiology, 37*, 1–7.
15. Marraffini, L. A. (2015). CRISPR-Cas immunity in prokaryotes. *Nature, 526*, 55–61.
16. Garneau, J. E., Dupuis, M. E., Villion, M., Romero, D. A., Barrangou, R., Boyaval, P., Fremaux, C., Horvath, P., Magadan, A. H., & Moineau, S. (2010). The CRISPR/Cas bacterial immune system cleaves bacteriophage and plasmid DNA. *Nature, 468*, 67–71.
17. Jinek, M., Chylinski, K., Fonfara, I., Hauer, M., Doudna, J. A., & Charpentier, E. (2012). A programmable dual-RNA-guided DNA endonuclease in adaptive bacterial immunity. *Science, 337*, 816–821.
18. Deltcheva, E., Chylinski, K., Sharma, C. M., Gonzales, K., Chao, Y., Pirzada, Z. A., Eckert, M. R., Vogel, J., & Charpentier, E. (2011). CRISPR RNA maturation by trans-encoded small RNA and host factor RNase III. *Nature, 471*, 602–607.
19. Sternberg, S. H., Redding, S., Jinek, M., Greene, E. C., & Doudna, J. A. (2014). DNA interrogation by the CRISPR RNA-guided endonuclease Cas9. *Nature, 507*, 62–67.
20. Mali, P., Yang, L., Esvelt, K. M., Aach, J., Guell, M., DiCarlo, J. E., Norville, J. E., & Church, G. M. (2013). RNA-guided human genome engineering via Cas9. *Science, 339*, 823–826.
21. Cho, S. W., Kim, S., Kim, J. M., & Kim, J. S. (2013). Targeted genome engineering in human cells with the Cas9 RNA-guided endonuclease. *Nature Biotechnology, 31*, 230–232.
22. Savic, N., & Schwank, G. (2016). Advances in therapeutic CRISPR/Cas9 genome editing. *Translational Research, 168*, 15–21.
23. Bannikov, A. V., & Lavrov, A. V. (2017). [CRISPR/CAS9, the King of Genome Editing Tools]. *Molecular Biology (Mosk), 51*, 582–594. 2017/09/14 ed.
24. Makarova, K. S., Wolf, Y. I., Alkhnbashi, O. S., Costa, F., Shah, S. A., Saunders, S. J., Barrangou, R., Brouns, S. J., Charpentier, E., Haft, D. H., et al. (2015). An updated evolutionary classification of CRISPR-Cas systems. *Nature Reviews. Microbiology, 13*, 722–736.
25. Mohanraju, P., Makarova, K. S., Zetsche, B., Zhang, F., Koonin, E. V., & van der Oost, J. (2016). Diverse evolutionary roots and mechanistic variations of the CRISPR-Cas systems. *Science, 353*, aad5147.
26. Doudna, J. A., & Charpentier, E. (2014). Genome editing. The new frontier of genome engineering with CRISPR-Cas9. *Science, 346*, 1258096.
27. Makarova, K. S., & Koonin, E. V. (2015). Annotation and classification of CRISPR-Cas Systems. *Methods in Molecular Biology, 1311*, 47–75.
28. Chylinski, K., Makarova, K. S., Charpentier, E., & Koonin, E. V. (2014). Classification and evolution of type II CRISPR-Cas systems. *Nucleic Acids Research, 42*, 6091–6105.
29. Jiang, F., & Doudna, J. A. (2017). CRISPR-Cas9 structures and mechanisms. *Annual Review of Biophysics, 46*, 505–529.
30. Chylinski, K., Le Rhun, A., & Charpentier, E. (2013). The tracrRNA and Cas9 families of type II CRISPR-Cas immunity systems. *RNA Biology, 10*, 726–737.
31. Hale, C. R., Zhao, P., Olson, S., Duff, M. O., Graveley, B. R., Wells, L., Terns, R. M., & Terns, M. P. (2009). RNA-guided RNA cleavage by a CRISPR RNA-Cas protein complex. *Cell, 139*, 945–956.
32. Gasiunas, G., Barrangou, R., Horvath, P., & Siksnys, V. (2012). Cas9-crRNA ribonucleoprotein complex mediates specific DNA cleavage for adaptive immunity in bacteria. *Proceedings of the National Academy of Sciences of the United States of America, 109*, E2579–E2586.
33. Nishimasu, H., Ran, F. A., Hsu, P. D., Konermann, S., Shehata, S. I., Dohmae, N., Ishitani, R., Zhang, F., & Nureki, O. (2014). Crystal structure of Cas9 in complex with guide RNA and target DNA. *Cell, 156*, 935–949.
34. Anders, C., Niewoehner, O., Duerst, A., & Jinek, M. (2014). Structural basis of PAM-dependent target DNA recognition by the Cas9 endonuclease. *Nature, 513*, 569–573.

35. Hsu, P. D., Scott, D. A., Weinstein, J. A., Ran, F. A., Konermann, S., Agarwala, V., Li, Y., Fine, E. J., Wu, X., Shalem, O., et al. (2013). DNA targeting specificity of RNA-guided Cas9 nucleases. *Nature Biotechnology, 31*, 827–832.
36. Tang, L., Yang, F., He, X., Xie, H., Liu, X., Fu, J., Xi, H., Lu, X., Liu, C., Song, Z., et al. (2019). Efficient cleavage resolves PAM preferences of CRISPR-Cas in human cells. *Cell Regeneration, 8*, 44–50.
37. Lieber, M. R. (2010). The mechanism of double-strand DNA break repair by the nonhomologous DNA end-joining pathway. *Annual Review of Biochemistry, 79*, 181–211.
38. San Filippo, J., Sung, P., & Klein, H. (2008). Mechanism of eukaryotic homologous recombination. *Annual Review of Biochemistry, 77*, 229–257.
39. Qi, L. S., Larson, M. H., Gilbert, L. A., Doudna, J. A., Weissman, J. S., Arkin, A. P., & Lim, W. A. (2013). Repurposing CRISPR as an RNA-guided platform for sequence-specific control of gene expression. *Cell, 152*, 1173–1183.
40. Perez-Pinera, P., Kocak, D. D., Vockley, C. M., Adler, A. F., Kabadi, A. M., Polstein, L. R., Thakore, P. I., Glass, K. A., Ousterout, D. G., Leong, K. W., et al. (2013). RNA-guided gene activation by CRISPR-Cas9-based transcription factors. *Nature Methods, 10*, 973–976.
41. Bikard, D., Jiang, W. Y., Samai, P., Hochschild, A., Zhang, F., & Marraffini, L. A. (2013). Programmable repression and activation of bacterial gene expression using an engineered CRISPR-Cas system. *Nucleic Acids Research, 41*, 7429–7437.
42. Chavez, A., Scheiman, J., Vora, S., Pruitt, B. W., Tuttle, M., Iyer, E. P. R., Lin, S. L., Kiani, S., Guzman, C. D., Wiegand, D. J., et al. (2015). Highly efficient Cas9-mediated transcriptional programming. *Nature Methods, 12*, 326–U365.
43. Konermann, S., Brigham, M. D., Trevino, A. E., Joung, J., Abudayyeh, O. O., Barcena, C., Hsu, P. D., Habib, N., Gootenberg, J. S., Nishimasu, H., et al. (2015). Genome-scale transcriptional activation by an engineered CRISPR-Cas9 complex. *Nature, 517*, 583–U332.
44. Tanenbaum, M. E., Gilbert, L. A., Qi, L. S., Weissman, J. S., & Vale, R. D. (2014). A protein-tagging system for signal amplification in gene expression and fluorescence imaging. *Cell, 159*, 635–646.
45. Gilbert, L. A., Horlbeck, M. A., Adamson, B., Villalta, J. E., Chen, Y., Whitehead, E. H., Guimaraes, C., Panning, B., Ploegh, H. L., Bassik, M. C., et al. (2014). Genome-scale CRISPR-mediated control of gene repression and activation. *Cell, 159*, 647–661.
46. Larson, M. H., Gilbert, L. A., Wang, X., Lim, W. A., Weissman, J. S., & Qi, L. S. (2013). CRISPR interference (CRISPRi) for sequence-specific control of gene expression. *Nature Protocols, 8*, 2180–2196.
47. Sampson, T. R., Saroj, S. D., Llewellyn, A. C., Tzeng, Y. L., & Weiss, D. S. (2013). A CRISPR/Cas system mediates bacterial innate immune evasion and virulence. *Nature, 497*, 254–257.
48. Price, A. A., Sampson, T. R., Ratner, H. K., Grakoui, A., & Weiss, D. S. (2015). Cas9-mediated targeting of viral RNA in eukaryotic cells. *Proceedings of the National Academy of Sciences of the United States of America, 112*, 6164–6169.
49. Moreno-Mateos, M. A., Fernandez, J. P., Rouet, R., Vejnar, C. E., Lane, M. A., Mis, E., Khokha, M. K., Doudna, J. A., & Giraldez, A. J. (2017). CRISPR-Cpf1 mediates efficient homology-directed repair and temperature-controlled genome editing. *Nature Communications, 8*, 2024.
50. Zetsche, B., Gootenberg, J. S., Abudayyeh, O. O., Slaymaker, I. M., Makarova, K. S., Essletzbichler, P., Volz, S. E., Joung, J., van der Oost, J., Regev, A., et al. (2015). Cpf1 is a single RNA-guided endonuclease of a class 2 CRISPR-Cas system. *Cell, 163*, 759–771.
51. Kleinstiver, B. P., Tsai, S. Q., Prew, M. S., Nguyen, N. T., Welch, M. M., Lopez, J. M., McCaw, Z. R., Aryee, M. J., & Joung, J. K. (2016). Genome-wide specificities of CRISPR-Cas Cpf1 nucleases in human cells. *Nature Biotechnology, 34*, 869–874.
52. Teng, F., Cui, T., Feng, G., Guo, L., Xu, K., Gao, Q., Li, T., Li, J., Zhou, Q., & Li, W. (2018). Repurposing CRISPR-Cas12b for mammalian genome engineering. *Cell Discovery, 4*, 63.

53. Strecker, J., Jones, S., Koopal, B., Schmid-Burgk, J., Zetsche, B., Gao, L., Makarova, K. S., Koonin, E. V., & Zhang, F. (2019). Engineering of CRISPR-Cas12b for human genome editing. *Nature Communications, 10*, 212.
54. Karvelis, T., Bigelyte, G., Young, J. K., Hou, Z., Zedaveinyte, R., Budre, K., Paulraj, S., Djukanovic, V., Gasior, S., Silanskas, A., et al. (2020). PAM recognition by miniature CRISPR-Cas12f nucleases triggers programmable double-stranded DNA target cleavage. *Nucleic Acids Research, 48*, 5016–5023.
55. Gao, Z., Fan, M., Das, A. T., Herrera-Carrillo, E., & Berkhout, B. (2020). Extinction of all infectious HIV in cell culture by the CRISPR-Cas12a system with only a single crRNA. *Nucleic Acids Research*. https://doi.org/10.1093/nar/gkaa226
56. Cox, D. B. T., Gootenberg, J. S., Abudayyeh, O. O., Franklin, B., Kellner, M. J., Joung, J., & Zhang, F. (2017). RNA editing with CRISPR-Cas13. *Science, 358*, 1019–1027.
57. Abudayyeh, O. O., Gootenberg, J. S., Essletzbichler, P., Han, S., Joung, J., Belanto, J. J., Verdine, V., Cox, D. B. T., Kellner, M. J., Regev, A., et al. (2017). RNA targeting with CRISPR-Cas13. *Nature, 550*, 280–284.
58. Zhang, C., Konermann, S., Brideau, N. J., Lotfy, P., Wu, X. B., Novick, S. J., Strutzenberg, T., Griffin, P. R., Hsu, P. D., & Lyumkis, D. (2018). Structural basis for the RNA-guided ribonuclease activity of CRISPR-Cas13d. *Cell, 175*, 212–223.e17.
59. Mahas, A., Aman, R., & Mahfouz, M. (2019). CRISPR-Cas13d mediates robust RNA virus interference in plants. *Genome Biology, 20*, 263.
60. Slaymaker, I. M., Mesa, P., Kellner, M. J., Kannan, S., Brignole, E., Koob, J., Feliciano, P. R., Stella, S., Abudayyeh, O. O., Gootenberg, J. S., et al. (2019). High-resolution structure of Cas13b and biochemical characterization of RNA targeting and cleavage. *Cell Reports, 26*, 3741–3751 e3745.
61. Xu, C., Zhou, Y., Xiao, Q., He, B., Geng, G., Wang, Z., Cao, B., Dong, X., Bai, W., Wang, Y., et al. (2021). Programmable RNA editing with compact CRISPR-Cas13 systems from uncultivated microbes. *Nature Methods, 18*, 499–506.
62. Abudayyeh, O. O., Gootenberg, J. S., Konermann, S., Joung, J., Slaymaker, I. M., Cox, D. B. T., Shmakov, S., Makarova, K. S., Semenova, E., Minakhin, L., et al. (2016). C2c2 is a single-component programmable RNA-guided RNA-targeting CRISPR effector. *Science, 353*, aaf5573.
63. East-Seletsky, A., O'Connell, M. R., Knight, S. C., Burstein, D., Cate, J. H. D., Tjian, R., & Doudna, J. A. (2016). Two distinct RNase activities of CRISPR-C2c2 enable guide-RNA processing and RNA detection. *Nature, 538*, 270–273.
64. Smargon, A. A., Cox, D. B. T., Pyzocha, N. K., Zheng, K. J., Slaymaker, I. M., Gootenberg, J. S., Abudayyeh, O. A., Essletzbichler, P., Shmakov, S., Makarova, K. S., et al. (2017). Cas13b is a type VI-B CRISPR-associated RNA-guided RNase differentially regulated by accessory proteins Csx27 and Csx28. *Molecular Cell, 65*, 618–630.e7.
65. Konermann, S., Lotfy, P., Brideau, N. J., Oki, J., Shokhirev, M. N., & Hsu, P. D. (2018). Transcriptome engineering with RNA-targeting Type VI-D CRISPR effectors. *Cell, 173*(665–676), e614.
66. Harrington, L. B., Burstein, D., Chen, J. S., Paez-Espino, D., Ma, E., Witte, I. P., Cofsky, J. C., Kyrpides, N. C., Banfield, J. F., & Doudna, J. A. (2018). Programmed DNA destruction by miniature CRISPR-Cas14 enzymes. *Science, 362*, 839–842.
67. Zhang, B., Wang, Q., Xu, X., Xia, Q., Long, F., Li, W., Shui, Y., Xia, X., & Wang, J. (2018). Detection of target DNA with a novel Cas9/sgRNAs-associated reverse PCR (CARP) technique. *Analytical and Bioanalytical Chemistry, 410*, 2889–2900.
68. Pardee, K., Green, A. A., Takahashi, M. K., Braff, D., Lambert, G., Lee, J. W., Ferrante, T., Ma, D., Donghia, N., Fan, M., et al. (2016). Rapid, low-cost detection of Zika virus using programmable biomolecular components. *Cell, 165*, 1255–1266.
69. Huang, M. Q., Zhou, X. M., Wang, H. Y., & Xing, D. (2018). Clustered regularly interspaced short palindromic repeats/Cas9 triggered isothermal amplification for site-specific nucleic acid detection. *Analytical Chemistry, 90*, 2193–2200.

70. Teng, F., Guo, L., Cui, T. T., Wang, X. G., Xu, K., Gao, Q. Q., Zhou, Q., & Li, W. (2019). CDetection: CRISPR-Cas12b-based DNA detection with sub-attomolar sensitivity and single-base specificity. *Genome Biology, 20*, 132.
71. Chen, J. S., Ma, E. B., Harrington, L. B., Da Costa, M., Tian, X. R., Palefsky, J. M., & Doudna, J. A. (2018). CRISPR-Cas12a target binding unleashes indiscriminate single-stranded DNase activity. *Science, 360*, 436–439.
72. Li, S. Y., Cheng, Q. X., Liu, J. K., Nie, X. Q., Zhao, G. P., & Wang, J. (2018). CRISPR-Cas12a has both cis- and trans-cleavage activities on single-stranded DNA. *Cell Research, 28*, 491–493.
73. Kellner, M. J., Koob, J. G., Gootenberg, J. S., Abudayyeh, O. O., & Zhang, F. (2019). SHERLOCK: Nucleic acid detection with CRISPR nucleases. *Nature Protocols, 14*, 2986–3012.
74. Gootenberg, J. S., Abudayyeh, O. O., Lee, J. W., Essletzbichler, P., Dy, A. J., Joung, J., Verdine, V., Donghia, N., Daringer, N. M., Freije, C. A., et al. (2017). Nucleic acid detection with CRISPR-Cas13a/C2c2. *Science, 356*, 438–442.
75. Myhrvold, C., Freije, C. A., Gootenberg, J. S., Abudayyeh, O. O., Metsky, H. C., Durbin, A. F., Kellner, M. J., Tan, A. L., Paul, L. M., Parham, L. A., et al. (2018). Field-deployable viral diagnostics using CRISPR-Cas13. *Science, 360*, 444–448.
76. Liu, Y. F., Xu, H. P., Liu, C., Peng, L. J., Khan, H., Cui, L. B., Huang, R., Wu, C., Shen, S. S., Wang, S., et al. (2019). CRISPR-Cas13a nanomachine based simple technology for avian influenza A (H7N9) virus on-site detection. *Journal of Biomedical Nanotechnology, 15*, 790–798.
77. Wu, Y. T., Liu, S. X., Wang, F., & Zeng, M. S. (2019). Room temperature detection of plasma epstein-barr virus DNA with CRISPR-Cas13. *Clinical Chemistry, 65*, 591–592.
78. Joung, J., Ladha, A., Saito, M., Kim, N. G., Woolley, A. E., Segel, M., Barretto, R. P. J., Ranu, A., Macrae, R. K., Faure, G., et al. (2020). Detection of SARS-CoV-2 with SHERLOCK one-pot testing. *New England Journal of Medicine, 383*, 1492–1494.
79. Normandin, E., Solomon, I. H., Zamirpour, S., Lemieux, J., Freije, C. A., Mukerji, S. S., Tomkins-Tinch, C., Park, D., Sabeti, P. C., & Piantadosi, A. (2020). Powassan virus neuropathology and genomic diversity in patients with fatal encephalitis. *Open Forum Infect Dis, 7*, ofaa392.
80. Barnes, K. G., Lachenauer, A. E., Nitido, A., Siddiqui, S., Gross, R., Beitzel, B., Siddle, K. J., Freije, C. A., Dighero-Kemp, B., Mehta, S. B., et al. (2020). Deployable CRISPR-Cas13a diagnostic tools to detect and report Ebola and Lassa virus cases in real-time. *Nature Communications, 11*, 4131.
81. Curti, L. A., Pereyra-Bonnet, F., Repizo, G. D., Fay, J. V., Salvatierra, K., Blariza, M. J., Ibanez-Alegre, D., Rinflerch, A. R., Miretti, M., & Gimenez, C. A. (2020). CRISPR-based platform for carbapenemases and emerging viruses detection using Cas12a (Cpf1) effector nuclease. *Emerging Microbes and Infecection, 9*, 1140–1148.
82. Broughton, J. P., Deng, X., Yu, G., Fasching, C. L., Servellita, V., Singh, J., Miao, X., Streithorst, J. A., Granados, A., Sotomayor-Gonzalez, A., et al. (2020). CRISPR-Cas12-based detection of SARS-CoV-2. *Nature Biotechnology, 38*, 870–874.
83. Ali, Z., Aman, R., Mahas, A., Rao, G. S., Tehseen, M., Marsic, T., Salunke, R., Subudhi, A. K., Hala, S. M., Hamdan, S. M., et al. (2020). iSCAN: An RT-LAMP-coupled CRISPR-Cas12 module for rapid, sensitive detection of SARS-CoV-2. *Virus Research, 288*, 198129.
84. Ackerman, C. M., Myhrvold, C., Thakku, S. G., Freije, C. A., Metsky, H. C., Yang, D. K., Ye, S. H., Boehm, C. K., Kosoko-Thoroddsen, T. S. F., Kehe, J., et al. (2020). Massively multiplexed nucleic acid detection with Cas13. *Nature, 582*, 277–282.
85. Ning, B., Yu, T., Zhang, S. W., Huang, Z., Tian, D., Lin, Z., Niu, A., Golden, N., Hensley, K., Threeton, B., et al. (2021). A smartphone-read ultrasensitive and quantitative saliva test for COVID-19. *Science Advances, 7*, eabe3703.
86. Fozouni, P., Son, S. M., Derby, M. D. D., Knott, G. J., Gray, C. N., D'Ambrosio, M. V., Zhao, C. Y., Switz, N. A., Kumar, G. R., Stephens, S. I., et al. (2021). Amplification-free

detection of SARS-CoV-2 with CRISPR-Cas13a and mobile phone microscopy. *Cell, 184*, 323–333.e9.
87. Chen, Y. J., Shi, Y., Chen, Y., Yang, Z., Wu, H., Zhou, Z. H., Li, J., Ping, J. F., He, L. P., Shen, H., et al. (2020). Contamination-free visual detection of SARS-CoV-2 with CRISPR/Cas12a: A promising method in the point-of-care detection. *Biosensors & Bioelectronics, 169*, 112642.
88. Arizti-Sanz, J., Freije, C. A., Stanton, A. C., Petros, B. A., Boehm, C. K., Siddiqui, S., Shaw, B. M., Adams, G., Kosoko-Thoroddsen, T. S. F., Kemball, M. E., et al. (2020). Streamlined inactivation, amplification, and Cas13-based detection of SARS-CoV-2. *Nature Communications, 11*, 5921.
89. Ramachandran, A., Huyke, D. A., Sharma, E., Sahoo, M. K., Huang, C., Banaei, N., Pinsky, B. A., & Santiago, J. G. (2020). Electric field-driven microfluidics for rapid CRISPR-based diagnostics and its application to detection of SARS-CoV-2. *Proceedings of the National Academy of Sciences of the United States of America, 117*, 29518–29525.
90. Huang, Z., Tian, D., Liu, Y., Lin, Z., Lyon, C. J., Lai, W. H., Fusco, D., Drouin, A., Yin, X. M., Hu, T., et al. (2020). Ultra-sensitive and high-throughput CRISPR-p owered COVID-19 diagnosis. *Biosensors & Bioelectronics, 164*, 112316.
91. Hou, T. Y., Zeng, W. Q., Yang, M. L., Chen, W. J., Ren, L. L., Ai, J. W., Wu, J., Liao, Y. L., Gou, X. J., Li, Y. J., et al. (2020). Development and evaluation of a rapid CRISPR-based diagnostic for COVID-19. *PLoS Pathogens, 16*, e1008705.
92. Patchsung, M., Jantarug, K., Pattama, A., Aphicho, K., Suraritdechachai, S., Meesawat, P., Sappakhaw, K., Leelahakorn, N., Ruenkam, T., Wongsatit, T., et al. (2020). Clinical validation of a Cas13-based assay for the detection of SARS-CoV-2 RNA. *Nature Biomedical Engineering, 4*, 1140–1149.
93. Guo, L., Sun, X., Wang, X., Liang, C., Jiang, H., Gao, Q., Dai, M., Qu, B., Fang, S., Mao, Y., et al. (2020). SARS-CoV-2 detection with CRISPR diagnostics. *Cell Discovery, 6*, 34.
94. Aquino-Jarquin, G. (2019). CRISPR-Cas14 is now part of the artillery for gene editing and molecular diagnostic. *Nanomedicine: Nanotechnology, Biology and Medicine, 18*, 428–431.
95. Khan, M. Z., Haider, S., Mansoor, S., & Amin, I. (2019). Targeting plant ssDNA viruses with engineered miniature CRISPR-Cas14a. *Trends in Biotechnology, 37*, 800–804.
96. Colomer-Lluch, M., Ruiz, A., Moris, A., & Prado, J. G. (2018). Restriction factors: From intrinsic viral restriction to shaping cellular immunity against HIV-1. *Frontiers in Immunology, 9*, 2876.
97. Zhang, F., Wen, Y., & Guo, X. (2014). CRISPR/Cas9 for genome editing: Progress, implications and challenges. *Human Molecular Genetics, 23*, R40–R46.
98. Check, E. (2002). A tragic setback. *Nature, 420*, 116–118.
99. Lino, C. A., Harper, J. C., Carney, J. P., & Timlin, J. A. (2018). Delivering CRISPR: A review of the challenges and approaches. *Drug Delivery, 25*, 1234–1257.
100. Kim, C. W., & Chang, K. M. (2013). Hepatitis C virus: Virology and life cycle. *Clinical and Molecular Hepatology, 19*, 17–25.
101. Sesmero, E., & Thorpe, I. F. (2015). Using the hepatitis C virus RNA-dependent RNA polymerase as a model to understand viral polymerase structure, function and dynamics. *Viruses-Basel, 7*, 3974–3994.
102. Moyo, B., Bloom, K., Scott, T., Ely, A., & Arbuthnot, P. (2018). Advances with using CRISPR/Cas-mediated gene editing to treat infections with hepatitis B virus and hepatitis C virus. *Virus Research, 244*, 311–320.
103. Ashraf, M. U., Salman, H. M., Khalid, M. F., Khan, M. H. F., Anwar, S., Afzal, S., Idrees, M., & Chaudhary, S. U. (2021). CRISPR-Cas13a mediated targeting of hepatitis C virus internal-ribosomal entry site (IRES) as an effective antiviral strategy. *Biomedicine & Pharmacotherapy, 136*, 111239.
104. Randall, G., Panis, M., Cooper, J. D., Tellinghuisen, T. L., Sukhodolets, K. E., Pfeffer, S., Landthaler, M., Landgraf, P., Kan, S., Lindenbach, B. D., et al. (2007). Cellular cofactors

affecting hepatitis C virus infection and replication. *Proceedings of the National Academy of Sciences of the United States of America, 104*, 12884–12889.
105. Tai, A. W., Benita, Y., Peng, L. F., Kim, S. S., Sakamoto, N., Xavier, R. J., & Chung, R. T. (2009). A functional genomic screen identifies cellular cofactors of hepatitis C virus replication. *Cell Host & Microbe, 5*, 298–307.
106. Supekova, L., Supek, F., Lee, J., Chen, S., Gray, N., Pezacki, J. P., Schlapbach, A., & Schultz, P. G. (2008). Identification of human kinases involved in hepatitis C virus replication by small interference RNA library screening. *The Journal of Biological Chemistry, 283*, 29–36.
107. Hamamoto, I., Nishimura, Y., Okamoto, T., Aizaki, H., Liu, M. Y., Mori, Y., Abe, T., Suzuki, T., Lai, M. M. C., Miyamura, T., et al. (2005). Human VAP-B is involved in hepatitis C virus replication through interaction with NS5A and NS5B. *Journal of Virology, 79*, 13473–13482.
108. Zhao, Y. L., Ren, Y. S., Zhang, X. P., Zhao, P., Tao, W. Y., Zhong, J., Li, Q., & Zhang, X. L. (2014). Ficolin-2 inhibits hepatitis C virus infection, whereas apolipoprotein E3 mediates viral immune escape. *Journal of Immunology, 193*, 783–796.
109. Wilkins, C., Woodward, J., Lau, D. T. Y., Barnes, A., Joyce, M., McFarlane, N., McKeating, J. A., Tyrrell, D. L., & Gale, M. (2013). IFITM1 Is a tight junction protein that inhibits hepatitis C virus entry. *Hepatology, 57*, 461–469.
110. Bukong, T. N., Kodys, K., & Szabo, G. (2013). Human ezrin-moesin-radixin proteins modulate hepatitis C virus infection. *Hepatology, 58*, 1569–1579.
111. Metz, P., Dazert, E., Ruggieri, A., Mazur, J., Kaderali, L., Kaul, A., Zeuge, U., Windisch, M. P., Trippler, M., Lohmann, V., et al. (2012). Identification of type I and type II interferon-induced effectors controlling hepatitis C virus replication. *Hepatology, 56*, 2082–2093.
112. Amet, T., Byrd, D., Hu, N., Sun, Q., Li, F., Zhao, Y., Hu, S., Grantham, A., & Yu, Q. (2014). BST-2 expression in human hepatocytes is inducible by all three types of interferons and restricts production of hepatitis C virus. *Current Molecular Medicine, 14*, 349–360.
113. Pan, X. B., Qu, X. W., Jiang, D., Zhao, X. L., Han, J. C., & Wei, L. (2013). BST2/Tetherin inhibits hepatitis C virus production in human hepatoma cells. *Antiviral Research, 98*, 54–60.
114. German Advisory Committee Blood, SAoPTbB. (2016). Human immunodeficiency virus (HIV). *Transfusion Medicine and Hemotherapy, 43*, 203–222.
115. Pace, M., & Frater, J. (2014). A cure for HIV: Is it in sight? *Expert Review of Anti-Infective Therapy, 12*, 783–791.
116. Yin, L., Zhao, F., Sun, H., Wang, Z., Huang, Y., Zhu, W., Xu, F., Mei, S., Liu, X., Zhang, D., et al. (2020). CRISPR-Cas13a inhibits HIV-1 infection. *Molecular Therapy - Nucleic Acids, 21*, 147–155.
117. Westerhout, E. M., Ter Brake, O., & Berkhout, B. (2006). The virion-associated incoming HIV-1 RNA genome is not targeted by RNA interference. *Retrovirology, 3*, 57–65.
118. Liao, H. K., Gu, Y., Diaz, A., Marlett, J., Takahashi, Y., Li, M., Suzuki, K., Xu, R., Hishida, T., Chang, C. J., et al. (2015). Use of the CRISPR/Cas9 system as an intracellular defense against HIV-1 infection in human cells. *Nature Communications, 6*, 6413.
119. Hu, W. H., Kaminski, R., Yang, F., Zhang, Y. G., Cosentino, L., Li, F., Luo, B. A., Alvarez-Carbonell, D., Garcia-Mesa, Y., Karn, J., et al. (2014). RNA-directed gene editing specifically eradicates latent and prevents new HIV-1 infection. *Proceedings of the National Academy of Sciences United States of America, 111*, 11461–11466.
120. Ebina, H., Misawa, N., Kanemura, Y., & Koyanagi, Y. (2013). Harnessing the CRISPR/Cas9 system to disrupt latent HIV-1 provirus. *Scientific Reports (UK), 3*, 2510.
121. Ueda, S., Ebina, H., Kanemura, Y., Misawa, N., & Koyanagi, Y. (2016). Anti-HIV-1 potency of the CRISPR/Cas9 system insufficient to fully inhibit viral replication. *Microbiology and Immunology, 60*, 483–496.
122. Yoder, K. E., & Bundschuh, R. (2016). Host double strand break repair generates HIV-1 strains resistant to CRISPR/Cas9. *Scientific Reports, 6*, 29530.
123. Yin, C., Zhang, T., Qu, X., Zhang, Y., Putatunda, R., Xiao, X., Li, F., Xiao, W., Zhao, H., Dai, S., et al. (2017). In vivo excision of HIV-1 provirus by saCas9 and multiplex single-guide RNAs in animal models. *Molecular Therapy, 25*, 1168–1186.

124. Kaminski, R., Bella, R., Yin, C., Otte, J., Ferrante, P., Gendelman, H. E., Li, H., Booze, R., Gordon, J., Hu, W., et al. (2016). Excision of HIV-1 DNA by gene editing: A proof-of-concept in vivo study. *Gene Therapy, 23*, 690–695.
125. Kaminski, R., Chen, Y., Fischer, T., Tedaldi, E., Napoli, A., Zhang, Y., Karn, J., Hu, W., & Khalili, K. (2016). Elimination of HIV-1 genomes from human T-lymphoid cells by CRISPR/Cas9 gene editing. *Scientific Reports, 6*, 22555.
126. Zhu, W., Lei, R., Le Duff, Y., Li, J., Guo, F., Wainberg, M. A., & Liang, C. (2015). The CRISPR/Cas9 system inactivates latent HIV-1 proviral DNA. *Retrovirology, 12*, 22.
127. Wang, Z., Pan, Q. H., Gendron, P., Zhu, W. J., Guo, F., Cen, S., Wainberg, M. A., & Liang, C. (2016). CRISPR/Cas9-derived mutations both inhibit HIV-1 replication and accelerate viral escape. *Cell Reports, 15*, 481–489.
128. Wang, G., Zhao, N., Berkhout, B., & Das, A. T. (2016). CRISPR-Cas9 can inhibit HIV-1 replication but NHEJ repair facilitates virus escape. *Molecular Therapy, 24*, 522–526.
129. Wang, Z., Wang, W. Z., Cui, Y. C., Pan, Q. H., Zhu, W. J., Gendron, P., Guo, F., Cen, S., Witcher, M., & Liang, C. (2018). HIV-1 employs multiple mechanisms to resist Cas9/single guide RNA targeting the viral primer binding site. *Journal of Virology, 92*, e01135–e01118.
130. Lebbink, R. J., de Jong, D. C. M., Wolters, F., Kruse, E. M., van Ham, P. M., Wiertz, E. J. H. J., & Nijhuis, M. (2017). Combinational CRISPR/Cas9 gene-editing approach can halt HIV replication and prevent viral escape. *Scientific Reports (UK), 7*, 41968.
131. Wang, G., Zhao, N., Berkhout, B., & Das, A. T. (2016). A combinatorial CRISPR-Cas9 attack on HIV-1 DNA extinguishes all infectious provirus in infected T cell cultures. *Cell Reports, 17*, 2819–2826.
132. Das, A. T., Binda, C. S., & Berkhout, B. (2019). Elimination of infectious HIV DNA by CRISPR-Cas9. *Current Opinion in Virology, 38*, 81–88.
133. Binda, C. S., Klaver, B., Berkhout, B., & Das, A. T. (2020). CRISPR-Cas9 Dual-gRNA attack causes mutation, excision and inversion of the HIV-1 proviral DNA. *Viruses, 12*, 330.
134. Darcis, G., Binda, C. S., Klaver, B., Herrera-Carrillo, E., Berkhout, B., & Das, A. T. (2019). The impact of HIV-1 genetic diversity on CRISPR-Cas9 antiviral activity and viral escape. *Viruses, 11*, 255.
135. Fagerlund, R. D., Staals, R. H., & Fineran, P. C. (2015). The Cpf1 CRISPR-Cas protein expands genome-editing tools. *Genome Biology, 16*, 251.
136. Gao, Z., Herrera-Carrillo, E., & Berkhout, B. (2018). Improvement of the CRISPR-Cpf1 system with ribozyme-processed crRNA. *RNA Biology, 15*, 1458–1467.
137. Saayman, S. M., Lazar, D. C., Scott, T. A., Hart, J. R., Takahashi, M., Burnett, J. C., Planelles, V., Morris, K. V., & Weinberg, M. S. (2016). Potent and targeted activation of latent HIV-1 using the CRISPR/dCas9 activator complex. *Molecular Therapy, 24*, 488–498.
138. Zhang, Y. G., Yin, C. R., Zhang, T., Li, F., Yang, W. S., Kaminski, R., Fagan, P. R., Putatunda, R., Young, W. B., Khalili, K., et al. (2015). CRISPR/gRNA-directed synergistic activation mediator (SAM) induces specific, persistent and robust reactivation of the HIV-1 latent reservoirs. *Scientific Reports (UK), 5*, 16277.
139. Limsirichai, P., Gaj, T., & Schaffer, D. V. (2016). CRISPR-mediated activation of latent HIV-1 expression. *Molecular Therapy, 24*, 499–507.
140. Saayman, S. M., Lazar, D. C., Scott, T. A., Burnett, J., Takahashi, M., Weinberg, M. S., & Morris, K. V. (2015). Potent and targeted activation of latent HIV-1 using multiplexed guide RNAs and the CRISPR/dCas9 activator complex. *Molecular Therapy, 23*, S276–S276.
141. Cocchi, F., Devico, A. L., Garzinodemo, A., Arya, S. K., Gallo, R. C., & Lusso, P. (1995). Identification of rantes, Mip-1-Alpha, and Mip-1-Beta as the major hiv-suppressive factors produced by Cd8(+) T-cells. *Science, 270*, 1811–1815.
142. Allers, K., Hutter, G., Hofmann, J., Loddenkemper, C., Rieger, K., Thiel, E., & Schneider, T. (2010). Evidence for the cure of HIV infection by CCR5-delta32/delta32 stem cell transplantation. *Blood, 117*, 2791–2799.
143. Hutter, G., Nowak, D., Mossner, M., Ganepola, S., Mussig, A., Allers, K., Schneider, T., Hofmann, J., Kucherer, C., Blau, O., et al. (2009). Long-term control of HIV by CCR5

Delta32/Delta32 stem-cell transplantation. *The New England Journal of Medicine, 360*, 692–698.

144. Li, C., Guan, X. M., Du, T., Jin, W., Wu, B., Liu, Y. L., Wang, P., Hu, B. D., Griffin, G. E., Shattock, R. J., et al. (2015). Inhibition of HIV-1 infection of primary CD4(+) T-cells by gene editing of CCR5 using adenovirus-delivered CRISPR/Cas9. *Journal of General Virology, 96*, 2381–2393.
145. Ye, L., Wang, J. M., Beyer, A. I., Teque, F., Cradick, T. J., Qi, Z. X., Chang, J. C., Bao, G., Muench, M. O., Yu, J. W., et al. (2014). Seamless modification of wild-type induced pluripotent stem cells to the natural CCR5 Delta 32 mutation confers resistance to HIV infection. *Proceedings of the National Academy of Sciences of the United States of America, 111*, 9591–9596.
146. Xu, L., Yang, H., Gao, Y., Chen, Z., Xie, L., Liu, Y., Liu, Y., Wang, X., Li, H., Lai, W., et al. (2017). CRISPR/Cas9-mediated CCR5 ablation in human hematopoietic stem/progenitor cells confers HIV-1 resistance in vivo. *Molecular Therapy, 25*, 1782–1789.
147. Nedellec, R., Coetzer, M., Lederman, M. M., Offord, R. E., Hartley, O., & Mosier, D. E. (2011). Resistance to the CCR5 inhibitor 5P12-RANTES requires a difficult evolution from CCR5 to CXCR4 coreceptor use. *PLoS One, 6*, e22020.
148. Connor, R. I., Sheridan, K. E., Ceradini, D., Choe, S., & Landau, N. R. (1997). Change in coreceptor use correlates with disease progression in HIV-1-infected individuals. *The Journal of Experimental Medicine, 185*, 621–628.
149. Huang, W., Toma, J., Stawiski, E., Fransen, S., Wrin, T., Parkin, N., Whitcomb, J. M., Coakley, E., Hecht, F. M., Deeks, S. G., et al. (2009). Characterization of human immunodeficiency virus type 1 populations containing CXCR4-using variants from recently infected individuals. *AIDS Research and Human Retroviruses, 25*, 795–802.
150. Schumann, K., Lin, S., Boyer, E., Simeonov, D. R., Subramaniam, M., Gate, R. E., Haliburton, G. E., Yee, C. J., Bluestone, J. A., Doudna, J. A., et al. (2015). Generation of knock-in primary human T cells using Cas9 ribonucleoproteins. *Proceedings of the National Academy of Sciences of the United States of America, 112*, 10437–10442.
151. Wang, Q., Chen, S., Xiao, Q., Liu, Z., Liu, S., Hou, P., Zhou, L., Hou, W., Ho, W., Li, C., et al. (2017). Genome modification of CXCR4 by Staphylococcus aureus Cas9 renders cells resistance to HIV-1 infection. *Retrovirology, 14*, 51.
152. Liu, Z., Chen, S., Jin, X., Wang, Q., Yang, K., Li, C., Xiao, Q., Hou, P., Liu, S., Wu, S., et al. (2017). Genome editing of the HIV co-receptors CCR5 and CXCR4 by CRISPR-Cas9 protects CD4(+) T cells from HIV-1 infection. *Cell & Bioscience, 7*, 47.
153. Dar, A., Kollet, O., & Lapidot, T. (2006). Mutual, reciprocal SDF-1/CXCR4 interactions between hematopoietic and bone marrow stromal cells regulate human stem cell migration and development in NOD/SCID chimeric mice. *Experimental Hematology, 34*, 967–975.
154. Nagasawa, T., Nakajima, T., Tachibana, K., Iizasa, H., Bleul, C. C., Yoshie, O., Matsushima, K., Yoshida, N., Springer, T. A., & Kishimoto, T. (1996). Molecular cloning and characterization of a murine pre-B-cell growth-stimulating factor stromal cell-derived factor 1 receptor, a murine homolog of the human immunodeficiency virus 1 entry coreceptor fusin. *Proceedings of the National Academy of Sciences of the United States of America, 93*, 14726–14729.
155. Malim, M. H., & Bieniasz, P. D. (2012). HIV restriction factors and mechanisms of evasion. *Cold Spring Harbor Perspectives in Medicine, 2*, a006940.
156. Bogerd, H. P., Kornepati, A. V., Marshall, J. B., Kennedy, E. M., & Cullen, B. R. (2015). Specific induction of endogenous viral restriction factors using CRISPR/Cas-derived transcriptional activators. *Proceedings of the National Academy of Sciences of the United States of America, 112*, E7249–E7256.
157. Kruize, Z., van Nuenen, A. C., van Wijk, S. W., Girigorie, A. F., van Dort, K. A., Booiman, T., & Kootstra, N. A. (2021). Nef obtained from individuals with HIV-1 vary in their ability to antagonize SERINC3-and SERINC5-mediated HIV-1 restriction. *Viruses-Basel, 13*, 423.
158. Usami, Y., Wu, Y., & Gottlinger, H. G. (2015). SERINC3 and SERINC5 restrict HIV-1 infectivity and are counteracted by Nef. *Nature, 526*, 218–223.

159. Li, H., Wang, S., Dong, X., Li, Q., Li, M., Li, J., Guo, Y., Jin, X., Zhou, Y., Song, H., et al. (2020). CRISPR-Cas13a cleavage of dengue virus NS3 gene efficiently inhibits viral replication. *Molecular Therapy - Nucleic Acids, 19*, 1460–1469.
160. Singsuksawat, E., Onnome, S., Posiri, P., Suphatrakul, A., Srisuk, N., Nantachokchawapan, R., Praneechit, H., Sae-Kow, C., Chidpratum, P., Hongeng, S., et al. (2021). Potent programmable antiviral against dengue virus in primary human cells by Cas13b RNP with short spacer and delivery by virus-like particle. *Molecular Therapy - Methods & Clinical Development., 21*, 729–740.
161. Wang, K. Z., Zou, C. L., Wang, X. J., Huang, C. X., Feng, T. T., Pan, W., Wu, Q. H., Wang, P. H., & Dai, J. F. (2018). Interferon-stimulated TRIM69 interrupts dengue virus replication by ubiquitinating viral nonstructural protein 3. *PLoS Pathogens, 14*, e1007287.
162. Abbott, T. R., Dhamdhere, G., Liu, Y. X., Lin, X. Q., Goudy, L., Zeng, L. P., Chemparathy, A., Chmura, S., Heaton, N. S., Debs, R., et al. (2020). Development of CRISPR as an antiviral strategy to combat SARS-CoV-2 and influenza. *Cell, 181*, 865–876.e12.
163. Blanchard, E. L., Vanover, D., Bawage, S. S., Tiwari, P. M., Rotolo, L., Beyersdorf, J., Peck, H. E., Bruno, N. C., Hincapie, R., Michel, F., et al. (2021). Treatment of influenza and SARS-CoV-2 infections via mRNA-encoded Cas13a in rodents. *Nature Biotechnology*. https://doi.org/10.1038/s41587-021-00822-w
164. Han, J., Perez, J. T., Chen, C., Li, Y., Benitez, A., Kandasamy, M., Lee, Y., Andrade, J., tenOever, B., & Manicassamy, B. (2018). Genome-wide CRISPR/Cas9 screen identifies host factors essential for influenza virus replication. *Cell Reports, 23*, 596–607.
165. Li, B., Clohisey, S. M., Chia, B. S., Wang, B., Cui, A., Eisenhaure, T., Schweitzer, L. D., Hoover, P., Parkinson, N. J., Nachshon, A., et al. (2020). Genome-wide CRISPR screen identifies host dependency factors for influenza A virus infection. *Nature Communications, 11*, 164.
166. Westera, L., Jennings, A. M., Maamary, J., Schwemmle, M., Garcia-Sastre, A., & Bortz, E. (2019). Poly-ADP ribosyl polymerase 1 (PARP1) regulates influenza A virus polymerase. *Advances in Virology, 2019*, 8512363.
167. Gea-Mallorqui, E. (2021). Mapping host restriction of SARS-CoV-2. *Nature Reviews. Immunology, 21*, 3.
168. Martin-Sancho, L., Lewinski, M. K., Pache, L., Stoneham, C. A., Yin, X., Becker, M. E., Pratt, D., Churas, C., Rosenthal, S. B., Liu, S., et al. (2021). Functional landscape of SARS-CoV-2 cellular restriction. *Molecular Cell*. https://doi.org/10.1101/2020.09.29.319566
169. Freije, C. A., Myhrvold, C., Boehm, C. K., Lin, A. E., Welch, N. L., Carter, A., Metsky, H. C., Luo, C. Y., Abudayyeh, O. O., Gootenberg, J. S., et al. (2019). Programmable inhibition and detection of RNA viruses using Cas13. *Molecular Cell, 76*, 826–837.e11.
170. Davison, A. J., Eberle, R., Ehlers, B., Hayward, G. S., McGeoch, D. J., Minson, A. C., Pellett, P. E., Roizman, B., Studdert, M. J., & Thiry, E. (2009). The order herpesvirales. *Archives of Virology, 154*, 171–177.
171. Lan, K., & Luo, M. H. (2017). Herpesviruses: Epidemiology, pathogenesis, and interventions. *Virologica Sinica, 32*, 347–348.
172. Lin, S. R., Yang, H. C., Kuo, Y. T., Liu, C. J., Yang, T. Y., Sung, K. C., Lin, Y. Y., Wang, H. Y., Wang, C. C., Shen, Y. C., et al. (2014). The CRISPR/Cas9 system facilitates clearance of the intrahepatic HBV templates in vivo. *Molecular Therapy - Nucleic Acids, 3*, e186.
173. Yuen, K. S., Wang, Z. M., Wong, N. M., Zhang, Z. Q., Cheng, T. F., Lui, W. Y., Chan, C. P., & Jin, D. Y. (2018). Suppression of Epstein-Barr virus DNA load in latently infected nasopharyngeal carcinoma cells by CRISPR/Cas9. *Virus Research, 244*, 296–303.
174. Wang, J., & Quake, S. R. (2014). RNA-guided endonuclease provides a therapeutic strategy to cure latent herpesviridae infection. *Proceedings of the National Academy of Sciences of the United States of America, 111*, 13157–13162.
175. Haddad, C. O., Kalt, I., Shovman, Y., Xia, L., Schlesinger, Y., Sarid, R., & Parnas, O. (2021). Targeting the Kaposi's sarcoma-associated herpesvirus genome with the CRISPR-Cas9 platform in latently infected cells. *Virology Journal, 18*, 56.

176. van Diemen, F. R., Kruse, E. M., Hooykaas, M. J., Bruggeling, C. E., Schurch, A. C., van Ham, P. M., Imhof, S. M., Nijhuis, M., Wiertz, E. J., & Lebbink, R. J. (2016). CRISPR/Cas9-mediated genome editing of herpesviruses limits productive and latent infections. *PLoS Pathogens, 12*, e1005701.
177. Aubert, M., Strongin, D. E., Roychoudhury, P., Loprieno, M. A., Haick, A. K., Klouser, L. M., Stensland, L., Huang, M. L., Makhsous, N., Tait, A., et al. (2020). Gene editing and elimination of latent herpes simplex virus in vivo. *Nature Communications, 11*, 4148.
178. Bai, Y., Xuan, B., Liu, H., Zhong, J., Yu, D., & Qian, Z. (2015). Tuberous sclerosis complex protein 2-independent activation of mTORC1 by human cytomegalovirus pUL38. *Journal of Virology, 89*, 7625–7635.
179. Johnson, K. E., Bottero, V., Flaherty, S., Dutta, S., Singh, V. V., & Chandran, B. (2014). IFI16 restricts HSV-1 replication by accumulating on the hsv-1 genome, repressing HSV-1 gene expression, and directly or indirectly modulating histone modifications. *PLoS Pathogens, 10*, e1004503.
180. Diner, B. A., Lum, K. K., Toettcher, J. E., & Cristea, I. M. (2016). Viral DNA sensors IFI16 and cyclic GMP-AMP synthase possess distinct functions in regulating viral gene expression, immune defenses, and apoptotic responses during herpesvirus infection. *mBio, 7*, e01553–e01516.
181. Bi, Y., Sun, L., Gao, D., Ding, C., Li, Z., Li, Y., Cun, W., & Li, Q. (2014). High-efficiency targeted editing of large viral genomes by RNA-guided nucleases. *PLoS Pathogens, 10*, e1004090.
182. Avey, D., Tepper, S., Li, W., Turpin, Z., & Zhu, F. (2015). Phosphoproteomic analysis of KSHV-infected cells reveals roles of ORF45-activated RSK during lytic replication. *PLoS Pathogens, 11*, e1004993.
183. Kanda, T., Furuse, Y., Oshitani, H., & Kiyono, T. (2016). Highly efficient CRISPR/Cas9-mediated cloning and functional characterization of gastric cancer-derived Epstein-Barr virus strains. *Journal of Virology, 90*, 4383–4393.
184. Bierle, C. J., Anderholm, K. M., Ben Wang, J., McVoy, M. A., & Schleiss, M. R. (2016). Targeted mutagenesis of guinea pig cytomegalovirus using CRISPR/Cas9-mediated gene editing. *Journal of Virology, 90*, 6989–6998.
185. Murrell, I., Wilkie, G. S., Davison, A. J., Statkute, E., Fielding, C. A., Tomasec, P., Wilkinson, G. W., & Stanton, R. J. (2016). Genetic stability of bacterial artificial chromosome-derived human cytomegalovirus during culture in vitro. *Journal of Virology, 90*, 3929–3943.
186. Auer, T. O., Duroure, K., De Cian, A., Concordet, J. P., & Del Bene, F. (2014). Highly efficient CRISPR/Cas9-mediated knock-in in zebrafish by homology-independent DNA repair. *Genome Research, 24*, 142–153.
187. Maresca, M., Lin, V. G., Guo, N., & Yang, Y. (2013). Obligate Ligation-Gated Recombination (ObLiGaRe): Custom-designed nuclease-mediated targeted integration through nonhomologous end joining. *Genome Research, 23*, 539–546.
188. Suzuki, K., Tsunekawa, Y., Hernandez-Benitez, R., Wu, J., Zhu, J., Kim, E. J., Hatanaka, F., Yamamoto, M., Araoka, T., Li, Z., et al. (2016). In vivo genome editing via CRISPR/Cas9 mediated homology-independent targeted integration. *Nature, 540*, 144–149.
189. Gong, Y., Bi, Y. W., Li, Z. H., Li, Y. Z., Yao, Y. T., Long, Q., Pu, T., Chen, C., Liu, T. Y., Dong, S. Z., et al. (2020). High-efficiency nonhomologous insertion of a foreign gene into the herpes simplex virus genome. *Journal of General Virology, 101*, 982–996.
190. Erwood, S., & Gu, B. (2020). Embryo-based large fragment knock-in in mammals: Why, how and what's next. *Genes-Basel, 11*, 140.
191. Fehrmann, F., & Laimins, L. A. (2003). Human papillomaviruses: Targeting differentiating epithelial cells for malignant transformation. *Oncogene, 22*, 5201–5207.
192. Ghosh, I., Mittal, S., Banerjee, D., Singh, P., Dasgupta, S., Chatterjee, S., Biswas, J., Panda, C., & Basu, P. (2014). Study of accuracy of colposcopy in VIA and HPV detection-based cervical cancer screening program. *The Australian and New Zealand Journal of Obstetrics and Gynaecology, 54*, 570–575.

193. Munoz, N., Kjaer, S. K., Sigurdsson, K., Iversen, O. E., Hernandez-Avila, M., Wheeler, C. M., Perez, G., Brown, D. R., Koutsky, L. A., Tay, E. H., et al. (2010). Impact of human papillomavirus (HPV)-6/11/16/18 vaccine on all HPV-associated genital diseases in young women. *Journal of the National Cancer Institute, 102*, 325–339.
194. Kennedy, E. M., Kornepati, A. V. R., Goldstein, M., Bogerd, H. P., Poling, B. C., Whisnant, A. W., Kastan, M. B., & Cullen, B. R. (2014). Inactivation of the human papillomavirus E6 or E7 gene in cervical carcinoma cells by using a bacterial CRISPR/Cas RNA-guided endonuclease. *Journal of Virology, 88*, 11965–11972.
195. Zhen, S., Hua, L., Takahashi, Y., Narita, S., Liu, Y. H., & Li, Y. (2014). In vitro and in vivo growth suppression of human papillomavirus 16-positive cervical cancer cells by CRISPR/Cas9. *Biochemical and Biophysical Research Communications, 450*, 1422–1426.
196. Porter, S. S., Stepp, W. H., Stamos, J. D., & McBride, A. A. (2017). Host cell restriction factors that limit transcription and replication of human papillomavirus. *Virus Research, 231*, 10–20.
197. Das, D., Smith, N., Wang, X., & Morgan, I. M. (2017). The deacetylase SIRT1 regulates the replication properties of human papillomavirus 16 E1 and E2. *Journal of Virology, 91*, e00102–e00117.
198. Peta, E., Cappellesso, R., Masi, G., Sinigaglia, A., Trevisan, M., Grassi, A., Di Camillo, B., Vassarotto, E., Fassina, A., Palu, G., et al. (2017). Down-regulation of microRNA-146a is associated with high-risk human papillomavirus infection and epidermal growth factor receptor overexpression in penile squamous cell carcinoma. *Human Pathology, 61*, 33–40.
199. DeSmet, M., Kanginakudru, S., Rietz, A., Wu, W. H., Roden, R., & Androphy, E. J. (2016). The replicative consequences of papillomavirus E2 protein binding to the origin replication factor ORC2. *PLoS Pathogens, 12*, e1005934.
200. McKinney, C. C., Kim, M. J., Chen, D., & McBride, A. A. (2016). Brd4 activates early viral transcription upon human papillomavirus 18 infection of primary keratinocytes. *MBio, 7*, e01644–e01616.
201. Meuris, F., Carthagena, L., Jaracz-Ros, A., Gaudin, F., Cutolo, P., Deback, C., Xue, Y. Z., Thierry, F., Doorbar, J., & Bachelerie, F. (2016). The CXCL12/CXCR4 signaling pathway: A new susceptibility factor in human papillomavirus pathogenesis. *PLoS Pathogens, 12*, e1006039.
202. Trépo, C., Chan, H. L. Y., & Lok, A. (2014). Hepatitis B virus infection. *The Lancet, 384*, 2053–2063.
203. Riviere, L., Ducroux, A., & Buendia, M. A. (2014). The oncogenic role of hepatitis B virus. *Recent Results in Cancer Research, 193*, 59–74.
204. Zhen, S., Hua, L., Liu, Y. H., Gao, L. C., Fu, J., Wan, D. Y., Dong, L. H., Song, H. F., & Gao, X. (2015). Harnessing the clustered regularly interspaced short palindromic repeat (CRISPR)/CRISPR-associated Cas9 system to disrupt the hepatitis B virus. *Gene Therapy, 22*, 404–412.
205. Li, H., Sheng, C., Liu, H., Liu, G., Du, X., Du, J., Zhan, L., Li, P., Yang, C., Qi, L., et al. (2016). An effective molecular target site in hepatitis B virus S gene for Cas9 cleavage and mutational inactivation. *International Journal of Biological Sciences, 12*, 1104–1113.
206. Seeger, C., & Sohn, J. A. (2014). Targeting hepatitis B virus with CRISPR/Cas9. *Molecular Therapy - Nucleic Acids, 3*, e216.
207. Kennedy, E. M., Bassit, L. C., Mueller, H., Kornepati, A. V. R., Bogerd, H. P., Nie, T., Chatterjee, P., Javanbakht, H., Schinazi, R. F., & Cullen, B. R. (2015). Suppression of hepatitis B virus DNA accumulation in chronically infected cells using a bacterial CRISPR/Cas RNA-guided DNA endonuclease. *Virology, 476*, 196–205.
208. Ramanan, V., Shlomai, A., Cox, D. B., Schwartz, R. E., Michailidis, E., Bhatta, A., Scott, D. A., Zhang, F., Rice, C. M., & Bhatia, S. N. (2015). CRISPR/Cas9 cleavage of viral DNA efficiently suppresses hepatitis B virus. *Scientific Reports, 5*, 10833.

209. Kostyushev, D., Brezgin, S., Kostyusheva, A., Zarifyan, D., Goptar, I., & Chulanov, V. (2019). Orthologous CRISPR/Cas9 systems for specific and efficient degradation of covalently closed circular DNA of hepatitis B virus. *Cellular and Molecular Life Sciences, 76*, 1779–1794.
210. Dong, C., Qu, L., Wang, H., Wei, L., Dong, Y., & Xiong, S. (2015). Targeting hepatitis B virus cccDNA by CRISPR/Cas9 nuclease efficiently inhibits viral replication. *Antiviral Research, 118*, 110–117.
211. Stone, D., Long, K. R., Loprieno, M. A., De Silva Feelixge, H. S., Kenkel, E. J., Liley, R. M., Rapp, S., Roychoudhury, P., Nguyen, T., Stensland, L., et al. (2021). CRISPR-Cas9 gene editing of hepatitis B virus in chronically infected humanized mice. *Molecular Therapy - Methods & Clinical Development, 20*, 258–275.
212. Croagh, C. M., Desmond, P. V., & Bell, S. J. (2015). Genotypes and viral variants in chronic hepatitis B: A review of epidemiology and clinical relevance. *World Journal of Hepatology, 7*, 289–303.
213. Nkongolo, S., Ni, Y., Lempp, F. A., Kaufman, C., Lindner, T., Esser-Nobis, K., Lohmann, V., Mier, W., Mehrle, S., & Urban, S. (2014). Cyclosporin A inhibits hepatitis B and hepatitis D virus entry by cyclophilin-independent interference with the NTCP receptor. *Journal of Hepatology, 60*, 723–731.
214. Yan, H., Zhong, G. C., Xu, G. W., He, W. H., Jing, Z. Y., Gao, Z. C., Huang, Y., Qi, Y. H., Peng, B., Wang, H. M., et al. (2012). Sodium taurocholate cotransporting polypeptide is a functional receptor for human hepatitis B and D virus. *eLife, 1*, e00049.
215. Wang, Y. P., Liu, F., He, H. W., Han, Y. X., Peng, Z. G., Li, B. W., You, X. F., Song, D. Q., Li, Z. R., Yu, L. Y., et al. (2010). Heat stress cognate 70 host protein as a potential drug target against drug resistance in hepatitis B virus. *Antimicrobial Agents and Chemotherapy, 54*, 2070–2077.
216. Stahl, M., Retzlaff, M., Nassal, M., & Beck, J. R. (2007). Chaperone activation of the hepadnaviral reverse transcriptase for template RNA binding is established by the Hsp70 and stimulated by the Hsp90 system. *Nucleic Acids Research, 35*, 6124–6136.
217. Cui, X., McAllister, R., Boregowda, R., Sohn, J. A., Cortes Ledesma, F., Caldecott, K. W., Seeger, C., & Hu, J. (2015). Does tyrosyl DNA phosphodiesterase-2 play a role in hepatitis B virus genome repair? *PLoS One, 10*, e0128401.
218. Koniger, C., Wingert, I., Marsmann, M., Rosler, C., Beck, J., & Nassal, M. (2014). Involvement of the host DNA-repair enzyme TDP2 in formation of the covalently closed circular DNA persistence reservoir of hepatitis B viruses. *Proceedings of the National Academy of Sciences of the United States of America, 111*, E4244–E4253.
219. Rost, M., Mann, S., Lambert, C., Doring, T., Thome, N., & Prange, R. (2006). Gamma-adaptin, a novel ubiquitin-interacting adaptor, and Nedd4 ubiquitin ligase control hepatitis B virus maturation. *The Journal of Biological Chemistry, 281*, 29297–29308.
220. Hartmann-Stuhler, C., & Prange, R. (2001). Hepatitis B virus large envelope protein interacts with gamma2-adaptin, a clathrin adaptor-related protein. *Journal of Virology, 75*, 5343–5351.
221. Jeong, G. U., Park, I. H., Ahn, K., & Ahn, B. Y. (2016). Inhibition of hepatitis B virus replication by a dNTPase-dependent function of the host restriction factor SAMHD1. *Virology, 495*, 71–78.
222. Miyakawa, K., Matsunaga, S., Watashi, K., Sugiyama, M., Kimura, H., Yamamoto, N., Mizokami, M., Wakita, T., & Ryo, A. (2015). Molecular dissection of HBV evasion from restriction factor tetherin: A new perspective for antiviral cell therapy. *Oncotarget, 6*, 21840–21852.
223. Yan, R., Zhao, X. S., Cai, D. W., Liu, Y. J., Block, T. M., Guo, J. T., & Guo, H. T. (2015). The interferon-inducible protein tetherin inhibits hepatitis B virus virion secretion. *Journal of Virology, 89*, 9200–9212.
224. Lv, M. Y., Zhang, B., Shi, Y., Han, Z., Zhang, Y., Zhou, Y. L., Zhang, W. Y., Niu, J. Q., & Yu, X. F. (2015). Identification of BST-2/tetherin-induced hepatitis B virus restriction and hepatocyte-specific BST-2 inactivation. *Scientific Reports (UK), 5*, 11736.

225. Wang, Y. X., Niklasch, M., Liu, T. T., Wang, Y., Shi, B. S., Yuan, W. J., Baumert, T. F., Yuan, Z. H., Tong, S. P., Nassal, M., et al. (2020). Interferon-inducible MX2 is a host restriction factor of hepatitis B virus replication. *Journal of Hepatology, 72*, 865–876.
226. Hallez, C., Li, X. X., Suspene, R., Thiers, V., Bouzidi, M. S., Dorobantu, C. M., Lucansky, V., Wain-Hobson, S., Gaudin, R., & Vartanian, J. P. (2019). Hypoxia-induced human deoxyribonuclease I is a cellular restriction factor of hepatitis B virus. *Nature Microbiology, 4*, 1196–1207.
227. Luther, D. C., Lee, Y. W., Nagaraj, H., Scaletti, F., & Rotello, V. M. (2018). Delivery approaches for CRISPR/Cas9 therapeutics in vivo: Advances and challenges. *Expert Opinion on Drug Delivery, 15*, 905–913.
228. Wienert, B., Wyman, S. K., Yeh, C. L. D., Conklin, B. R., & Corn, J. E. (2020). CRISPR off-target detection with DISCOVER-seq. *Nature Protocols, 15*, 1775–1799.
229. Zuo, E. W., Sun, Y. D., Wei, W., Yuan, T. L., Ying, W. Q., Sun, H., Yuan, L. Y., Steinmetz, L. M., Li, Y. X., & Yang, H. (2019). Cytosine base editor generates substantial off-target single-nucleotide variants in mouse embryos. *Science, 364*, 289–292.
230. Li, J. J., Hong, S. Y., Chen, W. J., Zuo, E. W., & Yang, H. (2019). Advances in detecting and reducing off-target effects generated by CRISPR-mediated genome editing. *Journal of Genetics and Genomics, 46*, 513–521.
231. Wienert, B., Wyman, S. K., Richardson, C. D., Yeh, C. D., Akcakaya, P., Porritt, M. J., Morlock, M., Vu, J. T., Kazane, K. R., Watry, H. L., et al. (2019). Unbiased detection of CRISPR off-targets in vivo using DISCOVER-Seq. *Science, 364*, 286–289.
232. Ryan, D. E., Taussig, D., Steinfeld, I., Phadnis, S. M., Lunstad, B. D., Singh, M., Vuong, X., Okochi, K. D., McCaffrey, R., Olesiak, M., et al. (2018). Improving CRISPR-Cas specificity with chemical modifications in single-guide RNAs. *Nucleic Acids Research, 46*, 792–803.
233. Ren, X. J., Yang, Z. H., Xu, J., Sun, J., Mao, D. C., Hu, Y. H., Yang, S. J., Qiao, H. H., Wang, X., Hu, Q., et al. (2014). Enhanced specificity and efficiency of the CRISPR/Cas9 system with optimized sgRNA parameters in drosophila. *Cell Reports, 9*, 1151–1162.
234. Yin, H., Song, C. Q., Suresh, S., Kwan, S. Y., Wu, Q. Q., Walsh, S., Ding, J. M., Bogorad, R. L., Zhu, L. J., Wolfe, S. A., et al. (2018). Partial DNA-guided Cas9 enables genome editing with reduced off-target activity. *Nature Chemical Biology, 14*, 311–316.
235. Wang, Y. F., Wang, B., Xie, H. H., Ren, Q. W., Liu, X. X., Li, F. F., Lv, X. J., He, X. B., Cheng, C. S., Deng, R. Z., et al. (2019). Efficient human genome editing using SaCas9 ribonucleoprotein complexes. *Biotechnology Journal, 14*, e1800689.
236. Kleinstiver, B. P., Pattanayak, V., Prew, M. S., Tsai, S. Q., Nguyen, N. T., & Joung, J. K. (2016). High-fidelity CRISPR-Cas9 nucleases with no detectable genome-wide off-target effects. *Molecular Therapy, 24*, S288–S288.
237. Tan, Y., Chu, A. H. Y., Bao, S., Hoang, D. A., Kebede, F. T., Xiong, W., Ji, M., Shi, J., & Zheng, Z. (2019). Rationally engineered Staphylococcus aureus Cas9 nucleases with high genome-wide specificity. *Proceedings of the National Academy of Sciences of the United States of America, 116*, 20969–20976.
238. Chen, P., Zhou, J., Wan, Y., Liu, H., Li, Y. Z., Liu, Z. X., Wang, H. J., Lei, J., Zhao, K., Zhang, Y. L., et al. (2020). A Cas12a ortholog with stringent PAM recognition followed by low off-target editing rates for genome editing. *Genome Biology, 21*, 18.
239. Kim, S., Kim, D., Cho, S. W., Kim, J., & Kim, J. S. (2014). Highly efficient RNA-guided genome editing in human cells via delivery of purified Cas9 ribonucleoproteins. *Genome Research, 24*, 1012–1019.
240. Shen, Y. M., Hirschhorn, R. R., Mercer, W. E., Surmacz, E., Tsutsui, Y., Soprano, K. J., & Baserga, R. (1982). Gene transfer: DNA microinjection compared with DNA transfection with a very high efficiency. *Molecular and Cellular Biology, 2*, 1145–1154.
241. Meaking, W. S., Edgerton, J., Wharton, C. W., & Meldrum, R. A. (1995). Electroporation-induced damage in mammalian cell DNA. *Biochimica et Biophysica Acta (BBA), 1264*, 357–362.

242. Alsaiari, S. K., Patil, S., Alyami, M., Alamoudi, K. O., Aleisa, F. A., Merzaban, J. S., Li, M., & Khashab, N. M. (2018). Endosomal escape and delivery of CRISPR/Cas9 genome editing machinery enabled by nanoscale zeolitic imidazolate framework. *Journal of the American Chemical Society, 140*, 143–146.
243. Lee, K., Conboy, M., Park, H. M., Jiang, F. G., Kim, H. J., Dewitt, M. A., Mackley, V. A., Chang, K., Rao, A., Skinner, C., et al. (2017). Nanoparticle delivery of Cas9 ribonucleoprotein and donor DNA in vivo induces homology-directed DNA repair. *Nature Biomedical Engineering, 1*, 889–901.
244. Zhang, S., Shen, J. T., Li, D. L., & Cheng, Y. Y. (2021). Strategies in the delivery of Cas9 ribonucleoprotein for CRISPR/Cas9 genome editing. *Theranostics, 11*, 614–648.
245. Xu, C. L., Ruan, M. Z. C., Mahajan, V. B., & Tsang, S. H. (2019). Viral delivery systems for CRISPR. *Viruses, 11*, 28.
246. Naso, M. F., Tomkowicz, B., Perry, W. L., & Strohl, W. R. (2017). Adeno-associated virus (AAV) as a vector for gene therapy. *BioDrugs, 31*, 317–334.
247. Cebrian-Serrano, A., & Davies, B. (2017). CRISPR-Cas orthologues and variants: Optimizing the repertoire, specificity and delivery of genome engineering tools. *Mammalian Genome, 28*, 247–261.
248. Herrera-Carrillo, E., & Berkhout, B. (2015). Potential mechanisms for cell-based gene therapy to treat HIV/AIDS. *Expert Opinion on Therapeutic Targets, 19*, 245–263.
249. Herrera-Carrillo, E., & Berkhout, B. (2015). Bone marrow gene therapy for HIV/AIDS. *Viruses, 7*, 3910–3936.
250. Laufs, S., Guenechea, G., Gonzalez-Murillo, A., Zsuzsanna Nagy, K., Luz Lozano, M., del Val, C., Jonnakuty, S., Hotz-Wagenblatt, A., Jens Zeller, W., Bueren, J. A., et al. (2006). Lentiviral vector integration sites in human NOD/SCID repopulating cells. *The Journal of Gene Medicine, 8*, 1197–1207.
251. Montini, E., Cesana, D., Schmidt, M., Sanvito, F., Ponzoni, M., Bartholomae, C., Sergi Sergi, L., Benedicenti, F., Ambrosi, A., Di Serio, C., et al. (2006). Hematopoietic stem cell gene transfer in a tumor-prone mouse model uncovers low genotoxicity of lentiviral vector integration. *Nature Biotechnology, 24*, 687–696.
252. Montini, E., Cesana, D., Schmidt, M., Sanvito, F., Bartholomae, C. C., Ranzani, M., Benedicenti, F., Sergi, L. S., Ambrosi, A., Ponzoni, M., et al. (2009). The genotoxic potential of retroviral vectors is strongly modulated by vector design and integration site selection in a mouse model of HSC gene therapy. *The Journal of Clinical Investigation, 119*, 964–975.
253. Wang, G. P., Levine, B. L., Binder, G. K., Berry, C. C., Malani, N., McGarrity, G., Tebas, P., June, C. H., & Bushman, F. D. (2009). Analysis of lentiviral vector integration in HIV+ study subjects receiving autologous infusions of gene modified CD4+ T cells. *Molecular Therapy, 17*, 844–850.
254. Wang, D., Mou, H., Li, S., Li, Y., Hough, S., Tran, K., Li, J., Yin, H., Anderson, D. G., Sontheimer, E. J., et al. (2015). Adenovirus-mediated somatic genome editing of Pten by CRISPR/Cas9 in mouse liver in spite of cas9-specific immune responses. *Human Gene Therapy, 26*, 432–442.
255. Annoni, A., Gregori, S., Naldini, L., & Cantore, A. (2019). Modulation of immune responses in lentiviral vector-mediated gene transfer. *Cellular Immunology, 342*, 103802.
256. Abordo-Adesida, E., Follenzi, A., Barcia, C., Sciascia, S., Castro, M. G., Naldini, L., & Lowenstein, P. R. (2005). Stability of lentiviral vector-mediated transgene expression in the brain in the presence of systemic antivector immune responses. *Human Gene Therapy, 16*, 741–751.
257. Mangeot, P. E., Risson, V., Fusil, F., Marnef, A., Laurent, E., Blin, J., Mournetas, V., Massourides, E., Sohier, T. J. M., Corbin, A., et al. (2019). Genome editing in primary cells and in vivo using viral-derived Nanoblades loaded with Cas9-sgRNA ribonucleoproteins. *Nature Communications, 10*, 45.
258. Wang, X. J., Zhong, M. T., Liu, Y., Ma, P. X., Dang, L., Meng, Q. Z., Wan, W. W., Ma, X. D., Liu, J., Yang, G., et al. (2020). Rapid and sensitive detection of COVID-19 using CRISPR/

Cas12a-based detection with naked eye readout, CRISPR/Cas12a-NER. *Scientific Bulletin, 65*, 1436–1439.

259. Joung, J., Ladha, A., Saito, M., Segel, M., Bruneau, R., Huang, M. W., Kim, N. G., Yu, X., Li, J., & Walker, B.D., et al. (2020). Point-of-care testing for COVID-19 using SHERLOCK diagnostics. *medRxiv*.
260. Arizti-Sanz, J., Freije, C. A., Stanton, A. C., Boehm, C. K., Petros, B. A., Siddiqui, S., Shaw, B. M., Adams, G., Kosoko-Thoroddsen, T. F., & Kemball, M. E., et al. (2020). Integrated sample inactivation, amplification, and Cas13-based detection of SARS-CoV-2. *bioRxiv*.

CRISPR Targeting the Integrated HTLV-1 Virus

Tasha Wilkie and Amanda R. Panfil

Abstract Human T-cell leukemia virus type 1 (HTLV-1) is a complex deltaretrovirus. As the only oncogenic human retrovirus known to date, HTLV-1 is the causative agent of adult T-cell leukemia/lymphoma (ATL), the immune-mediated neurodegenerative disease, HTLV-1-associated myelopathy/tropical spastic paraparesis (HAM/TSP), and a number of other inflammatory disorders. These HTLV-1-related diseases develop in a portion of infected individuals after a prolonged clinical latency period of several decades. Unlike other oncogenic retroviruses, HTLV-1 does not encode a host-derived oncogene or cause insertional cis-activation of a cellular oncogene. Instead, the virus encodes key regulatory and accessory genes, which mediate cellular transformation and proliferation. A large body of research has found that *Tax* and *Hbz* are essential players in this process, with other accessory genes likely contributing to viral pathogenesis. Over the past several decades, the development of animal models to study HTLV-1 infection and transformation has provided a useful tool to study the complex questions surrounding viral-mediated mechanisms of disease and potential therapeutic interventions. Genome editing of either *Tax* or *Hbz* may disable HTLV-1-infected cell growth and survival. Such disruption could also prevent immune modulatory effects and ultimately disrupt HTLV-1-mediated diseases. CRISPR targeting of the integrated HTLV-1 virus is an understudied, yet intriguing, technology that has the potential to be used as an HTLV-1 disease treatment. This chapter will briefly describe the HTLV-1 lifecycle,

T. Wilkie
Department of Veterinary Biosciences, College of Veterinary Medicine,
The Ohio State University, Columbus, OH, USA

Center for Retrovirus Research, The Ohio State University, Columbus, OH, USA

A. R. Panfil (✉)
Department of Veterinary Biosciences, College of Veterinary Medicine,
The Ohio State University, Columbus, OH, USA

Center for Retrovirus Research, The Ohio State University, Columbus, OH, USA

Comprehensive Cancer Center, The Ohio State University, Columbus, OH, USA
e-mail: panfil.6@osu.edu

Y. H. Yun, K. E. Yoder (eds.), *Biotechnologies for Gene Therapy*,
https://doi.org/10.1007/978-3-030-93333-3_6

how this virus mediates disease, the relevant in vitro and in vivo models used to study HTLV-1 pathogenesis, and, finally, the potential for gene editing as a promising HTLV-1 therapeutic.

1 Genome Organization

HTLV-1 belongs to the family *Retroviridae*, in the genus *Deltaretrovirus* [1]. Like other members of this family, HTLV-1 is an enveloped virus with a diameter of approximately 100 nanometers. The proteolipid envelope bilayer derived from host cell membrane is comprised of both viral glycoproteins (envelope; Env) and host proteins. The viral Env protein helps the virus bind to host cell receptors and enter new target cells. Matrix (MA) structural proteins line the inner membrane where it assists in particle assembly during viral spread. An icosahedral capsid (CA) encases the two copies of positive sense viral RNA, the functional viral enzymes protease (PR), reverse transcriptase (RT), and integrase (IN) together within the nucleocapsid.

HTLV-1 is a complex retrovirus that encodes the standard retroviral structural and enzymatic genes *gag, pro, pol,* and *env* (Fig. 1). In addition, there is a unique *pX* region located downstream of the envelope gene, which encodes several regulatory and accessory genes on both the sense and antisense genomic strands. The regulatory/accessory genes encoded within the sense genomic strand include *Tax*, *Rex*, *p30*, *p12/p8*, *p30*, and *p13*, while only one gene—*Hbz*—is derived from the antisense genomic strand. The HTLV-1 proviral genome is roughly 9 kb in length and flanked by characteristic long terminal repeats (LTRs) on both the 5′ and 3′ ends of the viral genome. These genomic regions are exact duplicates, which consist of a U3, R, and U5 region. These regions contain promoter elements, polyadenylation signal sequences, and other cis-acting regulatory sequences essential for proper viral transcription. Unlike other retroviruses, such as HIV-1, the HTLV-1 viral genome is highly conserved with sequence homogeneity both within the

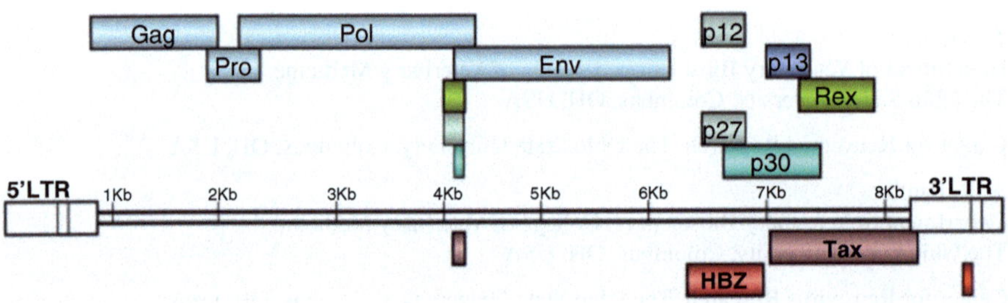

Fig. 1 Brief schematic of the HTLV-1 proviral genome. The approximate locations of the common retroviral structural and enzymatic genes, as well as the regulatory and accessory genes transcribed within the pX region are depicted

same host and between different HTLV-1 isolates. This is somewhat surprising given the error-prone nature of reverse transcriptase. The genetic stability of HTLV-1 is largely due to the clonal expansion of infected cells versus viral replication and infection of new cells, which is more common for many other viruses [2–5].

2 Geographical Distribution and Viral Transmission

Recent epidemiological studies estimate approximately 5–10 million people are infected with HTLV-1 worldwide [6]. However, these studies were based on known endemic regions where there are reliable epidemiological data. The actual number of infected individuals is suspected to be much higher, since other high cluster and populated areas such as China, India, and East Africa are less well documented. In addition to worldwide distribution, HTLV-1 infection is also highly endemic in several areas worldwide including Southwestern Japan, sub-Saharan Africa, the Caribbean area, South America, and foci in the Middle East and Australo-Melanesia. The USA and Europe (with the exception of Romania) show low incidence of HTLV-1 [6–8]. Of those seropositive individuals, most are immigrants or their children who originate from highly endemic regions.

HTLV-1 infection of new cells is heavily dependent on cell-to-cell transmission, as opposed to cell-free transmission [9–11]. Therefore, successful new infection requires the transfer of virus-infected cells. There are three main routes of viral transmission described for HTLV-1: mother-to-child (vertical transmission), infected blood products, and sexual transmission. Vertical transmission of HTLV-1 occurs through breast feeding, and there is limited evidence to suggest HTLV-1 is able to infect a fetus across the placental barrier [12]. The risk of transmission from mother to child increases with the duration of breastfeeding (>6 months), as protective maternal antibodies within the infant wane [13–15]. Studies show that the frequency of HTLV-1 transmission through breast feeding ranges from 4% to 25% in highly endemic regions [16]. HTLV-1 provirus is detected in both the semen and cervical secretion of HTLV-1-infected individuals [17]. Transmission can occur in both hetero- and homosexual intercourse; however, transmission from male-to-female is most frequent [18, 19]. As with vertical transmission, the incidence of infection is increased by proviral load and duration of exposure/sexual contact [20, 21]. Blood transfusions were a major route of HTLV-1 transmission prior to the implementation of routine serological screening of blood donors in the early 1990s in several countries [22]. As a result, this route of transmission has been almost eliminated in these countries. Despite the risk of transmission via blood, the screening of HTLV-1 in solid organ transplant donors is sporadic or nonexistent in most countries. Recently, this has resulted in several well-documented cases of HTLV-1 seroconversion and subsequent disease progression in transplant recipients [23–28].

3 Viral Lifecycle

HTLV-1 mediates cell entry through binding of the retroviral envelope protein (Env) to the primary receptor glucose transporter type 1 (GLUT1) [29]. Two additional cellular factors, neuropilin-1 (NRP1) and heparan sulfate proteoglycan (HSPG), function as coreceptors to facilitate viral entry [30, 31]. HTLV-1 predominantly immortalizes CD4$^+$ T-cells both in vitro and in vivo; however, provirus can also be detected in a variety of hematopoietic cells including CD8$^+$ T-cells, B cells, dendritic cells, monocytes, and macrophages [32].

Upon entry into the cell, the viral enzyme reverse transcriptase converts the viral RNA genome into double-stranded DNA [33, 34]. The viral DNA is then translocated to the nucleus where it is integrated into the host cell DNA via the viral enzyme integrase. The integrated form of the viral genome is referred to as the provirus. HTLV-1 integration occurs at random sites throughout the cellular genome, occasionally contained within transcriptionally active regions [35, 36]. HTLV-1 can be propagated in the infected cell through infectious replication where the integrated genome is expressed, packaged, and disseminated as free virus for infection of a target cell. More commonly, viral amplification occurs through clonal expansion of infected cells, or mitotic spread. On average, >99.9% of infected cells arise through mitotic spread within an infected host [37, 38]. Several of the viral regulatory and accessory genes play important roles within the viral lifecycle. However, this work will focus specifically on *Tax* and *Hbz* as these two viral genes play a critical role in infection, persistence, and disease development.

The viral protein Tax is a pleiotropic, oncogenic, transactivator protein essential for de novo infection and immortalization of infected cells [39–43]. Tax is generated from doubly spliced mRNA from the pX region and encodes a 353 amino acid protein (Fig. 2). Although Tax protein has been found in both the nuclear and cytoplasmic compartments of infected cells, it localizes primarily to the nucleus [44, 45]. Most notably, this protein functions as a viral transcriptional activator of the 5′ LTR and also various cellular signaling pathways including the CREB, NF-κB, and AP1 pathways [46–55]. Activation of these pathways helps stimulate clonal proliferation and survival of infected cells. In addition to its role as a transcriptional activator, Tax has also been shown to induce DNA damage, deregulate the cell cycle, inhibit apoptosis, and modulate miRNA expression [56–58]. The consequence of these actions leads to uncontrolled cell division, which leads to cell

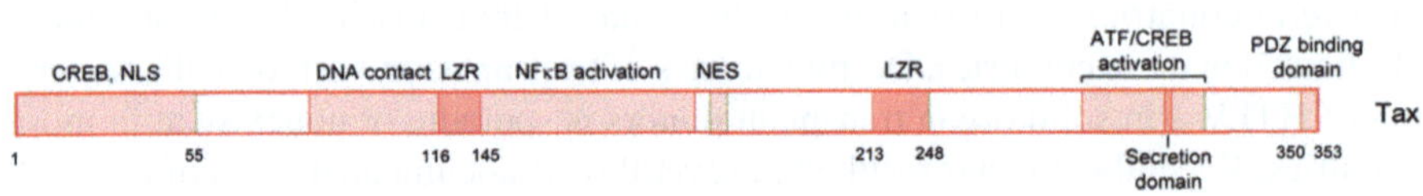

Fig. 2 Schematic of the Tax protein with relevant functional domains highlighted

proliferation and eventual transformation. Given its defined role in viral pathobiology, several important domains have been mapped within the Tax protein [59]:

- N-terminus: a CREB-binding domain and nuclear localization signal.
- Central portion: two leucine zipper-like motifs required for DNA interactions, protein dimerization, and NF-κB activation.
- C-terminus: ATF/CREB activation domain, p300/CBP-binding domain, and PDZ-binding domain.

Several of the transcriptional effects of Tax are counteracted by the viral protein Hbz. *Hbz* transcription initiates in the 3′ LTR and is regulated by viral CREs and several SP1-binding sites [60–62]. There are both spliced and unspliced *Hbz* transcript variants, and the proteins encoded by these transcripts have nearly identical amino acid sequences and functions [62]. However, spliced *Hbz* is more abundant in infected cells [63] and therefore has been well-studied. Subsequent discussion of *Hbz* refers to the major spliced variant. Hbz plays important roles in regulating genomic integrity, apoptosis, autophagy, and escape from the host immune system [64]. Interestingly, Hbz is able to stimulate proliferation of infected cells in both its protein and mRNA forms [61, 65]—making it a particularly intriguing genome editing target. The Hbz protein consists of 206 amino acids (Fig. 3) and is comprised of three functional domains: an N-terminal activation domain, a central basic region, and a C-terminal bZIP domain. The activation domain contains two well-studied LXXLL-like motifs, which enable Hbz to bind CBP/p300 [66]. The LXXLL motifs are also required for Hbz to activate TGF-β signaling [67]. Through interactions with CBP/p300, Hbz is able to sequester these factors away from Tax and repress Tax-mediated LTR-activation [66, 68, 69]. The bZIP domain allows Hbz to heterodimerize with cellular bZIP proteins of the AP1 superfamily (CREB2, c-Jun, JunB, JunD, CREB, MafB, and ATF3) and affects their binding to cellular DNA recognition sites [70–80].

The viral proteins Tax and Hbz play central roles in the HTLV-1-mediated disease process. Experimental loss of Tax through mutation or deletion prevents de novo spread of the virus and also HTLV-1-mediated T-cell immortalization in vitro [39, 43, 49]. Mutually, Tax overexpression can induce cellular transformation in vitro in the absence of other viral proteins and induces ATL-like disease in mice [40–42, 81]. These early seminal discoveries visibly identified Tax as an oncogenic viral protein. Identified several decades after the discovery of HTLV-1, Hbz also influences viral persistence and transformation. Loss of Hbz has no effect on T-cell immortalization in vitro, but is required for efficient viral infection and persistence in vivo [82]. In addition, shRNA-mediated knockdown of Hbz decreases

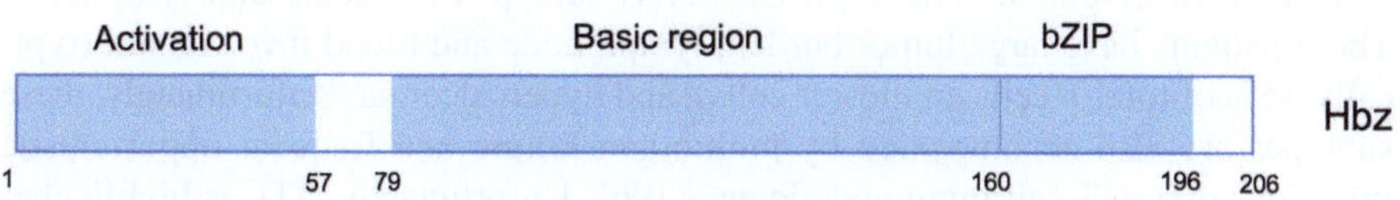

Fig. 3 Schematic of the Hbz protein with relevant functional domains highlighted

proliferation, tumor burden, and tumor infiltration to the surrounding tissues in vivo [83]. Recently, Hbz was shown to prevent Tax-induced transformation and senescence in a transgenic fly model [84]. The current theory within the field is that Tax is responsible for initiating transformation, while Hbz counteracts several Tax-induced effects and is responsible for the long-term proliferation of transformed cells.

4 HTLV-1-Associated Diseases

In HTLV-1-positive individuals, the virus persists within approximately 10^3 to 10^6 individual infected T-cell clones that survive for the lifetime of the host [85]. Until recently, viral transcription driven from the 5′ LTR (sense or plus-strand transcription) was believed to be silent, since plus-strand transcription from freshly isolated infected lymphocytes is absent. However, recent evidence has shown that Tax is transcribed in "intense intermittent bursts" in infected lymphocytes [86]. The factors responsible for driving these bursts are cellular stress, controlled by hypoxia, and glycolysis [87]. These studies also found Hbz, driven by minus-strand transcription, is transcribed at lower, more consistent levels in infected cells. Interestingly, these studies also found Hbz is silent in a portion of the cells at various times. These studies confirm Tax and Hbz are expressed, and their expression is tightly regulated in asymptomatic HTLV-1-infected individuals. DNA analyses of asymptomatic HTLV-1-infected individuals show little variation in proviral sequencing [88], confirming HTLV-1 exists predominantly in a latent form where the genome is maintained by mitotic host cell division followed by occasional reactivation and de novo infection.

HTLV-1 is the cause of an aggressive non-Hodgkin's peripheral T-cell malignancy called adult T-cell leukemia/lymphoma (ATL) [89–91]. This disease is most commonly a $CD4^+$ T-cell malignancy; however, there are rare examples of $CD8^+$ T-cell origin [92]. ATL is characterized by lymphadenopathy, skin lesions, increased abnormal lymphocytes often called "flower" cells, frequent blood and bone marrow involvement, hypercalcemia, and lytic bone lesions [93]. Very little is known why only a fraction (~5%) of infected individuals develop ATL; however, one risk factor is the patient proviral load. HTLV-1-infected patients with a higher proviral load (>4 copies/100 peripheral blood mononuclear cells; PBMCs) are at an increased risk for developing disease [94].

ATL is a heterogeneous disease divided into four clinical subtypes (acute, lymphoma, smoldering, chronic) based on several criteria [95]. The smoldering or chronic ATL is considered less aggressive and often involves a skin rash and minimal blood involvement. The aggressive ATL subtypes are acute and lymphoma. These patients have large tumor burden, lymph node and blood involvement (typically >5% of total T-cells are flower cells), and hypercalcemia. Unfortunately, these subtypes are also accompanied by multiorgan failure and frequent opportunistic infections due to T-cell immunodeficiency [96]. Unfortunately, ATL is highly chemotherapy and multidrug resistant, leading to poor prognosis. The median survival rate of the least aggressive subtypes of ATL (smoldering, chronic) is 32–55 months, while the survival rate of aggressive subtypes of ATL (lymphoma, acute) is 8–11 months [97].

Tax expression is typically quite low or undetectable in ATL patient cells. This arises through epigenetic silencing, 5′ LTR deletion, or abortive protein mutations within the Tax gene [98–100]. A recent study using an HTLV-1-transformed cell line, MT-1, discovered transient Tax expression, and this expression was critical for maintaining the infected cell population [101]. This transient Tax expression is believed to activate the antiapoptotic machinery even after Tax expression is again silenced or lost. Inversely, Hbz is the only viral gene that is consistently found in all ATL cases and is intact [61]. This evidence supports the theory that Hbz supports infected cell survival and the development of leukemia/lymphoma.

HTLV-1-associated myelopathy/tropical spastic paraparesis (HAM/TSP) is a slow progressing, chronic neurodegenerative disease occurring in ~5% of HTLV-1-infected individuals [102–105]. Characterized by painful stiffness and weakness of the legs (lower limb spasticity) and bowel and bladder disturbances, this disease manifests slowly with a steady progression over several years. HAM/TSP arises from inflammation and subsequent swelling that occurs in the spinal cord. HTLV-1-infected cells infiltrate the central nervous system and cause persistent immune activation against proliferating HTLV-1-infected CD4+ T-cells. HAM/TSP patients frequently display high proviral loads and high levels of HTLV-1-specific antibodies in the cerebrospinal fluid (CSF) [106–111]. Many symptoms of HAM/TSP are also found in patients with MS. In fact, the clinical courses of HAM/TSP and MS are similar with steady worsening of neurological function without any distinct relapse or period of remission.

The main reservoir for HTLV-1 in HAM/TSP patients are CD4+CD25+ T-cells [102, 107]. Like ATL, elevated patient proviral loads are strongly correlated with disease pathogenesis. Tax protein is detected in the CSF cells of HAM/TSP patients, while Tax mRNA is detected in cells isolated from spinal cord and cerebellar sections [108, 110, 111]. The persistent expression of the highly immunogenic Tax protein in an immune-privileged area like the CNS leads to direct cell damage (i.e., axonal degeneration) and contributes to lymphocyte activation. Although Hbz mRNA is detected in PBMCs from HAM/TSP patients, the transcript level is lower than in ATL patient PBMCs and correlates with the patient proviral load and disease severity [112]. Hbz transgenic mice present similar immunological features of HAM/TSP [113], again, supporting the role of Hbz in HAM/TSP disease pathology.

5 In Vitro and In Vivo Models of HTLV-1 Disease

HTLV-1 infects and immortalizes primary CD4+ T-cells in vitro [114]—identical to what is observed in HTLV-1-infected asymptomatic individuals and in the HTLV-1-mediated diseases, such as ATL and HAM/TSP. Cell-free infection by HTLV-1 is extremely inefficient [9–11]; therefore, infection requires the coculture of lethally irradiated HTLV-1 producer (infected) cells with naïve primary PBMCs from a healthy donor. Immortalization of recipient cells becomes apparent 5–6 weeks after coculture and can be detected by the expansion of infected cells from the peripheral blood lymphocyte mixed cell population. Using this technique, researchers have been readily able to examine the requirements of specific gene products on HTLV-1-mediated immortalization using proviral clones [82, 115–120].

Several in vivo models exist to examine various aspects of viral pathogenesis from infection to tumorigenesis. A brief description of the most relevant animal models is described below.

Early HTLV-1 infection events can be modeled in New Zealand white (NZW) rabbits. NZW rabbits inoculated with irradiated HTLV-1-infected cells become persistently infected and mimic early infection events in humans such as the antibody response against Gag and Env [82, 115, 119]. HTLV-1-infected rabbits do not develop disease related to HTLV-1, likely due to several factors such as lifespan length and immune pressure. However, this system enables the study of early viral infection events in the presence of a functional immune system. Long-term HTLV-1 latency is mediated in part by the host immune response, making this an attractive model to study viral persistence. Similar to the in vitro immortalization assay in primary cells, this model also allows for the in vivo functional study of various HTLV-1 proteins.

Immune-deficient mice (NOD.Cg-PrkdcSCIDIL2rgtm1Wjl/SzJ; NOG) inoculated subcutaneously with various HTLV-1-infected cell lines (Hut-102, SLB-1, ATL-ED, TL-Om1) will develop tumors [83, 121, 122]. The transplanted HTLV-1 tumor cell lines also secrete human IL-2Rα, which can be used as a biomarker for cellular proliferation in vivo. Previously, this mouse model has been successfully used to measure drug efficacy against ATL [121, 122] and the functional significance of various HTLV-1 proteins on tumorigenesis. Using this model system, shRNA knockdown of Hbz in HTLV-1-infected cells was found to decrease proliferation, tumor size, and infiltration of tumor cells to the surrounding environment [83].

Humanized immune system (HIS) mice develop T-cell tropism and lymphoproliferative disease after HTLV-1 infection. These mice are created by intrahepatic injection of human umbilical-cord stem cells into neonatal NSG mice. Within a period of several weeks, this will result in the development of phenotypically normal human lymphocytes [123, 124]. HIS mice inoculated with irradiated HTLV-1-producing cells consistently reproduce several key stages (persistent infection, chronic proliferation of $CD4^{+}$ T-cells, and lymphoproliferative disease) of HTLV-1-mediated tumorigenesis in a relatively compact time frame of 4–5 weeks. Like NZW rabbits, infection in these mice can also be induced using an infectious molecular clone of HTLV-1 and thus enables the in vivo functional study of various HTLV-1 proteins on HTLV-1-mediated tumorigenesis.

6 CRISPR/Cas9 Targeting of HTLV-1

Clustered regularly interspersed short palindromic repeat (CRISPR)/Cas9 genome editing is a relatively new technology that utilizes a guide RNA (gRNA) to target a site-specific DNA double strand break (DSB) by the Cas9 endonuclease. These DSBs are most often repaired by the nonhomologous end-joining pathway. This type of repair is error prone and typically introduces insertions and deletions (indels) at the repair site. CRISPR gene editing can lead to disruptions within genomic reading frames, DNA regulatory motifs, or the structure of encoded RNA elements. The use of CRISPR/Cas9 is rapidly moving into clinical trials and has been mostly focused on *ex vivo* modification of cells followed by transplant back into patients. This approach has been most beneficial for diseases including cancer and sickle cell disease [125, 126].

Future treatment of HTLV-1 diseased patients will likely involve in vivo delivery of CRISPR/Cas9. There are several well-known technical challenges associated with this type of approach including type of packaging vector, systemic versus targeted delivery, editing efficiency, off-target effects, and immunogenicity [127]. However, several points of evidence exist to support the development of this innovative and powerful genome editing technology to treat HTLV-1-mediated diseases:

1. **Previous studies demonstrated genome editing can disrupt HTLV-1-mediated proliferation.** Zinc finger nucleases (ZFNs) were reported to specifically recognize the HTLV-1 LTR and disrupt promoter function, ultimately inhibiting proliferation of HTLV-1-positive cells in cell culture [128]. Additional support for CRISPR targeting the HTLV-1 genome was recently reported with two different gRNAs targeting Hbz that disrupted ATL cell proliferation in vitro [129].
2. **HTLV-1 offers focused gRNA targeting.** The viral genome is highly conserved with sequence homogeneity within the same host and among different viral isolates [130, 131]. This is uncommon for a retrovirus given the error-prone nature of RT. However, HTLV-1 mostly persists within an infected host through mitotic host cell division [37, 38] and thus relies less on active replication compared to other retroviruses such as HIV-1.
3. **There is a strong need for innovative therapeutic approaches to treat HTLV-1-mediated diseases.** Current treatment strategies for HTLV-1-infected patients involve a "watch and wait" approach, monitoring proviral load, and other predictive risk factors for disease development. ATL caused by infectious HTLV-1 is an aggressive, chemotherapy-resistant, and highly fatal malignancy. The median survival time of patients who develop aggressive ATL subtypes is less than 1 year after diagnosis. Unfortunately, ATL patients who experience initial treatment success will consistently relapse [132, 133]. HAM/TSP represents a challenging disease with limited treatment options [134], and many treatments simply target pain or inflammation [102, 135]. Therapies that control the expression of HTLV-1 gene products represent an effective treatment for preventing and treating HAM/TSP.

4. **The most important drivers of HTLV-1-mediated transformation, proliferation, and immunological inducing effects are the *Tax* and *Hbz* genes.** Tax is the major driver of viral transcription, transformation, and immune dysregulation [136, 137], while HBZ supports proliferation and survival of HTLV-1-transformed cells [82, 83]. HTLV-1 ATL tumor cells typically do not express *Tax*, while Tax expression directly contributes to lymphocyte activation and immunopathogenesis in HAM/TSP patients [109]. Conversely, *Hbz* is always expressed in HTLV-1-infected cells, PBMCs of HTLV-1-infected individuals, and ATL tumor cells [138]. Together, these two viral proteins are essential to the pathophysiology of both HAM/TSP and ATL. Disrupting the function or expression of *Tax* and/or *Hbz* through genome editing could disturb HTLV-1-infected cell growth and survival and thus prevent immune modulatory effects and ultimately HTLV-1-associated diseases.
5. **Several in vitro and in vivo models exist to study HTLV-1 infection, immortalization, persistence, and tumorigenesis.** Well-established HTLV-1 animal models (rabbits, NOG mice, HIS mice) and in vitro cell immortalization studies represent several useful tools to study the various aspects of HTLV-1 infection and disease development. Most importantly, as CRISPR gene editing continues to move into the clinic, these in vivo models represent a system to eventually test delivery of CRISRP/Cas9 gene editing components against measurable gRNA-viral targets. CRISPR gene editing faces several significant obstacles including off-target editing, inefficient or off-target delivery, and stimulation of the host immune response. In vivo models could help to overcome these significant obstacles facing clinical application of CRISPR gene editing against HTLV-1-mediated diseases.

7 Concluding Remarks

HTLV-1 is an oncogenic human retrovirus and the infectious cause of both a neurodegenerative disease called HAM/TSP and an aggressive $CD4^+$ T-cell malignancy called ATL. Interestingly, both HAM/TSP and ATL develop in a portion of infected individuals after a prolonged clinical latency period. There is a strong need for innovative treatment approaches within the HTLV field, as both of these diseases have limited treatment options. Within the HTLV-1 genome, the viral genes *Tax* and *Hbz* offer well-defined molecular targets for CRISPR gene editing. HTLV-1 is also an excellent model to advance retroviral genome editing technologies due to the wide variety of animal models available within the field. Although there are several technical challenges to in vivo delivery of CRISPR/Cas9, there are several points of evidence, which suggest gene editing is a promising future HTLV-1 therapeutic.

References

1. Fields, B. N., Knipe, D. M., Howley, P. M., & Griffin, D. E. (2001). *Fields virology*. Lippincott Williams & Wilkins.
2. Furukawa, Y., et al. (1992). Frequent clonal proliferation of human T-cell leukemia virus type 1 (HTLV-1)-infected T cells in HTLV-1-associated myelopathy (HAM-TSP). *Blood, 80*, 1012–1016.
3. Wattel, E., Vartanian, J. P., Pannetier, C., & Wain-Hobson, S. (1995). Clonal expansion of human T-cell leukemia virus type I-infected cells in asymptomatic and symptomatic carriers without malignancy. *Journal of Virology, 69*, 2863–2868.
4. Gillet, N. A., et al. (2011). The host genomic environment of the provirus determines the abundance of HTLV-1-infected T-cell clones. *Blood, 117*, 3113–3122. https://doi.org/10.1182/blood-2010-10-312926
5. Kulkarni, A., & Bangham, C. R. M. (2018). HTLV-1: Regulating the balance between proviral latency and reactivation. *Frontiers in Microbiology, 9*, 449. https://doi.org/10.3389/fmicb.2018.00449
6. Gessain, A., & Cassar, O. (2012). Epidemiological aspects and world distribution of HTLV-1 infection. *Frontiers in Microbiology, 3*, 388. https://doi.org/10.3389/fmicb.2012.00388
7. Sarkis, S., et al. (2019). Role of HTLV-1 orf-I encoded proteins in viral transmission and persistence. *Retrovirology, 16*, 43. https://doi.org/10.1186/s12977-019-0502-1
8. Anyanwu, N. C. J., Ella, E. E., Ohwofasa, A., & Aminu, M. (2018). Re-emergence of human T-lymphotropic viruses in West Africa. *The Brazilian Journal of Infectious Diseases, 22*, 224–234. https://doi.org/10.1016/j.bjid.2018.05.003
9. Fan, N., et al. (1992). Infection of peripheral blood mononuclear cells and cell lines by cell-free human T-cell lymphoma/leukemia virus type I. *Journal of Clinical Microbiology, 30*, 905–910.
10. Pique, C., & Jones, K. S. (2012). Pathways of cell-cell transmission of HTLV-1. *Frontiers in Microbiology, 3*, 378. https://doi.org/10.3389/fmicb.2012.00378
11. Alais, S., Mahieux, R., & Dutartre, H. (2015). Viral source-independent high susceptibility of dendritic cells to human T-cell leukemia virus type 1 infection compared to that of T lymphocytes. *Journal of Virology, 89*, 10580–10590. https://doi.org/10.1128/JVI.01799-15
12. Fujino, T., & Nagata, Y. (2000). HTLV-I transmission from mother to child. *Journal of Reproductive Immunology, 47*, 197–206. https://doi.org/10.1016/s0165-0378(00)00054-1
13. Takahashi, K., et al. (1991). Inhibitory effect of maternal antibody on mother-to-child transmission of human T-lymphotropic virus type I. The mother-to-child transmission study group. *International Journal of Cancer, 49*, 673–677. https://doi.org/10.1002/ijc.2910490508
14. Hino, S., et al. (1994). Intervention of maternal transmission of HTLV-1 in Nagasaki, Japan. *Leukemia, 8*(Suppl 1), S68–S70.
15. Ureta-Vidal, A., et al. (1999). Mother-to-child transmission of human T-cell-leukemia/lymphoma virus type I: Implication of high antiviral antibody titer and high proviral load in carrier mothers. *International Journal of Cancer, 82*, 832–836. https://doi.org/10.1002/(sici)1097-0215(19990909)82:6<832::aid-ijc11>3.0.co;2-p
16. Ando, Y., et al. (2003). Long-term follow-up study of HTLV-I infection in bottle-fed children born to seropositive mothers. *The Journal of Infection, 46*, 9–11. https://doi.org/10.1053/jinf.2002.1081
17. Mansouri Torshizi, M., et al. (2014). Effect of human T-cell lymphotrophic virus type 1 (HTLV-1) in seropositive infertile women on intracytoplasmic sperm injection (ICSI) outcome. *The Iranian Journal of Reproductive Medicine, 12*, 15–18.
18. Kajiyama, W., et al. (1986). Intrafamilial transmission of adult T cell leukemia virus. *The Journal of Infectious Diseases, 154*, 851–857. https://doi.org/10.1093/infdis/154.5.851
19. Roucoux, D. F., et al. (2005). A prospective study of sexual transmission of human T lymphotropic virus (HTLV)-I and HTLV-II. *The Journal of Infectious Diseases, 191*, 1490–1497. https://doi.org/10.1086/429410

20. Kaplan, J. E., et al. (1996). Male-to-female transmission of human T-cell lymphotropic virus types I and II: Association with viral load. The retrovirus epidemiology donor study group. *Journal of Acquired Immune Deficiency Syndromes and Human Retrovirology, 12*, 193–201.
21. Shirley, M. H., Sayeed, S., Barnes, I., Finlayson, A., & Ali, R. (2013). Incidence of haematological malignancies by ethnic group in England, 2001-7. *British Journal of Haematology, 163*, 465–477. https://doi.org/10.1111/bjh.12562
22. Sato, H., & Okochi, K. (1986). Transmission of human T-cell leukemia virus (HTLV-I) by blood transfusion: Demonstration of proviral DNA in recipients' blood lymphocytes. *International Journal of Cancer, 37*, 395–400.
23. Glowacka, I., et al. (2013). Delayed seroconversion and rapid onset of lymphoproliferative disease after transmission of human T-cell lymphotropic virus type 1 from a multiorgan donor. *Clinical Infectious Diseases, 57*, 1417–1424. https://doi.org/10.1093/cid/cit545
24. Cook, L. B., et al. (2016). Rapid dissemination of human T-lymphotropic virus type 1 during primary infection in transplant recipients. *Retrovirology, 13*, 3. https://doi.org/10.1186/s12977-015-0236-7
25. de Mendoza, C., et al. (2019). HTLV-1 infection in solid organ transplant donors and recipients in Spain. *BMC Infectious Diseases, 19*, 706. https://doi.org/10.1186/s12879-019-4346-z
26. Ramanan, P., et al. (2014). Donor-transmitted HTLV-1-associated myelopathy in a kidney transplant recipient--case report and literature review. *American Journal of Transplantation, 14*, 2417–2421. https://doi.org/10.1111/ajt.12849
27. Taylor, G. P. (2018). Human T-lymphotropic virus type 1 infection and solid organ transplantation. *Reviews in Medical Virology, 28*. https://doi.org/10.1002/rmv.1970
28. Roc, L., et al. (2019). Rapid subacute myelopathy following kidney transplantation from HTLV-1 donors: Role of immunosuppresors and failure of antiretrovirals. *Therapeutic Advances in Infectious Disease, 6*, 2049936119868028. https://doi.org/10.1177/2049936119868028
29. Manel, N., et al. (2003). The ubiquitous glucose transporter GLUT-1 is a receptor for HTLV. *Cell, 115*, 449–459. https://doi.org/10.1016/s0092-8674(03)00881-x
30. Ghez, D., et al. (2006). Neuropilin-1 is involved in human T-cell lymphotropic virus type 1 entry. *Journal of Virology, 80*, 6844–6854. https://doi.org/10.1128/JVI.02719-05
31. Pinon, J. D., et al. (2003). Human T-cell leukemia virus type 1 envelope glycoprotein gp46 interacts with cell surface heparan sulfate proteoglycans. *Journal of Virology, 77*, 9922–9930. https://doi.org/10.1128/jvi.77.18.9922-9930.2003
32. Furuta, R., et al. (2017). Human T-cell leukemia virus type 1 infects multiple lineage hematopoietic cells in vivo. *PLoS Pathogens, 13*, e1006722. https://doi.org/10.1371/journal.ppat.1006722
33. Baltimore, D. (1970). RNA-dependent DNA polymerase in virions of RNA tumour viruses. *Nature, 226*, 1209–1211. https://doi.org/10.1038/2261209a0
34. Coffin, J. (1990). In B. N. Fields (Ed.), *Virology* (Vol. 2, pp. 1437–1500). Raven Press.
35. Meekings, K. N., Leipzig, J., Bushman, F. D., Taylor, G. P., & Bangham, C. R. (2008). HTLV-1 integration into transcriptionally active genomic regions is associated with proviral expression and with HAM/TSP. *PLoS Pathogens, 4*, e1000027. https://doi.org/10.1371/journal.ppat.1000027
36. Leclercq, I., et al. (2000). Host sequences flanking the human T-cell leukemia virus type 1 provirus in vivo. *Journal of Virology, 74*, 2305–2312. https://doi.org/10.1128/jvi.74.5.2305-2312.2000
37. Gessain, A., Gallo, R. C., & Franchini, G. (1992). Low degree of human T-cell leukemia/lymphoma virus type I genetic drift in vivo as a means of monitoring viral transmission and movement of ancient human populations. *Journal of Virology, 66*, 2288–2295.
38. Laydon, D. J., Sunkara, V., Boelen, L., Bangham, C. R. M., & Asquith, B. (2020). The relative contributions of infectious and mitotic spread to HTLV-1 persistence. *PLoS Computational Biology, 16*, e1007470. https://doi.org/10.1371/journal.pcbi.1007470
39. Robek, M. D., & Ratner, L. (1999). Immortalization of CD4(+) and CD8(+) T lymphocytes by human T-cell leukemia virus type 1 Tax mutants expressed in a functional molecular clone. *Journal of Virology, 73*, 4856–4865. https://doi.org/10.1128/JVI.73.6.4856-4865.1999

40. Fujita, M., & Shiku, H. (1995). Differences in sensitivity to induction of apoptosis among rat fibroblast cells transformed by HTLV-I tax gene or cellular nuclear oncogenes. *Oncogene, 11*, 15–20.
41. Tanaka, A., et al. (1990). Oncogenic transformation by the tax gene of human T-cell leukemia virus type I in vitro. *Proceedings of the National Academy of Sciences of the United States of America, 87*, 1071–1075. https://doi.org/10.1073/pnas.87.3.1071
42. Pozzatti, R., Vogel, J., & Jay, G. (1990). The human T-lymphotropic virus type I tax gene can cooperate with the ras oncogene to induce neoplastic transformation of cells. *Molecular and Cellular Biology, 10*, 413–417. https://doi.org/10.1128/mcb.10.1.413
43. Grassmann, R., et al. (1992). Role of human T-cell leukemia virus type 1 X region proteins in immortalization of primary human lymphocytes in culture. *Journal of Virology, 66*, 4570–4575. https://doi.org/10.1128/JVI.66.7.4570-4575.1992
44. Meertens, L., Chevalier, S., Weil, R., Gessain, A., & Mahieux, R. (2004). A 10-amino acid domain within human T-cell leukemia virus type 1 and type 2 tax protein sequences is responsible for their divergent subcellular distribution. *The Journal of Biological Chemistry, 279*, 43307–43320.
45. Turci, M., Romanelli, M. G., Lorenzi, P., Righi, P., & Bertazzoni, U. (2006). Localization of human T-cell lymphotropic virus type II Tax protein is dependent upon a nuclear localization determinant in the N-terminal region. *Gene, 365*, 119–124.
46. Zhao, L. J., & Giam, C. Z. (1991). Interaction of the human T-cell lymphotrophic virus type I (HTLV-I) transcriptional activator Tax with cellular factors that bind specifically to the 21-base-pair repeats in the HTLV-I enhancer. *Proceedings of the National Academy of Sciences of the United States of America, 88*, 11445–11449. https://doi.org/10.1073/pnas.88.24.11445
47. Yoshimura, T., Fujisawa, J., & Yoshida, M. (1990). Multiple cDNA clones encoding nuclear proteins that bind to the tax-dependent enhancer of HTLV-1: All contain a leucine zipper structure and basic amino acid domain. *The EMBO Journal, 9*, 2537–2542.
48. Zhao, L. J., & Giam, C. Z. (1992). Human T-cell lymphotropic virus type I (HTLV-I) transcriptional activator, Tax, enhances CREB binding to HTLV-I 21-base-pair repeats by protein-protein interaction. *Proceedings of the National Academy of Sciences of the United States of America, 89*, 7070–7074. https://doi.org/10.1073/pnas.89.15.7070
49. Felber, B. K., Paskalis, H., Kleinman-Ewing, C., Wong-Staal, F., & Pavlakis, G. N. (1985). The pX protein of HTLV-I is a transcriptional activator of its long terminal repeats. *Science, 229*, 675–679. https://doi.org/10.1126/science.2992082
50. Chu, Z. L., DiDonato, J. A., Hawiger, J., & Ballard, D. W. (1998). The tax oncoprotein of human T-cell leukemia virus type 1 associates with and persistently activates IkappaB kinases containing IKKalpha and IKKbeta. *The Journal of Biological Chemistry, 273*, 15891–15894. https://doi.org/10.1074/jbc.273.26.15891
51. Geleziunas, R., et al. (1998). Human T-cell leukemia virus type 1 Tax induction of NF-kappaB involves activation of the IkappaB kinase alpha (IKKalpha) and IKKbeta cellular kinases. *Molecular and Cellular Biology, 18*, 5157–5165. https://doi.org/10.1128/mcb.18.9.5157
52. Li, X. H., Murphy, K. M., Palka, K. T., Surabhi, R. M., & Gaynor, R. B. (1999). The human T-cell leukemia virus type-1 Tax protein regulates the activity of the IkappaB kinase complex. *The Journal of Biological Chemistry, 274*, 34417–34424. https://doi.org/10.1074/jbc.274.48.34417
53. Jin, D. Y., Giordano, V., Kibler, K. V., Nakano, H., & Jeang, K. T. (1999). Role of adapter function in oncoprotein-mediated activation of NF-kappaB. Human T-cell leukemia virus type I Tax interacts directly with IkappaB kinase gamma. *Journal of Biological Chemistry, 274*, 17402–17405. https://doi.org/10.1074/jbc.274.25.17402
54. Harhaj, E. W., & Sun, S. C. (1999). IKKgamma serves as a docking subunit of the IkappaB kinase (IKK) and mediates interaction of IKK with the human T-cell leukemia virus Tax protein. *The Journal of Biological Chemistry, 274*, 22911–22914. https://doi.org/10.1074/jbc.274.33.22911

55. Scoggin, K. E., Ulloa, A., & Nyborg, J. K. (2001). The oncoprotein Tax binds the SRC-1-interacting domain of CBP/p300 to mediate transcriptional activation. *Molecular and Cellular Biology, 21*, 5520–5530. https://doi.org/10.1128/MCB.21.16.5520-5530.2001
56. Nicot, C. (2015). HTLV-I tax-mediated inactivation of cell cycle checkpoints and DNA repair pathways contribute to cellular transformation: "A random mutagenesis model". *Journal of Cancer Science, 2*. https://doi.org/10.13188/2377-9292.1000009
57. Muhleisen, A., Giaisi, M., Kohler, R., Krammer, P. H., & Li-Weber, M. (2014). Tax contributes apoptosis resistance to HTLV-1-infected T cells via suppression of Bid and Bim expression. *Cell Death & Disease, 5*, e1575. https://doi.org/10.1038/cddis.2014.536
58. Fochi, S., et al. (2019). NF-kappaB and MicroRNA Deregulation Mediated by HTLV-1 Tax and HBZ. *Pathogens, 8*. https://doi.org/10.3390/pathogens8040290
59. Martinez, M. P., Al-Saleem, J., & Green, P. L. (2019). Comparative virology of HTLV-1 and HTLV-2. *Retrovirology, 16*, 21. https://doi.org/10.1186/s12977-019-0483-0
60. Larocca, D., Chao, L. A., Seto, M. H., & Brunck, T. K. (1989). Human T-cell leukemia virus minus strand transcription in infected cells. *Biochemical and Biophysical Research Communications, 163*, 1006–1013.
61. Satou, Y., Yasunaga, J., Yoshida, M., & Matsuoka, M. (2006). HTLV-I basic leucine zipper factor gene mRNA supports proliferation of adult T cell leukemia cells. *Proceedings of the National Academy of Sciences of the United States of America, 103*, 720–725. https://doi.org/10.1073/pnas.0507631103
62. Yoshida, M., Satou, Y., Yasunaga, J., Fujisawa, J., & Matsuoka, M. (2008). Transcriptional control of spliced and unspliced human T-cell leukemia virus type 1 bZIP factor (HBZ) gene. *Journal of Virology, 82*, 9359–9368.
63. Usui, T., et al. (2008). Characteristic expression of HTLV-1 basic zipper factor (HBZ) transcripts in HTLV-1 provirus-positive cells. *Retrovirology, 5*, 34.
64. Giam, C. Z., & Semmes, O. J. (2016). HTLV-1 infection and adult T-cell leukemia/lymphoma-A tale of two proteins: Tax and HBZ. *Viruses, 8*. https://doi.org/10.3390/v8060161
65. Mitobe, Y., Yasunaga, J., Furuta, R., & Matsuoka, M. (2015). HTLV-1 bZIP factor RNA and protein impart distinct functions on T-cell proliferation and survival. *Cancer Research, 75*, 4143–4152. https://doi.org/10.1158/0008-5472.CAN-15-0942
66. Clerc, I., et al. (2008). An interaction between the human T cell leukemia virus type 1 basic leucine zipper factor (HBZ) and the KIX domain of p300/CBP contributes to the down-regulation of tax-dependent viral transcription by HBZ. *The Journal of Biological Chemistry, 283*, 23903–23913. https://doi.org/10.1074/jbc.M803116200
67. Zhao, T., et al. (1865-1876, 2011). HTLV-1 bZIP factor enhances TGF-beta signaling through p300 coactivator. *Blood, 118*. https://doi.org/10.1182/blood-2010-12-326199. blood-2010-12-326199 [pii].
68. Gaudray, G., et al. (2002). The complementary strand of the human T-cell leukemia virus type 1 RNA genome encodes a bZIP transcription factor that down-regulates viral transcription. *Journal of Virology, 76*, 12813–12822.
69. Cook, P. R., Polakowski, N., & Lemasson, I. (2011). HTLV-1 HBZ protein deregulates interactions between cellular factors and the KIX domain of p300/CBP. *Journal of Molecular Biology, 409*, 384–398. https://doi.org/10.1016/j.jmb.2011.04.003
70. Raval, G. U., et al. (2015). Localization, quantification and interaction with host factors of endogenous HTLV-1 HBZ protein in infected cells and ATL. *Retrovirology, 12*, 59. https://doi.org/10.1186/s12977-015-0186-0
71. Basbous, J., et al. (2003). The HBZ factor of human T-cell leukemia virus type I dimerizes with transcription factors JunB and c-Jun and modulates their transcriptional activity. *The Journal of Biological Chemistry, 278*, 43620–43627. https://doi.org/10.1074/jbc.M307275200
72. Matsumoto, J., Ohshima, T., Isono, O., & Shimotohno, K. (2005). HTLV-1 HBZ suppresses AP-1 activity by impairing both the DNA-binding ability and the stability of c-Jun protein. *Oncogene, 24*, 1001–1010.

73. Hivin, P., et al. (2007). The HBZ-SP1 isoform of human T-cell leukemia virus type I represses JunB activity by sequestration into nuclear bodies. *Retrovirology, 4*, 14.
74. Thebault, S., Basbous, J., Hivin, P., Devaux, C., & Mesnard, J. M. (2004). HBZ interacts with JunD and stimulates its transcriptional activity. *FEBS Letters, 562*, 165–170.
75. Kuhlmann, A. S., et al. (2007). HTLV-1 HBZ cooperates with JunD to enhance transcription of the human telomerase reverse transcriptase gene (hTERT). *Retrovirology, 4*, 92.
76. Gazon, H., et al. (2012). Human T-cell leukemia virus type 1 (HTLV-1) bZIP factor requires cellular transcription factor JunD to upregulate HTLV-1 antisense transcription from the 3′ long terminal repeat. *Journal of Virology, 86*, 9070–9078. https://doi.org/10.1128/JVI.00661-12
77. Ma, Y., et al. (2013). The HTLV-1 HBZ protein inhibits cyclin D1 expression through interacting with the cellular transcription factor CREB. *Molecular Biology Reports, 40*, 5967–5975. https://doi.org/10.1007/s11033-013-2706-0
78. Lemasson, I., et al. (2007). Human T-cell leukemia virus type 1 (HTLV-1) bZIP protein interacts with the cellular transcription factor CREB to inhibit HTLV-1 transcription. *Journal of Virology, 81*, 1543–1553. https://doi.org/10.1128/JVI.00480-06
79. Ohshima, T., et al. (2010). HTLV-1 basic leucine-zipper factor, HBZ, interacts with MafB and suppresses transcription through a Maf recognition element. *Journal of Cellular Biochemistry, 111*, 187–194.
80. Hagiya, K., Yasunaga, J., Satou, Y., Ohshima, K., & Matsuoka, M. (2011). ATF3, an HTLV-1 bZip factor binding protein, promotes proliferation of adult T-cell leukemia cells. *Retrovirology, 8*, 19. https://doi.org/10.1186/1742-4690-8-19
81. Hasegawa, H., et al. (2006). Thymus-derived leukemia-lymphoma in mice transgenic for the Tax gene of human T-lymphotropic virus type I. *Nature Medicine, 12*, 466–472. https://doi.org/10.1038/nm1389
82. Arnold, J., et al. (2006). Enhancement of infectivity and persistence in vivo by HBZ, a natural antisense coded protein of HTLV-1. *Blood, 107*, 3976–3982. https://doi.org/10.1182/blood-2005-11-4551
83. Arnold, J., Zimmerman, B., Li, M., Lairmore, M. D., & Green, P. L. (2008). Human T-cell leukemia virus type-1 antisense-encoded gene, Hbz, promotes T-lymphocyte proliferation. *Blood, 112*, 3788–3797. https://doi.org/10.1182/blood-2008-04-154286
84. Akkouche, A., et al. (2021). In vivo antagonistic role of the Human T-Cell Leukemia Virus Type 1 regulatory proteins Tax and HBZ. *PLoS Pathogens, 17*, e1009219. https://doi.org/10.1371/journal.ppat.1009219
85. Bangham, C. R. M., Miura, M., Kulkarni, A., & Matsuoka, M. (2019). Regulation of latency in the human T cell leukemia virus, HTLV-1. *Annual Review of Virology, 6*, 365–385. https://doi.org/10.1146/annurev-virology-092818-015501
86. Billman, M. R., Rueda, D., & Bangham, C. R. M. (2017). Single-cell heterogeneity and cell-cycle-related viral gene bursts in the human leukaemia virus HTLV-1. *Wellcome Open Research, 2*, 87. https://doi.org/10.12688/wellcomeopenres.12469.2
87. Kulkarni, A., et al. (2017). Glucose metabolism and oxygen availability govern reactivation of the latent human retrovirus HTLV-1. *Cell Chemical Biology, **24***, 1377–1387. e1373. https://doi.org/10.1016/j.chembiol.2017.08.016
88. Katsuya, H., et al. (2019). The nature of the HTLV-1 provirus in naturally infected individuals analyzed by the viral DNA-capture-seq approach. *Cell Reports., 29*, 724–735.e724. https://doi.org/10.1016/j.celrep.2019.09.016
89. Yoshida, M., Miyoshi, I., & Hinuma, Y. (1982). Isolation and characterization of retrovirus from cell lines of human adult T-cell leukemia and its implication in the disease. *Proceedings of the National Academy of Sciences of the United States of America, 79*, 2031–2035.
90. Uchiyama, T., Yodoi, J., Sagawa, K., Takatsuki, K., & Uchino, H. (1977). Adult T-cell leukemia: Clinical and hematologic features of 16 cases. *Blood, 50*, 81–492.

91. Poiesz, B. J., et al. (1980). Detection and isolation of type C retrovirus particles from fresh and cultured lymphocytes of a patient with cutaneous T-cell lymphoma. *Proceedings of the National Academy of Sciences of the United States of America, 77*, 7415–7419.
92. Ureshino, H., Komiya, N., Sueoka, E., Kimura, S., & Miyahara, M. (2014). Mogamulizumab for adult T-cell leukemia/lymphoma expressing atypical phenotype CD4-/CD8+/CCR4+. *Open Journal of Hematology*. https://doi.org/10.13055/ojhmt_5_1_5.140623
93. Ishitsuka, K., & Tamura, K. (2014). Human T-cell leukaemia virus type I and adult T-cell leukaemia-lymphoma. *The Lancet Oncology, 15*, e517–e526. https://doi.org/10.1016/S1470-2045(14)70202-5
94. Iwanaga, M., et al. (2010). Human T-cell leukemia virus type I (HTLV-1) proviral load and disease progression in asymptomatic HTLV-1 carriers: A nationwide prospective study in Japan. *Blood, 116*, 1211–1219. https://doi.org/10.1182/blood-2009-12-257410
95. Shimoyama, M. (1991). Diagnostic criteria and classification of clinical subtypes of adult T-cell leukaemia-lymphoma. A report from the Lymphoma Study Group (1984-87). *British Journal of Haematology, 79*, 428–437. https://doi.org/10.1111/j.1365-2141.1991.tb08051.x
96. Yano, H., et al. (2007). Regulatory T-cell function of adult T-cell leukemia/lymphoma cells. *International Journal of Cancer, 120*, 2052–2057. https://doi.org/10.1002/ijc.22536
97. Panfil, A. R., Martinez, M. P., Ratner, L., & Green, P. L. (2016). Human T-cell leukemia virus-associated malignancy. *Current Opinion in Virology, 20*, 40–46. https://doi.org/10.1016/j.coviro.2016.08.009
98. Tamiya, S., et al. (1996). Two types of defective human T-lymphotropic virus type I provirus in adult T-cell leukemia. *Blood, 88*, 3065–3073.
99. Furukawa, Y., Kubota, R., Tara, M., Izumo, S., & Osame, M. (2001). Existence of escape mutant in HTLV-I tax during the development of adult T-cell leukemia. *Blood, 97*, 987–993.
100. Takeda, S., et al. (2004). Genetic and epigenetic inactivation of tax gene in adult T-cell leukemia cells. *International Journal of Cancer, 109*, 559–567. https://doi.org/10.1002/ijc.20007
101. Mahgoub, M., et al. (2018). Sporadic on/off switching of HTLV-1 Tax expression is crucial to maintain the whole population of virus-induced leukemic cells. *Proceedings of the National Academy of Sciences of the United States of America, 115*, E1269–E1278. https://doi.org/10.1073/pnas.1715724115
102. Enose-Akahata, Y., Vellucci, A., & Jacobson, S. (2017). Role of HTLV-1 Tax and HBZ in the pathogenesis of HAM/TSP. *Frontiers in Microbiology, 8*, 2563. https://doi.org/10.3389/fmicb.2017.02563
103. Gessain, A., et al. (1985). Antibodies to human T-lymphotropic virus type-I in patients with tropical spastic paraparesis. *Lancet, 2*, 407–410. https://doi.org/10.1016/s0140-6736(85)92734-5
104. Osame, M., et al. (1986). HTLV-I associated myelopathy, a new clinical entity. *Lancet, 1*, 1031–1032. https://doi.org/10.1016/s0140-6736(86)91298-5
105. Matsuura, E., et al. (2016). HTLV-1 associated myelopathy/tropical spastic paraparesis (HAM/TSP): A comparative study to identify factors that influence disease progression. *Journal of the Neurological Sciences, 371*, 112–116. https://doi.org/10.1016/j.jns.2016.10.030
106. Olindo, S., et al. (2005). HTLV-1 proviral load in peripheral blood mononuclear cells quantified in 100 HAM/TSP patients: A marker of disease progression. *Journal of the Neurological Sciences, 237*, 53–59. https://doi.org/10.1016/j.jns.2005.05.010
107. Nagai, M., et al. (1998). Analysis of HTLV-I proviral load in 202 HAM/TSP patients and 243 asymptomatic HTLV-I carriers: High proviral load strongly predisposes to HAM/TSP. *Journal of Neurovirology, 4*, 586–593. https://doi.org/10.3109/13550289809114225
108. Lehky, T. J., et al. (1995). Detection of human T-lymphotropic virus type I (HTLV-I) tax RNA in the central nervous system of HTLV-I-associated myelopathy/tropical spastic paraparesis patients by in situ hybridization. *Annals of Neurology, 37*, 167–175. https://doi.org/10.1002/ana.410370206
109. Andrade, R. G., et al. (2013). Strong correlation between tax and HBZ mRNA expression in HAM/TSP patients: Distinct markers for the neurologic disease. *Journal of Clinical Virology, 56*, 135–140. https://doi.org/10.1016/j.jcv.2012.10.003

110. Cartier, L., & Ramirez, E. (2005). Presence of HTLV-I Tax protein in cerebrospinal fluid from HAM/TSP patients. *Archives of Virology, 150*, 743–753. https://doi.org/10.1007/s00705-004-0443-3
111. Moritoyo, T., et al. (1999). Detection of human T-lymphotropic virus type I p40tax protein in cerebrospinal fluid cells from patients with human T-lymphotropic virus type I-associated myelopathy/tropical spastic paraparesis. *Journal of Neurovirology, 5*, 241–248. https://doi.org/10.3109/13550289909015810
112. Saito, M., et al. (2009). In vivo expression of the HBZ gene of HTLV-1 correlates with proviral load, inflammatory markers and disease severity in HTLV-1 associated myelopathy/tropical spastic paraparesis (HAM/TSP). *Retrovirology, 6*, 19. https://doi.org/10.1186/1742-4690-6-19
113. Satou, Y., et al. (2011). HTLV-1 bZIP factor induces T-cell lymphoma and systemic inflammation in vivo. *PLoS Pathogens, 7*, e1001274. https://doi.org/10.1371/journal.ppat.1001274
114. Anderson, M. D., Ye, J., Xie, L., & Green, P. L. (2004). Transformation studies with a human T-cell leukemia virus type 1 molecular clone. *Journal of Virological Methods, 116*, 195–202. https://doi.org/10.1016/j.jviromet.2003.11.016
115. Kannian, P., et al. (2012). Distinct transformation tropism exhibited by human T lymphotropic virus type 1 (HTLV-1) and HTLV-2 is the result of postinfection T cell clonal expansion. *Journal of Virology, 86*, 3757–3766. https://doi.org/10.1128/JVI.06900-11
116. Ye, J., Silverman, L., Lairmore, M. D., & Green, P. L. (2003). HTLV-1 Rex is required for viral spread and persistence in vivo but is dispensable for cellular immortalization in vitro. *Blood, 102*, 3963–3969. https://doi.org/10.1182/blood-2003-05-1490
117. Xie, L., Yamamoto, B., Haoudi, A., Semmes, O. J., & Green, P. L. (2006). PDZ binding motif of HTLV-1 Tax promotes virus-mediated T-cell proliferation in vitro and persistence in vivo. *Blood, 107*, 1980–1988. https://doi.org/10.1182/blood-2005-03-1333
118. Younis, I., Yamamoto, B., Phipps, A., & Green, P. L. (2005). Human T-cell leukemia virus type 1 expressing nonoverlapping tax and rex genes replicates and immortalizes primary human T lymphocytes but fails to replicate and persist in vivo. *Journal of Virology, 79*, 14473–14481. https://doi.org/10.1128/JVI.79.23.14473-14481.2005
119. Martinez, M. P., et al. (2019). HTLV-1 CTCF-binding site is dispensable for in vitro immortalization and persistent infection in vivo. *Retrovirology, 16*, 44. https://doi.org/10.1186/s12977-019-0507-9
120. Yin, H., et al. (2012). Human T-cell leukemia virus type 2 antisense viral protein 2 is dispensable for in vitro immortalization but functions to repress early virus replication in vivo. *Journal of Virology, 86*, 8412–8421. https://doi.org/10.1128/JVI.00717-12
121. Dewan, M. Z., et al. (2003). Rapid tumor formation of human T-cell leukemia virus type 1-infected cell lines in novel NOD-SCID/gammac(null) mice: Suppression by an inhibitor against NF-kappaB. *Journal of Virology, 77*, 5286–5294. https://doi.org/10.1128/jvi.77.9.5286-5294.2003
122. Ohsugi, T., et al. (2005). In vivo antitumor activity of the NF-kappaB inhibitor dehydroxymethylepoxyquinomicin in a mouse model of adult T-cell leukemia. *Carcinogenesis, 26*, 1382–1388. https://doi.org/10.1093/carcin/bgi095
123. Huey, D. D., & Niewiesk, S. (2018). Production of humanized Mice through stem cell transfer. *Curremt Protocols in Mouse Biology, 8*, 17–27. https://doi.org/10.1002/cpmo.38
124. Huey, D. D., et al. (2018). Role of wild-type and recombinant human T-cell leukemia viruses in lymphoproliferative disease in humanized NSG mice. *Comparative Medicine, 68*, 4–14.
125. Hirakawa, M. P., Krishnakumar, R., Timlin, J. A., Carney, J. P., & Butler, K. S. (2020). Gene editing and CRISPR in the clinic: Current and future perspectives. *Bioscience Reports, 40*. https://doi.org/10.1042/BSR20200127
126. Frangoul, H., et al. (2021). CRISPR-Cas9 gene editing for sickle cell disease and beta-thalassemia. *The New England Journal of Medicine, 384*, 252–260. https://doi.org/10.1056/NEJMoa2031054

127. Mout, R., Ray, M., Lee, Y. W., Scaletti, F., & Rotello, V. M. (2017). In vivo delivery of CRISPR/Cas9 for therapeutic gene editing: Progress and challenges. *Bioconjugate Chemistry, 28*, 880–884. https://doi.org/10.1021/acs.bioconjchem.7b00057
128. Tanaka, A., et al. (2013). A novel therapeutic molecule against HTLV-1 infection targeting provirus. *Leukemia, 27*, 1621–1627. https://doi.org/10.1038/leu.2013.46
129. Nakagawa, M., et al. (2018). Targeting the HTLV-I-regulated BATF3/IRF4 transcriptional network in adult T cell leukemia/lymphoma. *Cancer Cell, **34***, 286–297.e210. https://doi.org/10.1016/j.ccell.2018.06.014
130. Seiki, M., Hattori, S., Hirayama, Y., & Yoshida, M. (1983). Human adult T-cell leukemia virus: Complete nucleotide sequence of the provirus genome integrated in leukemia cell DNA. *Proceedings of the National Academy of Sciences of the United States of America, 80*, 3618–3622. https://doi.org/10.1073/pnas.80.12.3618
131. Gray, G. S., White, M., Bartman, T., & Mann, D. (1990). Envelope gene sequence of HTLV-1 isolate MT-2 and its comparison with other HTLV-1 isolates. *Virology, 177*, 391–395. https://doi.org/10.1016/0042-6822(90)90498-g
132. Utsunomiya, A., Choi, I., Chihara, D., & Seto, M. (2015). Recent advances in the treatment of adult T-cell leukemia-lymphomas. *Cancer Science, 106*, 344–351. https://doi.org/10.1111/cas.12617
133. Yves, P., Stephane, M., Rishika, B., Christine, D., & Gerard, P. (2015). Characteristics of adult T-cell leukemia/lymphoma patients with long survival: Prognostic significance of skin lesions and possible beneficial role of valproic acid. *Leukemia Research and Treatment, 2015*, 476805. https://doi.org/10.1155/2015/476805
134. Sato, T., et al. (2018). Mogamulizumab (Anti-CCR4) in HTLV-1-associated myelopathy. *The New England Journal of Medicine, 378*, 529–538. https://doi.org/10.1056/NEJMoa1704827
135. Bangham, C. R., Araujo, A., Yamano, Y., & Taylor, G. P. (2015). HTLV-1-associated myelopathy/tropical spastic paraparesis. *Nature Reviews. Disease Primers, 1*, 15012. https://doi.org/10.1038/nrdp.2015.12
136. Grassmann, R., Aboud, M., & Jeang, K. T. (2005). Molecular mechanisms of cellular transformation by HTLV-1 Tax. *Oncogene, 24*, 5976–5985.
137. Bex, F., & Gaynor, R. B. (1998). Regulation of gene expression by HTLV-I Tax protein. *Methods, 16*, 83–94. https://doi.org/10.1006/meth.1998.0646
138. Yasunaga, J., & Matsuoka, M. (2007). Leukaemogenic mechanism of human T-cell leukaemia virus type I. *Reviews in Medical Virology, 17*, 301–311. https://doi.org/10.1002/rmv.548

Retroviral Vectors for Gene Therapy of Monogenic Diseases

Kristine E. Yoder, Anthony J. Rabe, and Ross C. Larue

Abstract Gene therapy vectors derived from retroviruses are attractive for treating monogenic diseases due to the stable integration of a transgene in patient cells. In addition, retroviruses do not activate the immune system, allowing for multiple doses with the same viral vector. Retroviral vectors are also readily pseudotyped expanding their tropism to a variety of cell types. In spite of these advantages, gene therapy trials with gammaretrovirus murine leukemia virus (MLV) vectors led to leukemia in five patients. This has led to important questions about the safety of retroviral vectors. Lentiviral vectors have not yet been shown to cause any oncogenic transformation and may be clinically relevant for some monogenic diseases. Other retroviral vectors derived from alpharetroviruses or spumaviruses are in development. Spumavirus vectors have been safely used for gene therapy in animals but have not yet been tested in humans. With continuing advances in safety, retroviral or lentiviral gene therapy vectors should continue to lead to cures of monogenic diseases.

1 Introduction

Retroviruses have been engineered to create retroviral gene therapy vectors, allowing stable expression of a transgene to cure monogenic diseases [1]. All retroviruses have two defining enzymes: reverse transcriptase and integrase [2]. The virus particles contain two genomic RNAs. Following entry into a host cell, reverse transcriptase copies the genomic RNA to a linear double-stranded cDNA (Fig. 1). The cDNA and integrase are part of a poorly defined preintegration complex (PIC) that includes an assemblage of viral and host proteins. Most retroviruses require cellular division for the PIC to gain access to the host chromatin, while lentiviruses can

K. E. Yoder (✉) · A. J. Rabe · R. C. Larue (✉)
Department of Cancer Biology and Genetics, College of Medicine,
The Ohio State University, Columbus, OH, USA
e-mail: yoder.176@osu.edu; larue.22@osu.edu

Y. H. Yun, K. E. Yoder (eds.), *Biotechnologies for Gene Therapy*,
https://doi.org/10.1007/978-3-030-93333-3_7

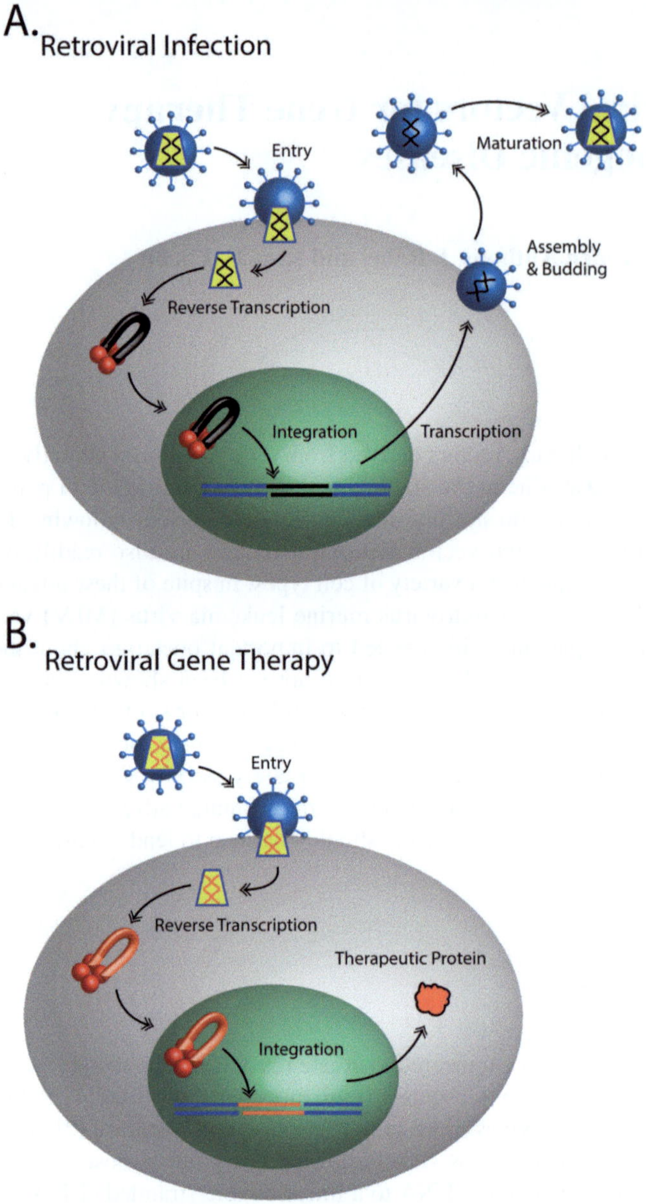

Fig. 1 The retroviral life cycle and transduction by retroviral vector particles. (**a**) The retroviral envelope protein determines the tropism of the virus and mediates entry to a target cell. Inside a cell, the viral genomic RNA (thin black lines) is reverse transcribed to a linear double-stranded cDNA. The viral cDNA (thick black lines) is bound by integrase, yielding a preintegration complex (PIC), which will mediate covalent joining to the host genome (blue). Lentiviruses have the unique ability to enter the nucleus without dissolution of the nuclear membrane during cellular division. Viral mRNAs and genomic RNA are transcribed from the integrated provirus by host transcription enzymes. Virus particles leaving the cell are not infectious and require the action of protease to cleave viral polyproteins, leading to mature, infectious virions. (**b**) Retroviral vector particles are generally pseudotyped with a membrane protein to allow altered tropism and entry to different cell types. Reverse transcription and integration are the same for retroviruses and retroviral vectors. The retroviral vector RNA (thin red lines) and cDNA (thick red lines) do not encode viral genes. Following integration, the provirus can only express the therapeutic protein of interest

cross an intact nuclear membrane [3, 4]. Integrase mediates the covalent joining of the viral cDNA to the host DNA, resulting in an integrated provirus. Integration site selection does not appear to be stochastic with each retrovirus displaying a unique preference for genomic features [5]. The proviral genome is transcribed by the host RNA Pol II transcription machinery to generate viral genomic RNA and mRNAs, which are translated in the cytoplasm. Progeny virions are assembled and released from the cell as immature uninfectious particles. The viral enzyme protease cleaves polyproteins in the virus particle, leading to an infectious virus. Retroviruses are classified as families: alpha (avian sarcoma and leukosis virus, ASLV), beta (mouse mammary tumor virus, MMTV), gamma (murine leukemia virus, MLV), delta (human T cell leukemia virus, HTLV-1), epsilon (walleye dermal sarcoma virus, WDSV), lentivirus (human immunodeficiency virus, HIV-1), and spumavirus (prototype foamy virus, PFV). The alpha through epsilon families are oncogenic. The lentivirus family causes immunodeficiency. The spumavirus family, also known as foamy viruses, does not appear to cause any disease [6].

Integration of retroviral gene therapy vector genomes offers the exciting possibility to cure monogenic diseases by stably introducing a corrective transgene. Retroviral gene therapy vectors recapitulate the early steps of the retroviral life cycle due to the inclusion of structural and enzymatic proteins (Fig. 1) [7–10]. The viral vectors are typically pseudotyped with an alternative envelope protein and the retroviral envelope gene is deleted. The engineered vector genome encodes a promotor driving the corrective transgene (Fig. 2). It also includes viral regulatory elements necessary for packaging the genomic RNA into vector particles and reverse transcription [13]. The vector genome does not encode any viral genes. With the exception of spumaviruses, retroviral vector particles may be pseudotyped with alternative membrane proteins changing the tropism and permitting transduction of different cell types [14, 15]. Retroviral-based vectors also offer the advantage of minimal innate and adaptive immunity responses, particularly in comparison with gene therapies derived from immunogenic adenoviruses [16–19]. Retroviral gene therapy vectors have functionally cured several monogenic disorders, including X-linked chronic granulomatous disease (X-CGD), Wiskott-Aldrich syndrome (WAS), X-linked adrenoleukodystrophy, and X-linked severe combined immune deficiency (X-SCID) [11, 20, 21]. Despite these initial successes, the field of retroviral gene therapy was suspended when an MLV-based gene therapy vector in clinical use treating X-SCID led to leukemia in several patients [22–24]. Genetic characterization of the tumors revealed MLV vector integration at the promoters of proto-oncogenes altering their expression and leading to oncogenic transformation. Interestingly, this is the same mechanism by which MLV infection causes leukemia in animals.

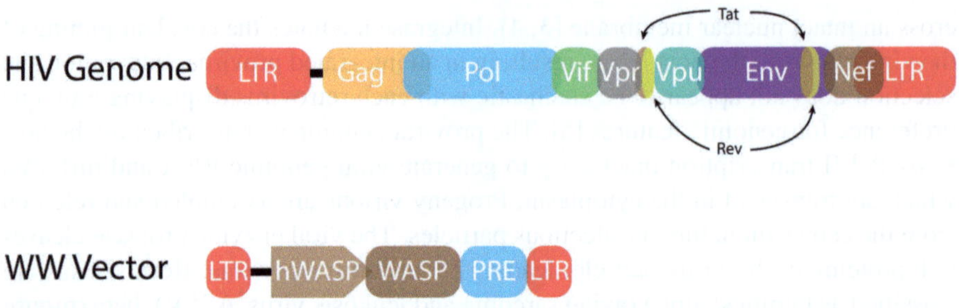

Fig. 2 HIV-1 viral genome and lentiviral vector genome. As in all retroviruses, the HIV-1 proviral genome (HIV Genome) is flanked by long terminal repeats (LTRs) that are generated during reverse transcription and includes binding sites for integrase and host transcription factors. All retroviral genomes include three genes. The *gag* gene encodes all structural proteins of the virus, *pol* encodes viral enzymes, and *env* encodes the envelope proteins. HIV-1 has six accessory genes not found in other retroviruses, including two spliced genes *tat* and *rev*. The genome of HIV-1-based lentiviral vectors (WW Vector) typically has deletions of the LTRs removing transcription factor–binding sites, termed self-inactivating (SIN). Only a gene of interest and its promoter are included in the vector genome. In this illustration, the *Wiskott-Aldrich syndrome protein* (*WASP*) gene is driven by the natural human WASP promoter (hWASP) [11]. A posttranscription regulatory element (PRE) is an RNA element that enhances the nuclear export of unspliced mRNA [12]

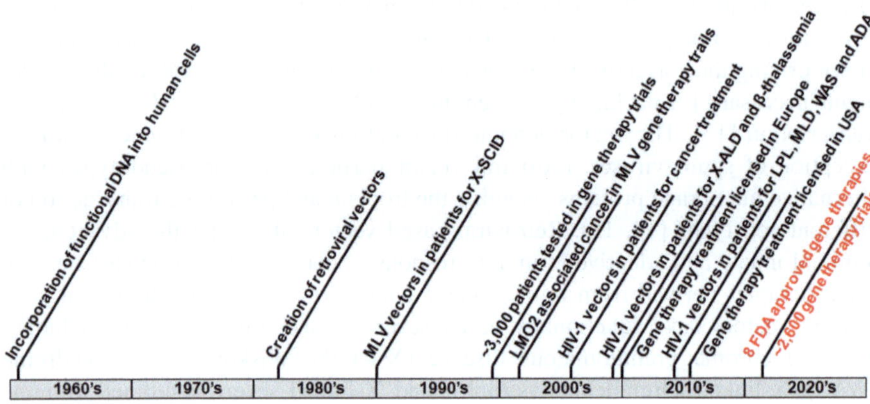

Fig. 3 Time line of developments in retroviral vectors for gene therapy

2 Retroviral Vectors Based on Gammaretrovirus MLV

The first retroviral gene therapy vectors were derived from MLV (Fig. 3). These vectors were used to treat X-SCID. Hematopoietic stem cells were transduced ex vivo with an MLV vector expressing the *interleukin 2 receptor common subunit gamma* (*IL-2RG*) gene and reintroduced to patients [25]. Between 1999 and 2009, twenty patients were treated with this gene therapy vector. Seventeen patients were cured of X-SCID, but five later developed leukemia [26, 27]. In these patients, the MLV vector proviral genome had integrated near known oncogenes *LMO-2* and

CCND2. The provirus dysregulated expression of these host genes, leading to oncogenic transformation. In addition, patients treated with MLV gene therapy vectors for WAS and X-CGD also developed leukemia [21, 28, 29]. These clinical trials demonstrated the ability of retroviral gene therapy to cure monogenic diseases, but also revealed significant risks.

MLV was shown to favor integration near transcription start sites (TSSs, 28% within 5 kb of a TSS compared to a matched random control value of 4.9%) [30]. This integration site preference was later discovered to be the result of MLV integrase binding to host BET proteins [31–33]. The family of BET proteins includes Brd2, 3, and 4, which are ubiquitously expressed. This protein family is involved in transcription, cell cycle regulation, and DNA replication. These proteins include two bromodomains near the amino terminus and an extra terminal (ET) domain near the carboxyl terminus. The bromodomains bind acetylated histone H3 and H4 tails, while the ET domain binds MLV integrase. Thus, the BET proteins are able to tether MLV integration complexes to chromatin at regions where H3 and H4 are acetylated including TSSs, enhancers, and super-enhancers. The MLV long terminal repeat (LTR) also includes strong transcriptional enhancers that can lead to inappropriate expression of host genes near the proviral vector genome [34]. MLV proviral genomes near oncogenes can lead to their dysregulation and ultimately transformation [35].

It is possible to engineer mutations in MLV integrase that prevent binding to BET proteins [36, 37]. MLV vectors with these integrase mutants retain infectivity and integration activity. These mutations reduced integration near TSSs, but not completely [36, 38]. Importantly, the MLV integrase mutants were characterized by infection in a mouse model [39]. Although there was less integration at TSSs and decreased rate of tumorigenesis, the tumors that did occur displayed integration near oncogenes. These data suggest that BET-independent MLV vectors may still lead to oncogenesis.

3 Lentiviral Vectors Derived from HIV-1

Retroviral gene therapy applications transitioned to lentiviral-based vectors instead of MLV-based vectors that led to leukemia. Lentiviruses, such as HIV-1, lead to immunosuppression but are not oncogenic. However, some safety concerns still exist. HIV-1 patients treated with antiretroviral therapy (ART) display a greater prevalence of non-AIDS-defining malignancies (NADM), including Hodgkin's lymphoma, oropharyngeal cancer, anal cancer, hepatocellular carcinoma, and non–small cell lung cancer [40]. Sequencing of integration sites in cells from ART treated patients revealed ~40% of HIV-1 latently infected cells were clonally expanded with the provirus at known oncogenes such as *MKL2* and *BACH2* [41, 42]. While the observation of clonal expansion by HIV-1 latently infected cells is concerning, there is no proven association with oncogenic transformation [43, 44]. Recent clinical trials with lentiviral gene therapy vectors to treat sickle cell disease were halted due to the development of acute myeloid leukemia (AML) in one patient and myelodysplastic syndrome (MDS) in another [45]. Genetic analysis of tumor cells from the AML patient revealed genomic rearrangements and integration at the *vesicle associate membrane protein 4* (*VAMP4*) gene, which is not associated with oncogenesis. It seems likely that genomic rearrangements were the cause of AML and were unrelated to the lentiviral vector. The integration site in cells from the MDS patient has not yet been reported.

One highly attractive feature of lentiviral vectors for gene therapy is their ability to transduce nondividing cells [46]. Resting cells, such as naïve T cells or hematopoietic stem cells, may be transduced ex vivo without activation. This may allow the cells to better retain their functionality when reinfused to patients [47]. Early versions of lentiviral-based vectors were associated with insertional mutagenesis in proliferative HSCs and tumors in mice [27, 48]. Subsequent development and engineering of lentiviral vectors has included multiple safety features. Virulence factors have been deleted, and the viral genome is encoded on multiple plasmids to reduce the possibility of recombination. Viral promoter elements present in the LTRs have been deleted, yielding self-inactivating (SIN) vectors [49, 50]. These safer lentiviral vectors have been used to treat WAS, metachromatic leukodystrophy, beta-thalassemia, X-linked adrenoleukodystrophy, and metachromatic leukodystrophy [11, 51–53]. Lentiviral gene therapy treatment of beta-thalassemia made transfusions unnecessary for several years, indicating that the transgene was successfully expressed on a long-term basis without any observed adverse reactions. Clonal expansion was observed in cells from one patient with integration at the proto-oncogene *high mobility group A2* (*HMGA2*) gene but did not lead to cancer [54]. Evaluation of cells from 22 additional patients revealed no evidence of clonal expansion [55].

The difference in oncogenic potential between MLV and HIV-1 derived gene therapy vectors is their integration site preference. While MLV integration prefers TSSs due to BET proteins (15–39% compared to 4.9% in matched randomized controls), HIV-1 prefers to integrate in the bodies of active transcribed genes (58%–86% compared to 45.7% in matched randomized controls) [30, 33, 56–60]. HIV-1 integration is directed by host protein lens epithelium–derived growth factor (LEDGF/p75), the first identified host cofactor for any retroviral integrase [61, 62]. LEDGF/p75 has a PWWP domain at the amino terminus and an integrase-binding domain (IBD) at the carboxyl terminus [63–65]. The PWWP domain binds to a tri-methylation posttranslational modification of histone H3 at lysine 36 (H3K36me3) and the IBD binds a groove created by two HIV-1 integrase protomers [66–71]. Binding of LEDGF/p75 helps to stabilize the HIV-1 integration complex and tethers it to nucleosomes decorated with H3K36me3. H3K36me3 is found in actively transcribed genes, explaining the preference for HIV-1 integration site selection [59, 72].

Other host proteins have been implicated in directing HIV-1 integration to genomic features without directly binding to integrase. The HIV-1 capsid core remains at least partially intact as the PIC enters the nucleus. Capsid binds several host proteins that appear to influence integration site preference including cleavage and polyadenylation specificity factor subunit 6 (CPSF6), nucleoporin protein 153

(Nup153), and E3 SUMO-protein ligase (RANBP2 or Nup358). CPSF6 is a chromatin-associated factor and part of several complexes in the nucleus including cleavage factor Im (CFIm) complex, paraspeckles, and nuclear speckles [73, 74]. The interaction of CPSF6 and HIV-1 capsid directs the PIC to gene dense regions [75–79]. Nup153 and RANBP2 are part of the nuclear pore complex [80–83]. Nup153 and RANBP2 participate in nuclear import of the capsid-encased PIC and depletion of either protein targets integration to less dense gene regions [84–89]. The interactions of capsid with nuclear pore proteins could explain the observation of HIV-1 integration events near the nuclear pore [90]. Depletion of the capsid-interacting proteins does not alter the preference for HIV-1 integration into active genes, but instead retargets integration away from gene dense regions. There have been no circumstances observed where HIV-1 integration occurs preferentially near TSSs, which correlates with oncogenesis.

4 Retroviral Vectors Based on Alpharetrovirus ASLV and Spumavirus PFV

Other retroviruses have been proposed as gene therapy vectors including ASLV and PFV. ASLV vectors are able to transduce macaque CD34+ hematopoietic progenitor cells with stable transgene expression up to 18 months [91]. There were no detectable integration events at enhancers or promoters [92]. In spite of these promising early results, ASLV vectors have not yet been used in human gene therapy trials.

Only the spumavirus subfamily (also known as foamy viruses) of retroviruses has been shown to not cause any disease in animal hosts or xenotropic human infections. There are also key differences between the life cycle of foamy viruses and other retroviruses (reviewed in [6]). In particular, the foamy viruses are not readily pseudotyped like other retroviruses. However, foamy virus vectors are able to transduce human primary macrophages, human and rhesus embryonic stem cells, human-induced pluripotent stem cells, and murine hematopoietic stem cells [9, 93–97]. Foamy viruses do have a slight preference for integration near TSSs, but significantly less than MLV [30, 98, 99]. Importantly, foamy viruses are not associated with oncogenesis.

Foamy virus vectors have not been used in human gene therapy trials but have treated five dogs deficient in CD18, the cause of canine leukocyte adhesion deficiency (CLAD) [100–102]. Foamy vector particles transduced CD34+ hematopoietic stem cells ex vivo, and the cells were infused back into the dogs [100]. Four dogs were disease free for at least 4–7 years [101]. Additionally, the animals did not develop any tumors [101, 102].

Retroviral gene therapy protocols typically include transduction of cells ex vivo, but foamy virus vectors have been given to dogs intravenously. Six dogs with X-SCID were treated with foamy virus vectors delivered intravenously [103, 104].

Two of the dogs survived 2.5 years with no evidence of oncogenesis in any of the animals [103]. These studies suggest that foamy virus vectors may be safe for human gene therapy trials with little risk of oncogenesis.

5 Conclusions

There have been efforts to engineer retroviral integration to target genetic "safe harbors" that will not lead to cancer in patients treated with retroviral gene therapy vectors [105]. However, these efforts have been largely unsuccessful. The goals of retroviral gene therapy are to eliminate the possibility of oncogenesis and to rescue gene expression for a functional cure. Of the retroviral gene therapies that have been used in humans, lentiviral vectors may be the closest to achieving these goals thus far [106, 107]. Importantly, there is no strong evidence that lentiviral vectors will lead to cancer in patients. The success of these gene therapies in treating several severe monogenic diseases suggests they will continue to be clinically relevant.FundingThis research was supported by NIH R01AI126742 (KEY), NIH R01AI150496 (KEY), and The Ohio State University Comprehensive Cancer Center (KEY).

References

1. Sinn, P. L., Sauter, S. L., & McCray, P. B., Jr. (2005). Gene therapy progress and prospects: Development of improved lentiviral and retroviral vectors - design, biosafety, and production. *Gene Therapy, 12*(14), 1089–1098. https://doi.org/10.1038/sj.gt.3302570
2. Coffin, J. M., Hughes, S. H., & Varmus, H. E. (1997). In J. M. Coffin, S. H. Hughes, & H. E. Varmus (Eds.), *Retroviruses*. Cold Spring Harbor Laboratory Press.
3. Kobiler, O., Drayman, N., Butin-Israeli, V., & Oppenheim, A. (2012). Virus strategies for passing the nuclear envelope barrier. *Nucleus, 3*(6), 526–539. https://doi.org/10.4161/nucl.21979
4. Matreyek, K. A., & Engelman, A. (2013). Viral and cellular requirements for the nuclear entry of retroviral preintegration nucleoprotein complexes. *Viruses, 5*(10), 2483–2511. https://doi.org/10.3390/v5102483
5. Shun, M. C., Raghavendra, N. K., Vandegraaff, N., Daigle, J. E., Hughes, S., Kellam, P., et al. (2007). LEDGF/p75 functions downstream from preintegration complex formation to effect gene-specific HIV-1 integration. *Genes & Development, 21*(14), 1767–1778.
6. Lindemann, D., & Rethwilm, A. (2011). Foamy virus biology and its application for vector development. *Viruses, 3*(5), 561–585. https://doi.org/10.3390/v3050561
7. Miller, A. D. (1992). Retroviral vectors. *Current Topics in Microbiology and Immunology, 158*, 1–24. https://doi.org/10.1007/978-3-642-75608-5_1
8. Naldini, L., Blomer, U., Gallay, P., Ory, D., Mulligan, R., Gage, F. H., et al. (1996). In vivo gene delivery and stable transduction of nondividing cells by a lentiviral vector. *Science, 272*(5259), 263–267. https://doi.org/10.1126/science.272.5259.263
9. Trobridge, G., Josephson, N., Vassilopoulos, G., Mac, J., & Russell, D. W. (2002a). Improved foamy virus vectors with minimal viral sequences. *Molecular Therapy, 6*(3), 321–328. https://doi.org/10.1006/mthe.2002.0672

10. Trobridge, G., Vassilopoulos, G., Josephson, N., & Russell, D. W. (2002b). Gene transfer with foamy virus vectors. *Methods in Enzymology, 346*, 628–648. https://doi.org/10.1016/s0076-6879(02)46082-x
11. Aiuti, A., Biasco, L., Scaramuzza, S., Ferrua, F., Cicalese, M. P., Baricordi, C., et al. (2013). Lentiviral hematopoietic stem cell gene therapy in patients with Wiskott-Aldrich syndrome. *Science, 341*(6148), 1233151. https://doi.org/10.1126/science.1233151
12. Zufferey, R., Donello, J. E., Trono, D., & Hope, T. J. (1999). Woodchuck hepatitis virus posttranscriptional regulatory element enhances expression of transgenes delivered by retroviral vectors. *Journal of Virology, 73*(4), 2886–2892. https://doi.org/10.1128/JVI.73.4.2886-2892.1999
13. Logan, A. C., Lutzko, C., & Kohn, D. B. (2002). Advances in lentiviral vector design for gene-modification of hematopoietic stem cells. *Current Opinion in Biotechnology, 13*(5), 429–436. https://doi.org/10.1016/s0958-1669(02)00346-4
14. Duverge, A., & Negroni, M. (2020). Pseudotyping lentiviral vectors: When the clothes make the virus. *Viruses, 12*(11). https://doi.org/10.3390/v12111311
15. Gutierrez-Guerrero, A., Cosset, F. L., & Verhoeyen, E. (2020). Lentiviral vector pseudotypes: Precious tools to improve gene modification of hematopoietic cells for research and gene therapy. *Viruses, 12*(9). https://doi.org/10.3390/v12091016
16. Mennechet, F. J. D., Paris, O., Ouoba, A. R., Salazar Arenas, S., Sirima, S. B., Takoudjou Dzomo, G. R., et al. (2019). A review of 65 years of human adenovirus seroprevalence. *Expert Review of Vaccines, 18*(6), 597–613. https://doi.org/10.1080/14760584.2019.1588113
17. Saez-Cirion, A., & Manel, N. (2018). Immune responses to retroviruses. *Annual Review of Immunology, 36*, 193–220. https://doi.org/10.1146/annurev-immunol-051116-052155
18. Sauter, D., & Kirchhoff, F. (2016). HIV replication: A game of hide and sense. *Current Opinion in HIV and AIDS, 11*(2), 173–181. https://doi.org/10.1097/COH.0000000000000233
19. Shirley, J. L., de Jong, Y. P., Terhorst, C., & Herzog, R. W. (2020). Immune responses to viral gene therapy vectors. *Molecular Therapy, 28*(3), 709–722. https://doi.org/10.1016/j.ymthe.2020.01.001
20. Eichler, F., Duncan, C., Musolino, P. L., Orchard, P. J., De Oliveira, S., Thrasher, A. J., et al. (2017). Hematopoietic stem-cell gene therapy for cerebral adrenoleukodystrophy. *The New England Journal of Medicine, 377*(17), 1630–1638. https://doi.org/10.1056/NEJMoa1700554
21. Ott, M. G., Schmidt, M., Schwarzwaelder, K., Stein, S., Siler, U., Koehl, U., et al. (2006). Correction of X-linked chronic granulomatous disease by gene therapy, augmented by insertional activation of MDS1-EVI1, PRDM16 or SETBP1. *Nature Medicine, 12*(4), 401–409. https://doi.org/10.1038/nm1393
22. Hacein-Bey-Abina, S., Garrigue, A., Wang, G. P., Soulier, J., Lim, A., Morillon, E., et al. (2008). Insertional oncogenesis in 4 patients after retrovirus-mediated gene therapy of SCID-X1. *The Journal of Clinical Investigation, 118*(9), 3132–3142. https://doi.org/10.1172/JCI35700
23. Hacein-Bey-Abina, S., Von Kalle, C., Schmidt, M., McCormack, M. P., Wulffraat, N., Leboulch, P., et al. (2003). LMO2-associated clonal T cell proliferation in two patients after gene therapy for SCID-X1. *Science, 302*(5644), 415–419. https://doi.org/10.1126/science.1088547
24. Howe, S. J., Mansour, M. R., Schwarzwaelder, K., Bartholomae, C., Hubank, M., Kempski, H., et al. (2008). Insertional mutagenesis combined with acquired somatic mutations causes leukemogenesis following gene therapy of SCID-X1 patients. *The Journal of Clinical Investigation, 118*(9), 3143–3150. https://doi.org/10.1172/JCI35798
25. Cavazzana-Calvo, M., Hacein-Bey, S., de Saint Basile, G., Gross, F., Yvon, E., Nusbaum, P., et al. (2000). Gene therapy of human severe combined immunodeficiency (SCID)-X1 disease. *Science, 288*(5466), 669–672. https://doi.org/10.1126/science.288.5466.669
26. Cavazzana, M., Six, E., Lagresle-Peyrou, C., Andre-Schmutz, I., & Hacein-Bey-Abina, S. (2016). Gene therapy for X-linked severe combined immunodeficiency: Where do we stand? *Human Gene Therapy, 27*(2), 108–116. https://doi.org/10.1089/hum.2015.137

27. Wu, C., & Dunbar, C. E. (2011). Stem cell gene therapy: The risks of insertional mutagenesis and approaches to minimize genotoxicity. *Frontiers in Medicine, 5*(4), 356–371. https://doi.org/10.1007/s11684-011-0159-1
28. Boztug, K., Schmidt, M., Schwarzer, A., Banerjee, P. P., Diez, I. A., Dewey, R. A., et al. (2010). Stem-cell gene therapy for the Wiskott-Aldrich syndrome. *The New England Journal of Medicine, 363*(20), 1918–1927. https://doi.org/10.1056/NEJMoa1003548
29. Stein, S., Ott, M. G., Schultze-Strasser, S., Jauch, A., Burwinkel, B., Kinner, A., et al. (2010). Genomic instability and myelodysplasia with monosomy 7 consequent to EVI1 activation after gene therapy for chronic granulomatous disease. *Nature Medicine, 16*(2), 198–204. https://doi.org/10.1038/nm.2088
30. Serrao, E., Ballandras-Colas, A., Cherepanov, P., Maertens, G. N., & Engelman, A. N. (2015). Key determinants of target DNA recognition by retroviral intasomes. *Retrovirology, 12*, 39. https://doi.org/10.1186/s12977-015-0167-3
31. De Rijck, J., de Kogel, C., Demeulemeester, J., Vets, S., El Ashkar, S., Malani, N., et al. (2013). The BET family of proteins targets moloney murine leukemia virus integration near transcription start sites. *Cell Reports, 5*(4), 886–894. https://doi.org/10.1016/j.celrep.2013.09.040
32. Gupta, S. S., Maetzig, T., Maertens, G. N., Sharif, A., Rothe, M., Weidner-Glunde, M., et al. (2013). Bromo- and extraterminal domain chromatin regulators serve as cofactors for murine leukemia virus integration. *Journal of Virology, 87*(23), 12721–12736. https://doi.org/10.1128/JVI.01942-13
33. Sharma, A., Larue, R. C., Plumb, M. R., Malani, N., Male, F., Slaughter, A., et al. (2013). BET proteins promote efficient murine leukemia virus integration at transcription start sites. *Proceedings of the National Academy of Sciences of the United States of America, 110*(29), 12036–12041. https://doi.org/10.1073/pnas.1307157110
34. Moiani, A., Miccio, A., Rizzi, E., Severgnini, M., Pellin, D., Suerth, J. D., et al. (2013). Deletion of the LTR enhancer/promoter has no impact on the integration profile of MLV vectors in human hematopoietic progenitors. *PLoS One, 8*(1), e55721. https://doi.org/10.1371/journal.pone.0055721
35. Sokol, M., Wabl, M., Ruiz, I. R., & Pedersen, F. S. (2014). Novel principles of gamma-retroviral insertional transcription activation in murine leukemia virus-induced end-stage tumors. *Retrovirology, 11*, 36. https://doi.org/10.1186/1742-4690-11-36
36. Aiyer, S., Swapna, G. V., Malani, N., Aramini, J. M., Schneider, W. M., Plumb, M. R., et al. (2014). Altering murine leukemia virus integration through disruption of the integrase and BET protein family interaction. *Nucleic Acids Research, 42*(9), 5917–5928. https://doi.org/10.1093/nar/gku175
37. Larue, R. C., Plumb, M. R., Crowe, B. L., Shkriabai, N., Sharma, A., DiFiore, J., et al. (2014). Bimodal high-affinity association of Brd4 with murine leukemia virus integrase and mononucleosomes. *Nucleic Acids Research, 42*(8), 4868–4881. https://doi.org/10.1093/nar/gku135
38. El Ashkar, S., De Rijck, J., Demeulemeester, J., Vets, S., Madlala, P., Cermakova, K., et al. (2014). BET-independent MLV-based vectors target away from promoters and regulatory elements. *Molecular Therapy - Nucleic Acids, 3*, e179. https://doi.org/10.1038/mtna.2014.33
39. Loyola, L., Achuthan, V., Gilroy, K., Borland, G., Kilbey, A., Mackay, N., et al. (2019). Disrupting MLV integrase:BET protein interaction biases integration into quiescent chromatin and delays but does not eliminate tumor activation in a MYC/Runx2 mouse model. *PLoS Pathogens, 15*(12), e1008154. https://doi.org/10.1371/journal.ppat.1008154
40. Lurain, K., Yarchoan, R., & Ramaswami, R. (2019). The changing face of HIV-associated malignancies: Advances, opportunities, and future directions. *American Society of Clinical Oncology Educational Book, 39*, 36–40. https://doi.org/10.1200/EDBK_100017
41. Maldarelli, F., Wu, X., Su, L., Simonetti, F. R., Shao, W., Hill, S., et al. (2014). HIV latency. Specific HIV integration sites are linked to clonal expansion and persistence of infected cells. *Science, 345*(6193), 179–183. https://doi.org/10.1126/science.1254194

42. Wagner, T. A., McLaughlin, S., Garg, K., Cheung, C. Y., Larsen, B. B., Styrchak, S., et al. (2014). HIV latency. Proliferation of cells with HIV integrated into cancer genes contributes to persistent infection. *Science, 345*(6196), 570–573. https://doi.org/10.1126/science.1256304
43. Gantner, P., Pagliuzza, A., Pardons, M., Ramgopal, M., Routy, J. P., Fromentin, R., et al. (2020). Single-cell TCR sequencing reveals phenotypically diverse clonally expanded cells harboring inducible HIV proviruses during ART. *Nature Communications, 11*(1), 4089. https://doi.org/10.1038/s41467-020-17898-8
44. Halvas, E. K., Joseph, K. W., Brandt, L. D., Guo, S., Sobolewski, M. D., Jacobs, J. L., et al. (2020). HIV-1 viremia not suppressible by antiretroviral therapy can originate from large T cell clones producing infectious virus. *The Journal of Clinical Investigation, 130*(11), 5847–5857. https://doi.org/10.1172/JCI138099
45. Kaiser, J. (2021). Gene therapy trials for sickle cell disease halted after two patients develop cancer. *Science*. https://doi.org/10.1126/science.abh1106
46. Swiggard, W. J., Baytop, C., Yu, J. J., Dai, J., Li, C., Schretzenmair, R., et al. (2005). Human immunodeficiency virus type 1 can establish latent infection in resting CD4+ T cells in the absence of activating stimuli. *Journal of Virology, 79*(22), 14179–14188. https://doi.org/10.1128/JVI.79.22.14179-14188.2005
47. McLean, A. R., & Michie, C. A. (1995). In vivo estimates of division and death rates of human T lymphocytes. *Proceedings of the National Academy of Sciences of the United States of America, 92*(9), 3707–3711. https://doi.org/10.1073/pnas.92.9.3707
48. Themis, M., Waddington, S. N., Schmidt, M., von Kalle, C., Wang, Y., Al-Allaf, F., et al. (2005). Oncogenesis following delivery of a nonprimate lentiviral gene therapy vector to fetal and neonatal mice. *Molecular Therapy, 12*(4), 763–771. https://doi.org/10.1016/j.ymthe.2005.07.358
49. Dull, T., Zufferey, R., Kelly, M., Mandel, R. J., Nguyen, M., Trono, D., et al. (1998). A third-generation lentivirus vector with a conditional packaging system. *Journal of Virology, 72*(11), 8463–8471.
50. Vannucci, L., Lai, M., Chiuppesi, F., Ceccherini-Nelli, L., & Pistello, M. (2013). Viral vectors: A look back and ahead on gene transfer technology. *New Microbiologica, 36*(1), 1–22.
51. Biffi, A., Montini, E., Lorioli, L., Cesani, M., Fumagalli, F., Plati, T., et al. (2013). Lentiviral hematopoietic stem cell gene therapy benefits metachromatic leukodystrophy. *Science, 341*(6148), 1233158. https://doi.org/10.1126/science.1233158
52. Cartier, N., Hacein-Bey-Abina, S., Bartholomae, C. C., Veres, G., Schmidt, M., Kutschera, I., et al. (2009). Hematopoietic stem cell gene therapy with a lentiviral vector in X-linked adrenoleukodystrophy. *Science, 326*(5954), 818–823. https://doi.org/10.1126/science.1171242
53. Milone, M. C., & O'Doherty, U. (2018). Clinical use of lentiviral vectors. *Leukemia, 32*(7), 1529–1541. https://doi.org/10.1038/s41375-018-0106-0
54. Cavazzana-Calvo, M., Payen, E., Negre, O., Wang, G., Hehir, K., Fusil, F., et al. (2010). Transfusion independence and HMGA2 activation after gene therapy of human beta-thalassaemia. *Nature, 467*(7313), 318–322. https://doi.org/10.1038/nature09328
55. Thompson, A. A., Walters, M. C., Kwiatkowski, J., Rasko, J. E. J., Ribeil, J. A., Hongeng, S., et al. (2018). Gene therapy in patients with transfusion-dependent beta-thalassemia. *The New England Journal of Medicine, 378*(16), 1479–1493. https://doi.org/10.1056/NEJMoa1705342
56. De Ravin, S. S., Su, L., Theobald, N., Choi, U., Macpherson, J. L., Poidinger, M., et al. (2014). Enhancers are major targets for murine leukemia virus vector integration. *Journal of Virology, 88*(8), 4504–4513. https://doi.org/10.1128/JVI.00011-14
57. Feng, L., Dharmarajan, V., Serrao, E., Hoyte, A., Larue, R. C., Slaughter, A., et al. (2016). The competitive interplay between allosteric HIV-1 integrase inhibitor BI/D and LEDGF/p75 during the early stage of HIV-1 replication adversely affects inhibitor potency. *ACS Chemical Biology*. https://doi.org/10.1021/acschembio.6b00167
58. Lafave, M. C., Varshney, G. K., Gildea, D. E., Wolfsberg, T. G., Baxevanis, A. D., & Burgess, S. M. (2014). MLV integration site selection is driven by strong enhancers and active promoters. *Nucleic Acids Research*. https://doi.org/10.1093/nar/gkt1399

59. Schroder, A. R., Shinn, P., Chen, H., Berry, C., Ecker, J. R., & Bushman, F. (2002). HIV-1 integration in the human genome favors active genes and local hotspots. *Cell, 110*(4), 521–529.
60. Wu, X., Li, Y., Crise, B., & Burgess, S. M. (2003). Transcription start regions in the human genome are favored targets for MLV integration. *Science, 300*(5626), 1749–1751. https://doi.org/10.1126/science.1083413
61. Cherepanov, P., Maertens, G., Proost, P., Devreese, B., Van Beeumen, J., Engelborghs, Y., et al. (2003). HIV-1 integrase forms stable tetramers and associates with LEDGF/p75 protein in human cells. *Journal of Biological Chemistry, 278*(1), 372–381.
62. Turlure, F., Devroe, E., Silver, P. A., & Engelman, A. (2004). Human cell proteins and human immunodeficiency virus DNA integration. *Frontiers in Bioscience, 9*, 3187–3208.
63. Cherepanov, P., Devroe, E., Silver, P. A., & Engelman, A. (2004). Identification of an evolutionarily conserved domain in human lens epithelium-derived growth factor/transcriptional co-activator p75 (LEDGF/p75) that binds HIV-1 integrase. *Journal of Biological Chemistry, 279*(47), 48883–48892.
64. Llano, M., Vanegas, M., Hutchins, N., Thompson, D., Delgado, S., & Poeschla, E. M. (2006). Identification and characterization of the chromatin-binding domains of the HIV-1 integrase interactor LEDGF/p75. *Journal of Molecular Biology, 360*(4), 760–773.
65. Turlure, F., Maertens, G., Rahman, S., Cherepanov, P., & Engelman, A. (2006). A tripartite DNA-binding element, comprised of the nuclear localization signal and two AT-hook motifs, mediates the association of LEDGF/p75 with chromatin in vivo. *Nucleic Acids Research, 34*(5), 1653–1675.
66. Cherepanov, P., Ambrosio, A. L., Rahman, S., Ellenberger, T., & Engelman, A. (2005). Structural basis for the recognition between HIV-1 integrase and transcriptional coactivator p75. *Proceedings of the National Academy of Sciences of the United States of America, 102*(48), 17308–17313.
67. Eidahl, J. O., Crowe, B. L., North, J. A., McKee, C. J., Shkriabai, N., Feng, L., et al. (2013). Structural basis for high-affinity binding of LEDGF PWWP to mononucleosomes. *Nucleic Acids Research.* https://doi.org/10.1093/nar/gkt074
68. Kessl, J. J., Li, M., Ignatov, M., Shkriabai, N., Eidahl, J. O., Feng, L., et al. (2011). FRET analysis reveals distinct conformations of IN tetramers in the presence of viral DNA or LEDGF/p75. *Nucleic Acids Research, 39*(20), 9009–9022.
69. McKee, C. J., Kessl, J. J., Shkriabai, N., Dar, M. J., Engelman, A., & Kvaratskhelia, M. (2008). Dynamic modulation of HIV-1 integrase structure and function by cellular lens epithelium-derived growth factor (LEDGF) protein. *Journal of Biological Chemistry, 283*(46), 31802–31812.
70. Pradeepa, M. M., Sutherland, H. G., Ule, J., Grimes, G. R., & Bickmore, W. A. (2012). Psip1/Ledgf p52 binds methylated histone H3K36 and splicing factors and contributes to the regulation of alternative splicing. *PLoS Genetics, 8*(5), e1002717. https://doi.org/10.1371/journal.pgen.1002717
71. van Nuland, R., van Schaik, F. M., Simonis, M., van Heesch, S., Cuppen, E., Boelens, R., et al. (2013). Nucleosomal DNA binding drives the recognition of H3K36-methylated nucleosomes by the PSIP1-PWWP domain. *Epigenetics & Chromatin, 6*(1), 12. https://doi.org/10.1186/1756-8935-6-12
72. De Rijck, J., Bartholomeeusen, K., Ceulemans, H., Debyser, Z., & Gijsbers, R. (2010). High-resolution profiling of the LEDGF/p75 chromatin interaction in the ENCODE region. *Nucleic Acids Research, 38*(18), 6135–6147. https://doi.org/10.1093/nar/gkq410
73. Cardinale, S., Cisterna, B., Bonetti, P., Aringhieri, C., Biggiogera, M., & Barabino, S. M. (2007). Subnuclear localization and dynamics of the Pre-mRNA 3′ end processing factor mammalian cleavage factor I 68-kDa subunit. *Molecular Biology of the Cell, 18*(4), 1282–1292. https://doi.org/10.1091/mbc.e06-09-0846
74. Ruepp, M. D., Aringhieri, C., Vivarelli, S., Cardinale, S., Paro, S., Schumperli, D., et al. (2009). Mammalian pre-mRNA 3′ end processing factor CF I m 68 functions in mRNA

export. *Molecular Biology of the Cell, 20*(24), 5211–5223. https://doi.org/10.1091/mbc.E09-05-0389
75. Achuthan, V., Perreira, J. M., Sowd, G. A., Puray-Chavez, M., McDougall, W. M., Paulucci-Holthauzen, A., et al. (2018). Capsid-CPSF6 interaction licenses nuclear HIV-1 trafficking to sites of viral DNA integration. *Cell Host & Microbe, 24*(3), 392–404.e398. https://doi.org/10.1016/j.chom.2018.08.002
76. Bejarano, D. A., Peng, K., Laketa, V., Borner, K., Jost, K. L., Lucic, B., et al. (2019). HIV-1 nuclear import in macrophages is regulated by CPSF6-capsid interactions at the nuclear pore complex. *Elife, 8*. https://doi.org/10.7554/eLife.41800
77. Lee, K., Ambrose, Z., Martin, T. D., Oztop, I., Mulky, A., Julias, J. G., et al. (2010). Flexible use of nuclear import pathways by HIV-1. *Cell Host and Microbe, 7*(3), 221–233.
78. Price, A. J., Fletcher, A. J., Schaller, T., Elliott, T., Lee, K., KewalRamani, V. N., et al. (2012). CPSF6 defines a conserved capsid interface that modulates HIV-1 replication. *PLoS Pathogens, 8*(8), e1002896. https://doi.org/10.1371/journal.ppat.1002896
79. Sowd, G. A., Serrao, E., Wang, H., Wang, W., Fadel, H. J., Poeschla, E. M., et al. (2016). A critical role for alternative polyadenylation factor CPSF6 in targeting HIV-1 integration to transcriptionally active chromatin. *Proceedings of the National Academy of Sciences of the United States of America, 113*(8), E1054–E1063. https://doi.org/10.1073/pnas.1524213113
80. Bichel, K., Price, A. J., Schaller, T., Towers, G. J., Freund, S. M., & James, L. C. (2013). HIV-1 capsid undergoes coupled binding and isomerization by the nuclear pore protein NUP358. *Retrovirology, 10*, 81. https://doi.org/10.1186/1742-4690-10-81
81. Frosst, P., Guan, T., Subauste, C., Hahn, K., & Gerace, L. (2002). Tpr is localized within the nuclear basket of the pore complex and has a role in nuclear protein export. *The Journal of Cell Biology, 156*(4), 617–630. https://doi.org/10.1083/jcb.200106046
82. Krull, S., Thyberg, J., Bjorkroth, B., Rackwitz, H. R., & Cordes, V. C. (2004). Nucleoporins as components of the nuclear pore complex core structure and Tpr as the architectural element of the nuclear basket. *Molecular Biology of the Cell, 15*(9), 4261–4277. https://doi.org/10.1091/mbc.e04-03-0165
83. Yokoyama, N., Hayashi, N., Seki, T., Pante, N., Ohba, T., Nishii, K., et al. (1995). A giant nucleopore protein that binds Ran/TC4. *Nature, 376*(6536), 184–188. https://doi.org/10.1038/376184a0
84. Buffone, C., Martinez-Lopez, A., Fricke, T., Opp, S., Severgnini, M., Cifola, I., et al. (2018). Nup153 unlocks the nuclear pore complex for HIV-1 nuclear translocation in nondividing cells. *Journal of Virology, 92*(19). https://doi.org/10.1128/JVI.00648-18
85. Koh, Y., Wu, X., Ferris, A. L., Matreyek, K. A., Smith, S. J., Lee, K., et al. (2013). Differential effects of human immunodeficiency virus type 1 capsid and cellular factors nucleoporin 153 and LEDGF/p75 on the efficiency and specificity of viral DNA integration. *Journal of Virology, 87*(1), 648–658. https://doi.org/10.1128/JVI.01148-12
86. Lelek, M., Casartelli, N., Pellin, D., Rizzi, E., Souque, P., Severgnini, M., et al. (2015). Chromatin organization at the nuclear pore favours HIV replication. *Nature Communications, 6*, 6483. https://doi.org/10.1038/ncomms7483
87. Matreyek, K. A., & Engelman, A. (2011). The requirement for nucleoporin NUP153 during human immunodeficiency virus type 1 infection is determined by the viral capsid. *Journal of Virology, 85*(15), 7818–7827.
88. Matreyek, K. A., Yucel, S. S., Li, X., & Engelman, A. (2013). Nucleoporin NUP153 phenylalanine-glycine motifs engage a common binding pocket within the HIV-1 capsid protein to mediate lentiviral infectivity. *PLoS Pathogens, 9*(10), e1003693. https://doi.org/10.1371/journal.ppat.1003693
89. Ocwieja, K. E., Brady, T. L., Ronen, K., Huegel, A., Roth, S. L., Schaller, T., et al. (2011). HIV integration targeting: A pathway involving Transportin-3 and the nuclear pore protein RanBP2. *PLoS Pathogens, 7*(3), e1001313.

90. Marini, B., Kertesz-Farkas, A., Ali, H., Lucic, B., Lisek, K., Manganaro, L., et al. (2015). Nuclear architecture dictates HIV-1 integration site selection. *Nature, 521*(7551), 227–231. https://doi.org/10.1038/nature14226
91. Hu, J., Ferris, A., Larochelle, A., Krouse, A. E., Metzger, M. E., Donahue, R. E., et al. (2007). Transduction of rhesus macaque hematopoietic stem and progenitor cells with avian sarcoma and leukosis virus vectors. *Human Gene Therapy, 18*(8), 691–700. https://doi.org/10.1089/hum.2006.175
92. Hu, J., Renaud, G., Gomes, T. J., Ferris, A., Hendrie, P. C., Donahue, R. E., et al. (2008). Reduced genotoxicity of avian sarcoma leukosis virus vectors in rhesus long-term repopulating cells compared to standard murine retrovirus vectors. *Molecular Therapy, 16*(9), 1617–1623. https://doi.org/10.1038/mt.2008.135
93. Deyle, D. R., Khan, I. F., Ren, G., & Russell, D. W. (2013). Lack of genotoxicity due to foamy virus vector integration in human iPSCs. *Gene Therapy, 20*(8), 868–873. https://doi.org/10.1038/gt.2013.6
94. Gharwan, H., Hirata, R. K., Wang, P., Richard, R. E., Wang, L., Olson, E., et al. (2007). Transduction of human embryonic stem cells by foamy virus vectors. *Molecular Therapy, 15*(10), 1827–1833. https://doi.org/10.1038/sj.mt.6300244
95. Nasimuzzaman, M., Lynn, D., Ernst, R., Beuerlein, M., Smith, R. H., Shrestha, A., et al. (2016). Production and purification of high-titer foamy virus vector for the treatment of leukocyte adhesion deficiency. *Molecular Therapy – Methods & Clinical Development, 3*, 16004. https://doi.org/10.1038/mtm.2016.4
96. Taylor, J. A., Vojtech, L., Bahner, I., Kohn, D. B., Laer, D. V., Russell, D. W., et al. (2008). Foamy virus vectors expressing anti-HIV transgenes efficiently block HIV-1 replication. *Molecular Therapy, 16*(1), 46–51. https://doi.org/10.1038/sj.mt.6300335
97. Vassilopoulos, G., Trobridge, G., Josephson, N. C., & Russell, D. W. (2001). Gene transfer into murine hematopoietic stem cells with helper-free foamy virus vectors. *Blood, 98*(3), 604–609. https://doi.org/10.1182/blood.v98.3.604
98. Nowrouzi, A., Dittrich, M., Klanke, C., Heinkelein, M., Rammling, M., Dandekar, T., et al. (2006). Genome-wide mapping of foamy virus vector integrations into a human cell line. *The Journal of General Virology, 87*(Pt 5), 1339–1347. https://doi.org/10.1099/vir.0.81554-0
99. Trobridge, G. D., Miller, D. G., Jacobs, M. A., Allen, J. M., Kiem, H. P., Kaul, R., et al. (2006). Foamy virus vector integration sites in normal human cells. *Proceedings of the National Academy of Sciences of the United States of America, 103*(5), 1498–1503. https://doi.org/10.1073/pnas.0510046103
100. Bauer, T. R., Jr., Allen, J. M., Hai, M., Tuschong, L. M., Khan, I. F., Olson, E. M., et al. (2008). Successful treatment of canine leukocyte adhesion deficiency by foamy virus vectors. *Nature Medicine, 14*(1), 93–97. https://doi.org/10.1038/nm1695
101. Bauer, T. R., Jr., Tuschong, L. M., Calvo, K. R., Shive, H. R., Burkholder, T. H., Karlsson, E. K., et al. (2013). Long-term follow-up of foamy viral vector-mediated gene therapy for canine leukocyte adhesion deficiency. *Molecular Therapy, 21*(5), 964–972. https://doi.org/10.1038/mt.2013.34
102. Ohmine, K., Li, Y., Bauer, T. R., Jr., Hickstein, D. D., & Russell, D. W. (2011). Tracking of specific integrant clones in dogs treated with foamy virus vectors. *Human Gene Therapy, 22*(2), 217–224. https://doi.org/10.1089/hum.2010.072
103. Burtner, C. R., Beard, B. C., Kennedy, D. R., Wohlfahrt, M. E., Adair, J. E., Trobridge, G. D., et al. (2014). Intravenous injection of a foamy virus vector to correct canine SCID-X1. *Blood, 123*(23), 3578–3584. https://doi.org/10.1182/blood-2013-11-538926
104. Humbert, O., Chan, F., Rajawat, Y. S., Torgerson, T. R., Burtner, C. R., Hubbard, N. W., et al. (2018). Rapid immune reconstitution of SCID-X1 canines after G-CSF/AMD3100 mobilization and in vivo gene therapy. *Blood Advances, 2*(9), 987–999. https://doi.org/10.1182/bloodadvances.2018016451

105. Papapetrou, E. P., Lee, G., Malani, N., Setty, M., Riviere, I., Tirunagari, L. M., et al. (2011). Genomic safe harbors permit high beta-globin transgene expression in thalassemia induced pluripotent stem cells. *Nature Biotechnology, 29*(1), 73–78. https://doi.org/10.1038/nbt.1717
106. El Ashkar, S., Van Looveren, D., Schenk, F., Vranckx, L. S., Demeulemeester, J., De Rijck, J., et al. (2017). Engineering next-generation BET-independent MLV vectors for safer gene therapy. *Molecular Therapy - Nucleic Acids, 7*, 231–245. https://doi.org/10.1016/j.omtn.2017.04.002
107. Hocum, J. D., Linde, I., Rae, D. T., Collins, C. P., Matern, L. K., & Trobridge, G. D. (2016). Retargeted foamy virus vectors integrate less frequently near proto-oncogenes. *Scientific Reports, 6*, 36610. https://doi.org/10.1038/srep36610

DNA Origami Nanodevices for Therapeutic Delivery Applications

Patrick D. Halley, Christopher R. Lucas, Nikša Roki, Nicholas J. Vantangoli, Kurtis P. Chenoweth, and Carlos E. Castro

Abstract Structural DNA Nanotechnology allows for the construction of complex, 2D, and 3D structures via molecular self-assembly. These versatile nanoscale structures combine several advantages, which make them well suited for numerous biomedical applications. Particular advantages of DNA nanostructures include unprecedented control over nanoscale geometry and precise functionalization with molecules including small molecule therapeutics, nucleic acids (e.g. antisense oligonucleotides or small interfering RNA), gene and gene editing molecules, and targeting molecules to create drug delivery devices that can target molecular payloads to a variety of cells. Furthermore, recent advances have led to new capabilities such as tunable drug release rates and multiple approaches to improve biological stability. Here we provide an overview of the DNA nanostructure design, manufacturing, and characterization processes; briefly describe nonclinical applications; and focus on preclinical development of DNA nanostructures as drug delivery devices with an emphasis on nanostructures engineered via the DNA origami fabrication method.

P. D. Halley · K. P. Chenoweth
Department of Mechanical and Aerospace Engineering, College of Engineering, The Ohio State University, Columbus, OH, USA

C. R. Lucas · N. Roki · N. J. Vantangoli
Department of Mechanical and Aerospace Engineering, College of Engineering, The Ohio State University, Columbus, OH, USA

The Ohio State Comprehensive Cancer Center, The Ohio State University, Columbus, OH, USA

C. E. Castro (✉)
Department of Mechanical and Aerospace Engineering, College of Engineering, The Ohio State University, Columbus, OH, USA

Department of Physics, The Ohio State University, Columbus, OH, USA
e-mail: castro.39@osu.edu

Y. H. Yun, K. E. Yoder (eds.), *Biotechnologies for Gene Therapy*,
https://doi.org/10.1007/978-3-030-93333-3_8

1 Introduction

Throughout the multiple life science disciplines, DNA (deoxyribonucleic acid) typically refers to the genetic and hereditary material of the living cell. Within the context of the nanoengineering discipline, DNA is not only considered the genetic and hereditary material of life, but also as construction material to build incredibly complex structures at the nanoscale. From its ideation of DNA nanotechnology in 1982 to its renaissance in 2006 with scaffolded DNA origami to 2021, the field has advanced from simple planar junctions to the ability to construct extremely complex static 2D DNA predefined nanoscale objects and further to the development of 3D nanoscale objects with increasing levels of complexity and curvature [1–5]. Throughout the 2010s, the development of a multitude of DNA nanostructure applications began to emerge including biosensing [6, 7], plasmonics [8], templating [9], DNA computing [10, 11], molecular machines [12, 13], and drug delivery [14–16]. Furthermore, the emergence of numerous, unique applications for DNA nanostructures suggests a powerful platform technology for commercial development. One of the most powerful properties of DNA nanostructures is the ability to add a controlled number of oligonucleotide molecular attachments with a customizable sequence in a site-specific manner. This enables complementary sequence-specific attachment of virtually any sequence of DNA, RNA, and microRNA (miR) oligonucleotides allowing for potential medical device development to diagnose diseases such as viral and microbial infection, and cancer. Furthermore, attachment of antibody molecules to oligonucleotides [17] allows for a platform to organize and detect specific peptide molecules with disease relevance. A key advantage of DNA nanostructure medical device platforms is the ability to achieve a high degree of functionalized nanostructure uniformity, which is critical for an effective diagnostic medical device, in addition to drug discovery.

DNA nanostructures hold great potential for multifunctional drug delivery device development [16], with an ability to attach small molecule intercalating agents [18–21], antisense oligonucleotide molecules [22, 23], genes and gene editing molecules [24, 25], and targeting aptamers [26, 27] or antibody fragments [28] to deliver a therapeutic payload to intended target cells. Here we provide an overview of the DNA nanostructure design, manufacturing, and characterization processes; briefly describe nonclinical applications; and focus on preclinical development of DNA nanostructures as drug delivery devices with an emphasis on nanostructures engineered via the DNA origami fabrication method.

2 Design and Fabrication

The ability to design synthetic DNA nanostructures was initially conceived in 1982 by Nadrian Seeman [1] (Fig. 1a). He proposed the idea of leveraging nucleic acid junctions, in particular Holliday junctions [29], to build up larger arrays and lattices.

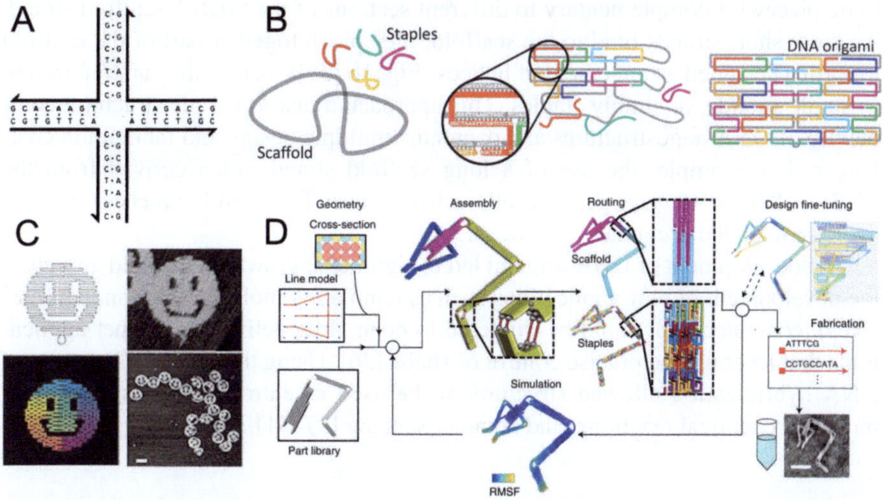

Fig. 1 DNA nanostructure design and concepts. (**a**) Seeman's original nucleic acid junction design [1]. (**b**) A schematic of the staples design and annealing process of scaffolded DNA origami nanostructures [35]. (**c**) Rothemund's smiley face design with AFM images of early DNA origami nanostructures [2]. (**d**) Design flow using MagicDNA from geometry conception to fabrication [55]. (**a**) is reprinted from *Journal of Theoretical Biology*, volume 99 Issue 2, N. Seeman, "Nucleic acid junctions and lattices", pages 237–247, copyright (1982), with permission from Elsevier. (**b**) is reprinted from *Trends in Molecular Medicine*, volume 24 issue 7, H. Singh, E. Kopperger, F. Simmel, "A DNA nanorobot uprises against cancer", pages 591–593, copyright (2018), with permission from Elsevier. (**c**) is reprinted from *Nature*, volume 440, P. Rothemund, "Folding DNA to create nanoscale shapes and patterns", pages 297–302, copyright (2006), with permission from Springer Nature. chapter, Pages No., Copyright (Year), with permission from Elsevier [OR APPLICABLE SOCIETY COPYRIGHT OWNER]. (**a**) Seeman, N. C. Nucleic acid junctions and lattices. Journal of Theoretical Biology 99, 237–247 (1982). (**b**) Singh, Kopperger, Simmel. "A DNA Nanorobot Uprises against Cancer." *Trends in Molecular Medicine* (2018). (**c**) Rothemund, P. W. K. Folding DNA to create nanoscale shapes and patterns. Nature 440, 297–302 (2006). (**d**) Huang, C.-M., Kucinic, A., Johnson, J. A., Su, H.-J. & Castro, C. E. Integrated computer-aided engineering and design for DNA assemblies. Nature Materials 1–8 (2021) doi: 10.1038/s41563-021-00978-5

This idea, which was originally conceived for applications in crystallography [1, 30], gave rise to the field of structural DNA nanotechnology [31]. Early structures were constructed both from DNA and expanded to RNA, which featured motifs inspired by natural RNA constructs [32]. The design of these structures was limited to a small nanoscale size, with the design process being ad hoc and the overall structure versatility was limited. Design capabilities expanded with the development of DNA tiles that could be assembled into sheets, ribbons, and lattices [33, 34]. The field took a major step in 2006 with the development of scaffolded DNA origami [2]. In the DNA origami approach, a long single-stranded DNA (ssDNA) scaffold strand, typically ~7000–8000 nucleotides (nt) long, is folded into a compact structure through base-pairing interactions with many (~150–200) shorter ssDNA oligonucleotides, typically ~25–60 bases long. The short oligo nucleotides are designed to be piecewise complementary to different sections of the ssDNA scaffold strand. As these short strands bind to the scaffold, they pinch together part of the scaffold to form a designed arrangement of helices (Fig. 1b [35]); hence, they are referred to as staple strands, or simply staples. This approach transformed the structural complexity of DNA nanostructures and overcame multiple design and fabrication challenges. For example, the use of a long scaffold strand, often derived from the M13mp18 bacteriophage genome, alleviated the need for careful sequence design to avoid unintended secondary structures.

The development of DNA origami led to significant growth in the field as well as the development of new applications such as templating molecules or nanoparticles (NPs), containers that could be triggered to open, drug delivery and other applications that leverage the precise control of shape [36]. Using the predictable nature of DNA hybridization allowed structures to be used as entropic springs, in single-molecule chemical reactions, and aptamer spacing [37–41].

2.1 Computer-Aided Design Approaches

Advances in structural and functional complexity of DNA nanostructures have proceeded hand-in-hand with computer-aided design approaches. Early design efforts relied on custom computer codes or other ad hoc structure and/or sequence design approaches (Fig. 1c) [42–46]. Leveraging the larger design space of scaffolded DNA origami methods quickly required more powerful software tools. The fundamental design steps in DNA origami include (1) conceiving of a target shape; (2) approximating that shape out of cylinders, which represent dsDNA helices; (3) routing the scaffold through the geometric design; (4) routing the staple strands throughout the structure; and (5) determining the sequence of the staple strands according to their piecewise binding to the scaffold. In 2009, Douglas et al., developed caDNAno, a computer-aided design (CAD) software tool that became the primary standard of use for DNA origami design for nearly a decade [12, 27, 47–50]. CaDNAno greatly facilitates the aforementioned key steps of DNA origami design in user-friendly software. The development of caDNAno [47] was critical to expansion of the field, since it provided an accessible graphical user interface that allowed for versatile design, which significantly lowered the barrier for new users to develop complex structures and new applications. As the design space continued to expand, interest in new geometries and capabilities drove a need for new design tools. In particular, the development of automated top-down design approaches such as DAEDALUS [51], vHelix [52], and others [53, 54] has been critical to facilitating rapid design of a wide range of wireframe structures; and the recently developed MagicDNA [55] integrates bottom-up and top-down approaches with convenient user interfaces to significantly expand the design space, especially enabling design of complex structures with many components (Fig. 1d).

Increasing design complexity led to a critical need for simulation tools to predict folded structure shapes, functional properties such as flexibility [56, 57], or even folding pathways [58]. Modeling approaches have included theoretical approaches to predict folding pathways [59] or deformation [60, 61] and computational approaches spanning continuum (e.g., finite element methods [62, 63]), to fully atomistic molecular dynamic simulations [64]. The CANDO (Computer-Aided eNgineering for Dna Origami) finite element approach [63] has been a very useful tool for structure prediction to guide the design process of static structures; however, this continuum approach lacks molecular details that are often important for describing local behavior or dynamic motion. Atomistic approaches provide these details [64], but the computational demands severely limit the timescale of simulations; hence, atomistic approaches have been limited to smaller structures and cannot predict dynamic properties such as large conformational changes. Over the last decade, the emergence of powerful coarse-grained models of DNA, especially the coarse-grained model developed by Doye and colleagues referred to as oxDNA [65, 66], has enabled detailed studies of the relation of molecular design parameters to structure and functional properties [56, 57, 67]. The advancement and development of new simulation tools [68] is likely to continue advancing design and function capabilities, for example, to directly test interactions of DNA structures with proteins or other biomolecules.

2.2 *Fabrication and Scaling Up Production*

After the design process, most origami structures fold robustly at high yields [2, 47, 62]. Basic DNA origami fabrication occurs by preparing a mixture of scaffold DNA (usually derived from the M13 bacteriophage genome [2, 69]), and ~150–200 short-oligonucleotides (25–70 bases long) referred to as staples, magnesium chloride (or other salt), and a water-based tris-EDTA buffer. To induce self-assembly, the solution is heated and then slowly cooled over a period of hours to days, though appropriate timing and temperatures can vary from structure to structure. Additionally, there are minor differences in specific protocols between laboratories, commonly used folding protocols are described in detail here [2, 46, 47, 62, 70]. These thermal annealing protocols allow for the breaking of internal secondary structure within the scaffold, and as the temperature cools, complementary segments between the staples and scaffold DNA form to facilitate correct placement of each staple strand.

In addition to the variability of times and temperatures, other method variations of folding have been shown including isothermal folding [71, 72], folding at a single temperature during a reduced thermal ramp [73], or in a biotechnological growth process [74]. In particular, the last two can be utilized to scale fabrication, as single temperature folding has been expanded to fold most structures in a matter of hours at mg scales [75], a 1500-fold increase over conventional methods. Furthermore, biotechnological mass production of DNA origami can create DNA origami on the scales required for clinical applications and drives the cost of fabrication to an estimated ~$0.20/mg when scaled to an 800 L volume pilot-scale fermentor [74]. M13mp18 bacteriophage DNA (or cloned versions) is the most commonly used scaffold for DNA origami fabrication, grown in competent E. coli cells, through well-established stirred flask batch reactions [62]. In addition to flask growth, groups have shown that growth of M13 phage DNA can be substantially increased by growing E. coli in high-density stirred tank bioreactors [76–78]. Although these processes are effective, they do leave behind a large amount of lipopolysaccharides, also known as endotoxins. In many clinical applications, these endotoxins can cause an inflammation response [79], but can be removed using detergents [49, 80] or commercially available kits during scaffold growth.

M13-based scaffolds are typically created in a variety of lengths (7000–9000 bases) [2, 32, 47, 62]. Additionally, custom scaffolds have been fabricated to be shorter or longer, using different phagemid bases such as pBluescript [80], excision from M13mp18 [81], or creating hybrid scaffold [82]. Furthermore, custom design-specific scaffolds can be created through the implementation of helper plasmids to control the desired sequence [83]. This process also allows for the ability to make more complex structures that use multiple scaffolds, which have different sequences [55, 83].

2.3 *Purification and Storage Methods*

Various methods for purification of DNA origami are utilized, with gel electrophoresis extraction [84] remaining the most common. The agarose gel readout using DNA stains and fluorescence verifies proper folding and the process of separating of free staples, misfolded, or aggregated structures from well-folded ones is relatively simple, though it is limited in scalability. More scalable options include ultrafiltration [85], ultracentrifugation [86], polyethylene glycol (PEG) precipitation [87], magnetic bead capture, FPLC [17]: each with strengths and weaknesses of their own.

Ultracentrifugation, or rate-zonal centrifugation, purifies a mixture of a DNA origami folding reaction by creating a gradient tube of various glycerol concentrations and loading the mixture of origami on top and a high-speed spin in an ultracentrifuge [17, 86]. After running ultracentrifugation, the origami separates into the microlayers of the gradient. Staple strands appear in lower percent glycerol layers near the top, while aggregated structures sink with higher percent glycerol. The desired origami would be present somewhere in the middle. Reconstituting the origami in folding buffer and running gel electrophoresis on various fractions would show the effectiveness of the gradient method. The final product contains highly purified nanostructures but at a cost of yield [86].

Stahl et al. showed that various shapes of origami can be purified through PEG precipitation, which leverages depletion forces from PEG to cause precipitation of DNA origami structures during centrifugation [87]. Most notably, the shapes include designs incorporating up to 100 helix bundles along with straight rods, bent rods, and even origami shaped like a robot. The lab was able to effectively purify the origami through ten cycles of PEG purification. This allowed for excess staple strands to be removed through the supernatant. Observation of the shapes under agarose gel electrophoresis showed significant purity when ran against the unpurified origami. By labelling the staples with fluorescent dye, the group demonstrated a near 100% success rate for removing staple strands with relatively little loss, although this method can exacerbate any aggregation in the sample [17, 87]. DNA can degrade over time, so it is vital to take appropriate precautions and extra care to prolong the lifetime of the structures and reagents. The company Integrated DNA Technologies (IDTs), which is widely used as a source for DNA strands for DNA origami folding, suggests that temperature and medium are the most vital factors to consider in DNA storage [88]. For optimum long-term storage, keeping the oligonucleotides at −20 °C will allow for stability for years in TE buffer or nuclease-free water. Oligos can also be stored at 4 °C or as dry pellets for months. With DNA origami, it is vital that staples are stored effectively. By storage conditions for staples, Kielar was able to observe triangle origami through atomic force microscopy (AFM) on how intact they remained [89]. These tests showed that staples stored in −20 °C for up to 43 months could produce up to 80% intact origami. Under the harsher conditions, these staples stored for longer could produce up to 66% intact origami.

Cryopreservation is also an option for the preservation of folded DNA origami. Xin studied this by putting 2D and 3D DNA structures through multiple freeze–thaw cycles [90]. After 32 cycles, 2D and 3D DNA structures were able to remain intact. However, ice crystal formation was noticed in the 2D structures and increased its sensitivity; 3D DNA under 1000 freeze cycles showed heavy damage to the structure. Cycles were tested at −20 °C and −196 °C. The inclusion of cryoprotectants such as glycerol and trehalose resulted in 85–90% of the origami being intact after many freeze–thaw cycles.

3 Characterization Methods

DNA origami is observed under various types of readouts. Most structures are first characterized by agarose gel electrophoresis, and then microscopy is utilized. Transmission electron microscopy (TEM) and AFM offer precise and detailed images and fluorescence microscopy can be utilized for biological interactions and real-time function of nanostructures. Additionally, functional DNA origami for clinical applications and interactions with biological materials can be observed through in vitro and in vivo experiments.

3.1 Gel Electrophoresis

Agarose gel electrophoresis (AGE) is typically the first step in evaluating a DNA origami self-assembly reaction, where folded DNA origami structures are made to run through a gel driven by an applied voltage [42]. The folded structure of the DNA origami influences the electrophoretic mobility in the gel. Testing a folded structure against references like the ssDNA scaffold and a 1 kb ladder provides a useful means to identify a band corresponding to successfully folded DNA origami. Oftentimes, a well-folded structure will run faster than the ssDNA scaffold, but this depends on the structure, and confirming proper folding using imaging methods like TEM or AFM is critical. Generally, increased gel mobility has correlated to better folded structure, as observed under microscopy [69]. AGE is particularly useful for optimizing assembly conditions for factors such as ion concentrations or folding temperatures by screening and identifying conditions that lead to maximal yield in the well-folded structure band. Gel electrophoresis is also useful to evaluate the presence of excess staple strands in solution after other purification methods, aggregation, in the form of dimers or larger aggregates, or attachment of functionalities such as aptamers or proteins to the origami that cause a gel shift [17].

3.2 Microscopy

Visual quantification of structure is a critical step in characterization that is carried out through TEM or AFM imaging. The most commonly used approaches are negative stain TEM and dry AFM (detailed protocols can be found here [62]). For TEM, briefly, structures are deposited on an electron microscopy grid and stained with a heavy metal compound, most often a solution of Uranyl Formate or Uranyl Acetate, which enhances contrast during imaging. TEM is particularly useful for evaluating multilayer 3D DNA origami structures that contain bundles of dsDNA helices as components. Recent work has led to advances in cryogenic TEM (cryoTEM) for DNA origami that allow high-resolution full reconstruction of complex 3D shapes [91, 92] or low-resolution reconstruction of many conformations of flexible structures [93]. Similarly, AFM is a highly useful tool to visualize folded DNA structures. Since it is a scanning probe approach, it is best suited for 2D, especially flat single-layer structures. Although it is lower throughput and not well suited for complex 3D structures compared to TEM, particular advantages include high contrast for single-layer structures, and possibilities for imaging in solution that allow for high resolution [2], dynamic binding or reconfigurations of molecules or motifs of interest [94, 95], or surface diffusion of structures [96].

In addition to TEM and AFM for visualization and quantification of the geometry, fluorescent microscopy leverages the site-specific incorporation of fluorophores for spatial and dynamic readouts. Early applications leveraged the nanometer control over placement of fluorescence markers to create calibration tools, or rulers, for various fluorescence methods [97–99]. DNA origami was also essential in the development of the super-resolution imaging method referred to as DNA-PAINT (Points Accumulation for Imaging in Nanoscale Topography) [100] and its applications [101–103]. In addition, time-resolved ensemble and single molecule fluorescence methods have been used to quantify the dynamic behavior of DNA origami devices, for example, to quantify kinetics of conformational changes [104–106]. Other approaches continue to emerge to characterize DNA origami properties such as small angle x-ray scattering [107, 108] and microwave microfluidics [109] that provide advantages like label-free detection.

One of the most widely studied applications of DNA origami is as a drug carrier. Fluorescence imaging has played a key role in quantifying several key aspects of drug delivery functions such as cell binding, uptake, and drug localization [18, 19, 26].

4 Nontherapeutic Uses for DNA Origami

The versatility in geometry and mechanical and dynamic properties in combination with substantial options of moiety attachment methods that are commercially available make DNA origami a useful tool for many different functions. Here, we briefly highlight a few nontherapeutic areas of research that DNA origami has been employed.

4.1 Biosensing

The precise nature of DNA origami and the well-characterized research of nucleic acids as a material allows for the usage of these nanostructures as a ruler or force sensor. The controllable spacing of DNA origami platforms has been utilized for nanoscopic ruler measurements in super-resolution microscopy [98, 110, 111]. Additionally, this feature has also been used in biological applications, such as quantifying effects of ligand spacing (Fig. 2a), used in both T-cell and B-cell activation studies [112–115]. Furthermore, dynamic DNA origami has been used as force sensors for the study of nucleosomes [116, 117] and real-time magnetic actuation with magnetic beads [118].

Origami nanostructures have also been used as biosensors, with controllable factors that could have uses in basic research sensing and therapy-based applications. Examples of the later include "nanorobots" that present conformation changes when responding to a target [26, 27] and origami with controllable lids that open with stimuli [44]. Additionally, there have been examples of different readouts used for sensing including fluorophores [119], enzyme–substrate interactions [120], and TEM angle readouts [116].

4.2 Templating Nanofabrication and Plasmonics

Due to the precise control design and functionalization of DNA origami, it provides a natural foundation for the template of other materials. DNA origami can be designed to have many shapes, which can be transferred to less programmable

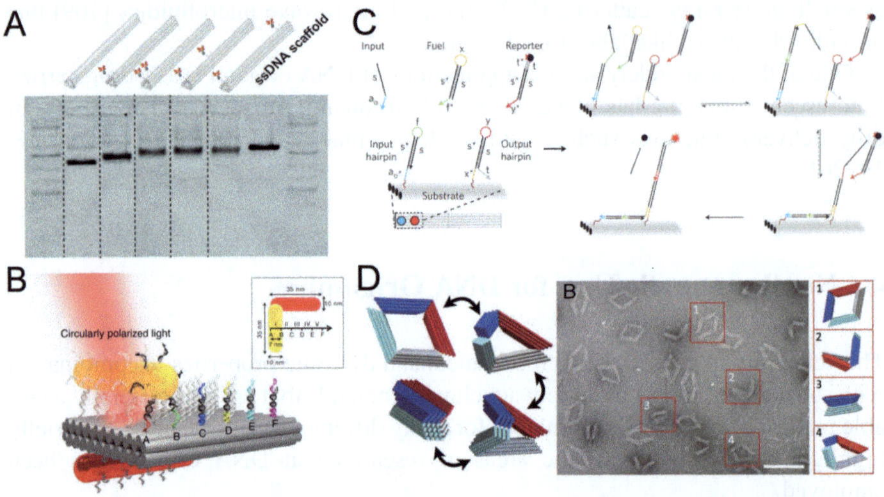

Fig. 2 Nontherapeutic used of DNA origami nanostructures. (**a**) Controlled spatial arrangement of EphA2 antibodies on a DNA origami rod [114]. (**b**) A schematic of a gold nanorod walker moving across a DNA origami raft using polarized light and energy-mediated strand displacement [128]. (**c**) A schematic of toehold-mediated strand displacement for modular DNA computing using DNA origami [138]. (**d**) A 4-sided hinge, or Bennett Linkage, that shows the controlled nature of dynamic DNA origami nanostructures and the possibility for nanomachine mechanisms [12]. (**c**) is reprinted from *Nature Nanotechnology*, volume 12, G. Chatterjee, N. Dalchau, R. Muscat, A. Phillips, G. Seelig, "A spatially localized architecture for fast and modular DNA computing", pages 920–927, copyright (2017), with permission from Springer Nature

materials. One example of this is protein patterning or lithography [121, 122] where the cavities of DNA origami were used to directly absorb proteins. Although direct attachment methods can also be used to fabricate complex protein structures [123]. Nanostructures can also be integrated with silica, strengthening chemical, thermal, and mechanical properties [124, 125]. Additionally, it has been used as a tool for silicon lithography allowing for a minimal feature size as low as 2 nm [126].

Metallic nanoparticles invite great research interest due to their unique optical properties. Due to the precise design and functionalization of DNA origami, metallic nanoparticles, and especially gold nanoparticles and nanorods, can be arranged on DNA origami nanoparticles to manipulate optical frequencies [127]. DNA oligo sequences are covalently linked to gold nanoparticles allowing for the precise alignment of gold rods. Additionally, Zhou et al. created a gold nanorod walker by using a combination of circularly polarized light and toehold-mediated DNA manipulation (Fig. 2b) [128]. Walking plasmonic nanorods act as translocation reporters, delivering their own location in real time via spectroscopy. This was accomplished by linking a gold nanoparticle to various DNA legs and having complement toehold strands placed on a 2D surface of DNA. The distance between toeholds, or step size, was 7 nm. Similar experiments with step sizes of 6 nm measured a speed of one step every 5 min [129]. Transportation of proteins along microtubules in cells has been the inspiration for nanorod walkers, but this natural phenomenon can move as fast as 1 μm/s. Li et al. postulated that the relatively slow speed for the man-made walkers is due to a lack of optimization rather than an inherent limitation to DNA walkers [130]. The fluorescent properties and Raman scattering properties of the nanoparticles allow immediate and precise locating of the walker, which may prove useful for evaluating future DNA locomotive devices.

4.3 DNA Computation

The density of information contained in DNA makes it appealing for large-scale data storage. It can store approximately 1.7×10^{19} bytes/g, approximately 8 orders of magnitudes denser than traditional hard drives [131]. The data storage required by 2025 has been estimated to be 1.75×10^{24} bytes, which will exceed the storage capacity of current methods [132]. Artist Joe Davis was the first to utilize DNA as a storage system in 1988, storing an image in black and white pixels [133]. In 2012, Church et al. demonstrated that large amounts of data can be stored in DNA form [134]. They converted various forms of media into a 5.27-megabit bitstream and synthesized the oligos using an ink-jet printed, high-fidelity DNA microchip system. This allowed for precise, rapid creation of DNA strands with excellent robustness (i.e., ten errors were found out of 5.27 million base pairs in the sequence). DNA has been shown to be a cryptographic tool to secure data with a high degree of confidentiality, integrity, and availability [135], and more recent work has demonstrated advances like the use of DNA origami for encrypted communication [136] and the development of approaches for random access memory [137]. Furthermore, Traditional computing concepts can be combined with aspects of DNA-based interactions to allow for a molecular computing system. DNA is particularly well suited for molecular computing applications because of its potential for a higher number of well-defined orthogonal interactions when compared with other molecules because of the sequence specificity. To facilitate molecular computing, single-stranded DNA can be used to fuel a molecular reaction, replacing one partially complement strand with a fully complementary one in a useful process called toehold-mediated strand displacement [139]. This allows for a scalable, complex computing system with DNA [140–142]; however, the time per computing step is limited by diffusion of DNA strands in solution. Recently, Chatterjee et al. demonstrated how a domino architecture of DNA strands immobilized on a DNA origami structure can form a series of logic gates using spatial separation. They created OR and AND gates with a system of spatially separated hairpins localized on a DNA origami platform (Fig. 2c) [138]. This capacity to perform logic computation on DNA origami has exciting potential for diagnostics that can evaluate several biomarkers [143].

4.4 Molecular Machines

In addition to designing precise and complex geometries at the molecular scale, DNA has emerged as a highly useful material to develop dynamic molecular devices for applications in molecular transport, biophysical measurement, controlling molecular interactions, and nanorobotics [144–147]. The first reconfigurable DNA structure was presented by Mao et al. in 1999 based on a transition of B-DNA (right-handed) to Z-DNA (left-handed) driven by changes in ion conditions to change the handedness of a DNA structure [148]. Soon after, Yurke et al. took a major step with development of toehold-mediated strand displacement to reconfigure a DNA-based molecular tweezer [139]. Later this concept was expanded to demonstrate DNA construct that can take successive steps along a DNA track [149], or DNA walkers, which continue to be of interest [150, 151]for applications like nanoscale transport [152] and molecular assembly [153]. The development of DNA origami [2] brought new opportunities to design more complex dynamic devices like containers with triggered opening [44] and structures with local reconfiguration of strands [154] or large-scale shape changes [155].

The ability to include single-stranded DNA (ssDNA) into DNA origami nanostructures provides a means to selectively incorporate flexible domains into DNA devices. This approach has been used to design nanomachines with well-defined rotational and translational motion similar to hinge and sliding joints in macroscopic machines (Fig. 2d) [12]. More recently, a number of approaches have been developed to actuate these dynamics devices including toehold-mediated strand displacement [141, 153], ion or temperature-based actuation [106, 156, 157], photoactuation [158, 159], or even rapid electrical or magnetic actuation approaches [160, 161]. These dynamic DNA origami devices have been demonstrated in applications like force sensing [104], probing molecular interactions or conformational changes [116, 117, 162], applying forces to biomolecules [163], and controlling molecular or nanoparticle assemblies [164, 165].

5 DNA Origami for Drug Delivery Applications

DNA origami-based nanoparticles have many inherent and designed properties that make them attractive candidates for drug delivery. For example, DNA polymers are inherently soluble, biocompatible, biodegradable, and their level of bioactivity can be controlled based on their sequence and chemical modifications. Their most unique property, the precise self-assembly based on DNA sequence complementarity, translates into reproducible and uniform DNA nanoparticle manufacturing with nanometer-level accuracy and down to a single nucleobase resolution. This in turn allows for the key features for drug delivery such as the precise control over nanoparticle geometry, composition, degradation, and the surface topography of therapeutic, targeting, and other molecules of interest [3, 166, 167]. In this manner, DNA origami–based nanoparticles provide unprecedented control over their design and represent a promising new strategy in the drug delivery field.

5.1 DNA Origami Stability in Physiological and High-Temperature Conditions

Structural stability of DNA nanostructures under physiological conditions is a critical component to optimize for drug delivery applications as the technology progresses toward the clinic [3, 16, 168, 169]. Within both in vitro and in vivo biological systems, DNA nanostructures are under constant assault by acidic conditions and molecules that cause structural degradation. In particular, in in vitro cell culture media containing 10% fetal bovine serum (FBS) supplement contains a high degree of nuclease activity (estimated at 256 U/L equivalents of DNase I) [49]. Furthermore, DNA nanostructures in vivo quickly encounter nucleases in the periphery upon i.v. systemic administration, as well as net positively charged proteins that bind to the polyanionic backbone of DNA double helices [170, 171]that may interfere with intended drug delivery applications. Furthermore, once DNA origami nanostructures enter the cell via the endolysosomal pathway [18, 19], they will encounter acidic conditions (pH 4–5) upon endosome–lysosome fusion, which may further degrade DNA nanostructure structural integrity. Thus, addressing physiological conditions and biomolecules that will degrade DNA nanostructures is critical for the success of drug delivery experiments both in vitro and in vivo.

Previous work by Hahn et al. studied the impact of the physiological cation concentrations on denaturation and the presence of FBS in cell culture media on DNA origami nanostructure digestion [172]. Findings from their study revealed DNA nanostructure denaturation due to cation depletion occurred in both design- and time-dependent manners [172] with a nanotube structure being more stable in lower magnesium concentrations compared to nanooctahedon and nanorod structures [172], suggesting that geometry plays a role in cation-dependent structural integrity. DNA origami nanostructure structural stability in vitro was improved by adjustment of cation concentration in cell culture medium with $MgSO_4$, which prevented denaturation [172]. This work also studied FBS-mediated digestion of DNA origami structures in the presence of heat inactivated (56 °C for 30 min) FBS added as a supplement to cell culture media (10%) in a time-dependent manner. In this study, digestion did not appear to be dependent on shape for the structures tested, and heat inactivation of FBS at 75 ° C and addition of actin to cell culture medium inhibited nuclease activity. Interestingly, other findings showed that dense, rod-shaped DNA origami nanostructures were stable when cultured with 20% FBS for 24 h [19], suggesting that multiple DNA helical layers and cross-sectional design (large vs. small or square lattice vs. honeycomb lattice) may influence nuclease activity. Together, these findings suggest that consideration of cation concentrations, geometric design of DNA origami nanostructures, and nuclease activity are all important when designing in vitro cell culture experiments with DNA origami nanostructures.

Since previous studies reported bare DNA nanostructures were susceptible to nuclease-mediated digestion alone [16, 49, 62, 136, 173], and in the presence of 10% FBS [49, 171], researchers have developed protection strategies to increase DNA nanostructure structural stability [171, 174]. Introduction of chemical modifications to DNA nanostructures can be advantageous to increase structural stability of DNA nanostructures in physiological in vitro experimental systems. These include the 2'-OMe modification on RNA molecules attached to DNA nanostructures [175], ultraviolet (UV) [176], or chemical crosslinking [177, 178], encapsulation in PEGylated liposomes [49], direct PEG attachment with oligolysine PEG conjugate molecules [179], and peptoid coating [173]. Although multiple previous studies have shown that chemically modifying DNA origami nanostructures increases structural stability in vitro, the influence of such modifications and geometrical designs in terms of in vivo stability remains less well understood. Zhang et al. previously performed biodistribution studies of quantum-dot labeled DNA origami nanostructures of varying geometries (triangle, rectangle, and rod) and showed that triangle-shaped DNA origami nanostructures preferentially accumulated in tumor tissue at 24 h post i.v. injection [20], suggesting that geometric shape, accessibility, and/or surface area may play a role with respect to structural stability in vivo.

In order to improve DNA origami structural stability in vivo, Perrault and Shih previously showed that encapsulating DNA origami nanostructures with PEGylated liposomes significantly extended pharmacokinetics (PK) half-life [49], and also avoided recognition by the immune system, suggesting an innovative protection approach for future DNA origami drug delivery applications. Furthermore, more recent findings by William Shih and colleagues revealed that electrostatic attachment of oligolysine–PEG conjugate molecules significantly improved PK elimination half-life of DNA origami nanostructures [179], suggesting a straightforward and effective nuclease protection strategy to increase DNA nanostructure structural stability in vivo. Although multiple DNA origami nanostructure nuclease protection strategies have been proposed [171, 174, 180], their impact on therapeutic payload in terms of delivery efficiency and PK circulation/elimination half-lives remains underexplored and will be important as this exciting drug delivery platform moves toward clinical applications, including the development of organic-specific therapies [230, 231].

5.2 *Loading and Release Methods*

A key advantage of DNA nanostructures is the relative ease of functionalization with a wide array of molecules or materials [4]. With respect to drug delivery applications, a plethora of nanomaterials and molecules, including imaging agents, drugs, and biologicals such as proteins and therapeutic nucleic acids (e.g., siRNA, mRNA) can be incorporated into DNA origami designs via diverse mechanisms [4, 169, 181, 182]. Among these mechanisms, the ones based on nucleic acid hybridization, intercalation, or programmable encapsulation/entrapment into the internal cavity of nanostructures are unique to DNA origami and nucleic acids (Fig. 3). In addition to these attractive properties, DNA is amenable for chemical conjugation, electrostatic interactions, enzymatic labeling, etc. [178, 183, 184].

Hybridization methods, based on DNA complementarity, offer site specificity and accuracy in the placement of molecules on or within DNA origami structures, wherein DNA origami can be thought of as a breadboard onto which molecules of interest are placed [98, 110]. Molecules can be incorporated either during or after DNA origami folding, depending on the functionalization of interest and the desired design and properties of the delivery system. Incorporating molecules during the DNA origami folding reaction generally results in a better yield of incorporation compared to postfolding incorporation. However, postfolding incorporation tends to

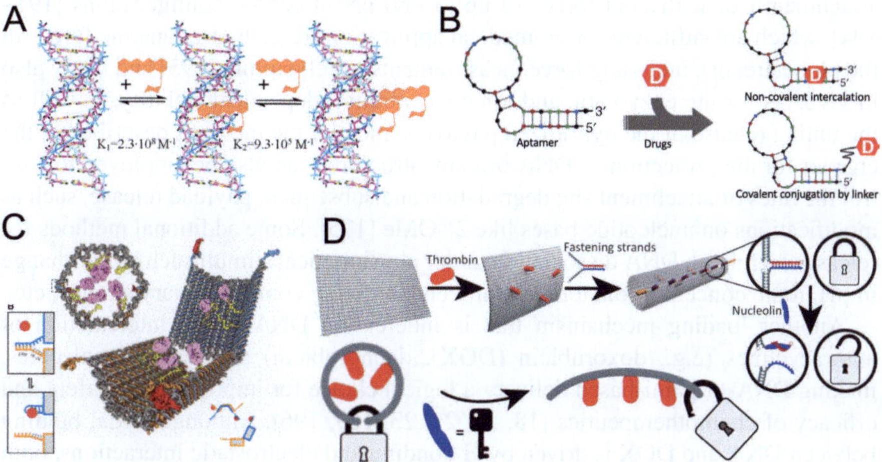

Fig. 3 DNA-enabled loading and release methods of DNA nanoparticles. (**a**) Mechanism and binding constant for intercalation of anthracyclines to dsDNA [199]. (**b**) Nucleic acid aptamer loaded with drug molecules either by intercalation or chemical conjugation to target drug to a specific cell [202]. (**c**) A DNA origami nanorobot for specific targeted delivery and logic-gated payload release based on DNA sequence complementarity and aptamer reconfiguration upon binding to its target [ref]. (**d**) Blood coagulating agent (thrombin) is enclosed into DNA origami tube using DNA complementarity in order to prevent nonspecific tissue coagulation. Upon targeting tumor vasculature, the coagulating agent is exposed, causing coagulation and necrosis of the tumor site [35]. (**a**) is reprinted from *The Journal of Physical Chemistry B*, volume 118, C. Perez-Arnaiz, N. Busto, J.M. Leal, B. Garcia, "New insights into the mechanism of the DNA/Doxorubicin interaction", pages 1288–1295, copyright (2014), with permission from the American Chemical Society. (**b**) is reprinted from *Molecular Therapy Nucleic Acids*, volume 3, H. Sun, X. Zhu, P. Lu, R. Rosato, W. Tan, Y. Zu, "Oligonucleotide aptamers: new tools for targeted Cancer therapy", page e182, copyright (2014), with permission from the American Society of Gene and Cell Therapy. (**c**) is reprinted from *Science*, volume 335, S. Douglas, I. Bachelet, G. Church, "A logic gated nanorobot for targeted transport of molecular payloads", pages 831–834, copyright (2012), with permission from the American Association for the Advancement of Science. (**d**) is reprinted from *Trends in Molecular Medicine*, volume 24 issue 7, H. Singh, E. Kopperger, F. Simmel, "A DNA nanorobot uprises against cancer", pages 591–593, copyright (2018), with permission from Elsevier

be more cost-effective (i.e., may require less material). To accomplish attachment during the folding reaction, molecules of interest are simply conjugated to one of the oligonucleotides that are used to fold the scaffold into the DNA origami shape [99, 185]. A wide variety of conjugation chemistries can be used to accomplish this, described elsewhere [184, 186–188]. For postfolding attachment, the molecules are usually first covalently attached to oligonucleotides of desired sequence. These oligonucleotides are then used to attach molecules to single-stranded complementary oligonucleotide extensions (also referred to as arms or overhangs), which are incorporated at specific locations on DNA origami. For example, researchers have used this strategy to attach antibodies [17]. Relevant for drug delivery, hybridization-based attachments, when properly designed, are sufficiently stable in the biological environment [189, 190]. Nucleic acid hybridization is driven by complementary hydrogen bonding and stabilization by pi-pi molecular bond stacking; this kind of attachment can withstand forces of up to ~60 pN in certain configurations [191–194], which are sufficient for biomedical applications (e.g., hydrodynamic forces in the bloodstream), including force measurements in cell culture [195] and likely also in vivo. As for the enzymatic and ion concentration-dependent stability, as well as the tuning release of the hybridized payloads, most of the methods described in the chapter for the protection of DNA origami structures can also be employed to control the rates of attachment site degradation and subsequent payload release, such as modifications on nucleotide bases like 2'-OMe [175]. Some additional methods for release can rely on DNA dissociation under physiological stimuli such as the change in pH, ionic concentration, temperature, encountering complementary strands, etc.

Another loading mechanism that is inherent in DNA is the intercalation of anthracyclines (e.g., doxorubicin (DOX), daunorubicin) and similar molecules, making DNA-origami-based delivery a logical choice for improving the safety and efficacy of chemotherapeutics [18, 20, 21, 23, 169, 196]. Although initial binding between DNA and DOX is driven by H bonding and electrostatic interactions, both nucleobases and anthracyclines can form pi bonds, allowing for stable intercalation and a complex stability constant between 32 and 930/km (equivalent to K_d of 3.13×10^{-5} and 1.08×10^{-6} M) [197–199] (Fig. 3a). Moreover, while intercalation complexes have favorable release profiles, their release is enhanced in acidic pH, as found in cell endosomes, lysosomes, and the tumor extracellular matrix. The loading capacity is another advantageous feature. For example, compared to the approximate 12.5% (w/w) loading capacity of the gold standard and clinically approved liposomal nanoparticles (e.g., Doxil), DNA-based nanoparticles can approximately achieve 17.9% (w/w) [200]. One of the first nucleic acid designs incorporating doxorubicin intercalation was a trojan aptamer directed against cells in culture expressing the prostate-specific membrane antigen [201]. This early work demonstrated the potential of DNA nanoparticles as drug carriers that can be specifically targeted to designated molecules and cells, for example, using drug-conjugated aptamers [202] (Fig. 3b). More complex DNA structures followed, including DNA origami, which demonstrated the therapeutic efficacy and improved properties for withstanding the biological environment [18, 20, 203, 204]. For example, Halley et al. demonstrated that daunorubicin-loaded DNA origami was stable and more effective than the free drug counterpart in cell culture [196], and other groups have shown efficacy in tumor models in vivo [20, 24].

Besides the drug delivery potential of DNA origami as static nanoparticles, they offer unique delivery approaches based on dynamic structures, which represent the next level of control over site-specific drug delivery. Current efforts are focused on testing programmable DNA origami designs with stimuli-responsive mechanisms [27] that can trigger the release or exposure of encapsulated or entrapped therapeutics (Fig. 3c). While most stimuli-responsive DNA nanodevice designs have not yet been utilized for cell culture or in vivo applications [205–207], the concept has been demonstrated in vivo using a DNA nanotube [26]. In this setup, a DNA nanotube loaded with thrombin, an enzyme that promotes blood coagulation, is specifically targeted to the tumor endothelium via a nucleolin-targeting aptamer. When the aptamer is bound to its target, the nucleolin, it causes the nanotube to open and expose thrombin in order to induce thrombosis in the tumor vasculature, cause subsequent tumor necrosis, and growth inhibition (Fig. 3d [35]).

5.3 *Targeting DNA Origami Nanostructures*

DNA origami nanotechnology holds potential for geometry-based and ligand-based targeting due to precision of its design and the ability to precisely functionalize with one or more targeting moieties [23, 27, 175]. The geometric features of NPs (size and shape) not only influence the nano-bio interface, which affects cell binding and intracellular transport, but also influence the hydrodynamics, diffusion, and interactions with anatomical features of organs and tumors [208]. Accordingly, the majority of studies using DNA-based NPs, including DNA origami, have focused on tumor targeting based on NP geometry, ranging from triangles, rods, etc. and using sizes predominantly in the range of ~2 to ~100 nm [209–212] relying primarily on the enhanced permeability and retention (EPR) effect for delivery. The exploration of DNA origami NPs for ligand-targeting applications is in the early stages of development. Ligand-based targeting holds potential to deliver of DNA NPs to specific targets at the organ, tissue, cell, and subcellular levels [27, 204, 213–218]. Achieving ligand-based targeting tends to not only enhance the therapeutic effects of drugs, but also reduces the side effects associated with a particular therapy [219, 220]. As mentioned earlier, DNA designs can incorporate targeting [26, 27, 221], peptides [222, 223], and nucleic acids [211, 223–225]in a very precise and versatile manner, allowing for the controlled density and spatial organization of the ligands [115, 226], as well as particular payload combinations [211].

Although ligand-targeted DNA designs are capable of specific targeting in vivo and in the absence of passive targeting associated with tumor biology, most studies have been conducted at the cell culture level or focus on in vivo efficacy studies [18, 20, 188, 196, 203, 210] without detailed studies of specific targeting. For example, various spherical, triangular, rod-like, tubular, branched, and wireframe DNA designs have been targeted via ligands to receptors such as nucleolin, folic acid,

HER2, and mucin 1, as reviewed in [227]. Furthermore, nucleic acids can be reformulated into targeting moieties themselves, as exemplified by DNA and RNA aptamers, some of which are at the clinical testing stages or have been approved for clinical use [228, 229]. For example, a doxorubicin-loaded aptamer can be viewed as the first nucleic-acid-made targeted nanocarrier that has demonstrated specific targeting [201]. These examples indicate applications primarily focused on the therapeutic delivery of anthracyclines for cancer treatment, but immunomodulation [222], vaccination applications [223], and the delivery of nucleic acids (siRNA, mRNA, etc.) [23, 175], and genes and gene editing molecules [24, 25] are additional avenues being explored. Overall, the flexibility of DNA designs and the precise combinations of ligand and therapeutic agents onto DNA origami represent some of the advantages of this technology and a potential step forward in personalized targeted therapy.

5.4 Cellular Activation

The unprecedented control over placement of targeting moieties [26, 27, 221], peptides [222, 223], and nucleic acids [211, 223, 224] enabled by DNA origami makes them ideally suited to present molecular ligands to cell surface receptors to study receptor-mediated signaling and biological responses. One of the first examples of employing DNA origami nanostructures to induce intracellular signaling and cellular activation was demonstrated by Schuller, et al. where the authors showed that attachment of 62 cytosine-phosphate-guanine (CpG) sequences to a hollow DNA origami nanotube induced strong elevations in IL-6 and IL-12 (p70) relative to equivalent amounts of CpG alone (or delivered via lipofectamine) to murine splenic mononuclear cells via TLR9 receptor recognition [85], suggesting DNA origami as a useful system to stimulate immune cells in culture. Interestingly, Schuller et al. also revealed DNA origami nanotubes (with no added CpG sequences) alone activated both dendritic cells and B cells, as indicated by elevation of the immune activation surface marker CD69, in both wild-type and TLR9-deficient cells, suggesting DNA origami alone can stimulate immune cell activation via a TLR9-independent mechanism [85].

Another early study by Douglas, et al. sought to selectively deliver molecular payloads to induce cellular signaling and activation via a logic-gated nanorobot [27] (Fig. 3c). The authors constructed a DNA nanorobot that resembled a “clam-shell” that was loaded with antibody fragments specific to cell surface receptors and were closed via aptamer locks. As proof-of-principle, the study went on to show that DNA nanorobots may employed to selectively deliver molecules to either suppress cellular signaling and growth in NKL leukemia cells or induce a cellular activation phenotype in T cells [27]. Collectively, these findings suggest that DNA origami nanostructures represent an effective molecular delivery system for ligand presentation to cell surface receptors to study molecular and cellular level biological responses in cell culture systems in vitro.

Taking advantage of the impeccable control over the number and spatial arrangement of ligand molecules presented on the surface to cell surface receptors, a previous study by Pedersen, et al. utilized a rectangle DNA origami nanostructure to induce surface receptor clustering to sensitize the Transforming Growth Factor-β (TGFβ) signaling pathway in NMuMG cells [226]. This was accomplished by incorporating 10 biotinylated staple overhangs into the central portion of the DNA origami rectangle, which would serve as streptavidin and subsequent biotinylated-peptide ligands to the TGFβ receptor. The authors showed that treating NMUmG cells with TGFβ peptides assembled on rectangle DNA origami nanostructures induced Smad2/3 nuclear localization and a significant increase in a TGFβ luciferase assay upon TGFβ stimulation relative to controls. The authors hypothesized that the effect was due to TGFβ receptor preclustering [226], suggesting that spatial arrangement of ligands or peptides on the DNA origami surface influences cell surface receptor signaling function.

In addition to the ability to cluster ligands on the DNA origami surface [226], previous studies studied how ligand spacing impacts cell surface receptor [77, 78, 114, 115]. Shaw et al. constructed DNA origami nanostructures, rod-shaped "nanocalipers" to present 2 ephrin ligands to EphA2 receptors on the surface of human breast cancer cells at precisely defined distances apart (40 and 100 nm) and showed that ligands spaced at 40 nm significantly increased the level of EphA2 activation (EphA2 receptor phosphorylation) and significantly modulated a downstream cellular response (decreased cellular migration) relative to ligands spaced at 100 nm and control conditions [115]. More recently to further support the claim that ligand spacing impacts cell surface receptor function, Verheyen, et al. constructed DNA origami devices with either a single or 2 ephrin ligands (ephrin A5-dimers) spaced 14, 40, or 100 nm apart to investigate EphA2 receptor activation and transcriptional response upon receptor ligation [114]. The authors showed that ligands spaced either 40 or 100 nm apart allowed for the highest level of EphA2 receptor activation and differentially expressed genes induced by EphA2 receptor signaling relative to the DNA origami nanocalipers alone that did not present ephrin-A5 [114]. Furthermore, Angelin, et al. developed a rectangle DNA origami device, termed multiscale origami structures as interface for cells (MOSAIC) capable of binding either 4, 5, 8, or 12 epidermal growth factor (EGF) molecules arranged at either "near" or "far" configurations to study the impact on EGF–receptor (EGFR) function on MCF-7 cells [231]. Interestingly, the authors showed that EGFR activation was dependent upon both number of ligands where 8 and 12 ligands increased activation relative to 4 and 5, and that ligands placed at "far" configurations allowed for higher EGFR activation relative to the "close" configuration [231], which interestingly reveals a different trend for EGFR versus EphA2. These differences could also be explained by solution (EphA5) versus surface (EGF) DNA origami bound ligand presentation and/or downstream distinct signaling mechanisms induced upon receptor ligation. In another study, Wang et al. demonstrated the ability to regulate death receptors clustering using a DNA origami, finding the critical ligand spacing for inducing cell death through apoptosis [232]. Veneziano et al. recently extended this concept to study the role of organization and spacing of an immunogenic protein domain on the activation of B cells [222] (Fig. 4a). These studies demonstrate DNA origami nanostructures are a useful platform to control both the number and spatial arrangement of biochemical ligand presentation to cell surface receptors to study receptor function and downstream cellular activation and responses.

In addition to studying how ligand spacing affects cell surface receptor function, the DNA origami platform also allows researchers to evaluate how B cell [222] and T cell [112, 233] antigen receptors respond to either precisely spaced antigen ligands or antigen receptors, as well as the impact of antigen binding on downstream cellular activation. Additionally, previous work by Shaw, et al. evaluated how antibody spacing affects antigen binding [221]. The authors leveraged spatial control over antigen molecules to show that antibodies bind to 2 antigens in a bivalent manner at separation distances between 3–17 nm, and that binding affinities

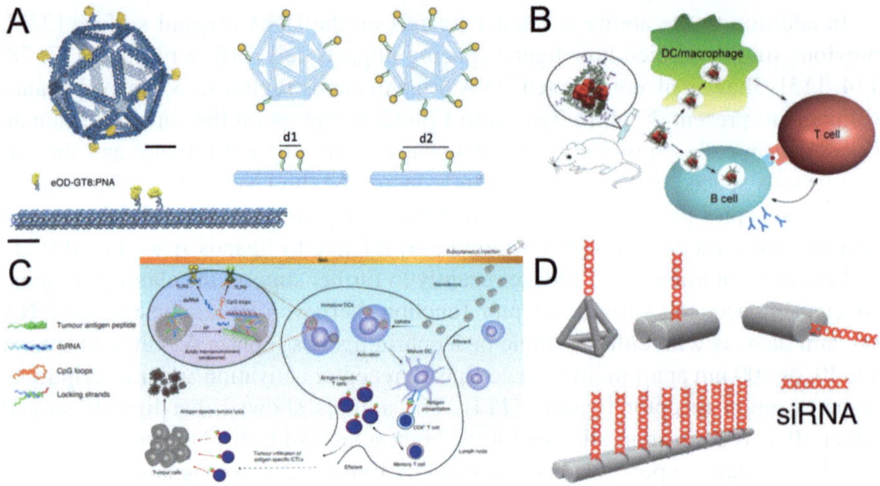

Fig. 4 DNA origami nanostructures for cellular activation, nucleic acid delivery, and vaccine development. (**a**) DNA origami for controlled spatial organization of immunogenic antigens to study the effects of number and organization on cell activation [222]. (**b**) A DNA nanostructure-based vaccine, containing a model protein as an antigen and CpG as an adjuvant, is processed by antigen-presenting cells, allowing for T cell activation and antibody production [239]. (**c**) A DNA origami tube vaccine, decorated with peptide antigens, as well as CpG and dsRNA adjuvants, and capable of inducing potent cytotoxic cells against tumor antigens [223]. (**d**) Control over spatial orientation and density of targeting ligands for the modulation of siRNA therapeutic efficacy as demonstrated for effective gene silencing in plants [224]. (**a**) is reprinted from *Nature Nanotechnology*, volume 15, R. Veneziano, T. Moyer, M. Stone, E. Wamhoff, B. Read, S. Mukherjee, T. Shepherd, J. Das, W. Schief, D. Irvine, M. Bathe, "Role of nanoscale antigen organization on B-cell activation probed using DNA origami", pages 716–723, copyright (2020), with permission from Springer Nature. (**b**) is reprinted from *Nano Letters*, volume 12, X. Liu, Y. Xu, T. Yu, C. Clifford, Y. Liu, H. Yan, Y. Chang, "A DNA nanostructure platform for directed assembly of synthetic vaccines", pages 4254–4259, copyright (2012), with permission from the American Chemical Society. (**c**) is reprinted from *Nature Materials*, volume 20, S. Liu, Q. Jiang, X. Zhao, R. Zhao, Y. Wang,Y. Wang, J. Liu, Y. Shang, S. Zhao, T. Wu, G. Nie, B. Ding, "A DNA nanodevice-based vaccine for cancer immunotherapy", pages 421–430, copyright (2020), with permission from Springer Nature

changed as a function of antigen distance with a distinct distance preference of ~16 nm. Interestingly, clear differences in spatial tolerance exist between IgM and IgG isotypes and between low- and high-affinity antibodies [221]. Veneziano, et al. employed DNA origami to arrange an engineered outer domain of the HIV-1 glycoprotein-120 (eOD-GT8), a clinical vaccine immunogen on the surface of an icosahedron-shaped structure and showed that B cell signaling is maximized by as few as 5 antigens spaced on the surface of a 40 nm viral-like particle [222]. The authors went on to show that antigen spacing up to ~25–30 nm monotonically increases B cell receptor activation and that scaffold rigidity is essential for robust B cell receptor triggering [222]. Lastly, very recently, Dong et al. utilized the DNA origami platform as a "peg board" to precisely bind up to 72 ssDNA "ligands" in a number and spatially arranged manner to engage a synthetic DNA chimeric antigen receptor (CAR) on the T cell surface to study the initiation and dynamics of synthetic T cell receptor signaling [112]. The authors showed that spatial arrangement of receptors determines the ligand density threshold for triggering and encodes the temporal kinetics of downstream signaling activities, and also that signaling sensitivity to a small cluster of high affinity ligands is enhanced by the presence of surrounding nonstimulating low-affinity ligands [112]. Collectively, these findings enabled by the extremely versatile DNA origami platform that precisely controls ligand number and spatial arrangement will guide the development of novel vaccine and immunotherapy technologies.

5.5 *Vaccine Development*

Vaccination is one of the most important prophylactic public measures to control, prevent, or eradicate infectious diseases. Despite being considered as one of the greatest technological achievements to better human health, only one human disease has been eradicated so far, necessitating the development of more effective and personalized vaccines. Moreover, rapidly customizable vaccine delivery technologies are paramount during emergency situations such as infectious disease pandemics. In the light of this and the overall success of vaccination strategy, the development of a new generation of vaccines has been inspired, including therapeutic vaccines and immunotherapies to treat cancers, autoimmune disorders, as well as drug abuse [234]. Many of these new avenues require development of subunit vaccines, which have improved safety profiles compared to live or attenuated vaccines, but usually lack proper immunogenicity unless combined with potent adjuvants. A particularly promising strategy toward addressing the challenge of suboptimal immunogenicity is to incorporate antigen subunits and adjuvants into nanoparticles [235–237]. Since these nanoparticle-based vaccines resemble natural pathogens such as bacteria and viruses, they can enhance the immunostimulatory effects of adjuvants and antigens [238]. Hence, the advantage of nanoparticle-mediated vaccine delivery strategy does not only involve an enhanced payload of therapeutic molecules and nanoparticle geometry-mediated delivery, particularly the enhanced recognition and uptake of nanoparticles by the immune system, but also the multivalent display of antigens and adjuvants along with the adjuvant properties of nanoparticles that are beneficial for immunotherapy applications [238].

Although DNA can be used as a therapeutic material in vaccine formulations, it can also be used as a structural material to precisely and accurately assemble nanoparticulate vaccine forms that could integrate adjuvants, antigens, and other immunostimulatory agents such as check-point inhibitors, all of which can be placed on the same nanoparticle and with precisely controlled geometry (Fig. 4b) [223, 239]. Similar to targeting applications, which benefit from the precise and accurate surface organization of molecules such as targeting ligands, vaccine formulation efficacy can be modulated by the geometry and density of antigenic and adjuvant molecules on the nanoparticle surface [222]. Another attractive property of DNA-based delivery systems is that they can be designed for the enhanced intracellular delivery of vaccine components [223]. Moreover, unlike the majority of nanoparticle systems that are limited to the endolysosomal pathway, DNA origami is capable of delivering therapeutics to the endolysosomal pathway and subsequently the cell cytosol [18, 196, 225]. The intracellular fate of adjuvants and antigens is particularly important in controlling the types of immune responses desired for each specific application (e.g., antibody-mediated immunity and cytotoxic T cell immunity) [240]. In general, cytosolic delivery capability is a requirement for the delivery of most biologics, including gene regulating agents (e.g., siRNA, mRNA, etc.) useful in potential future designs of DNA origami platforms for the delivery of mRNA and DNA vaccines [241].

Early studies toward the development of DNA-origami-based vaccines focused on demonstrating the enhanced immunostimulatory properties of adjuvants assembled on DNA origami nanoparticles compared to free adjuvants [85, 242–245]. Soon after, a complete DNA-origami-based vaccine formulation, consisting of model protein antigen and a CpG oligonucleotide adjuvant, demonstrated enhanced antibody production against the antigen without eliciting antibodies against the carrier itself (i.e., against DNA origami) (Fig. 4c) [223]. By utilizing advantages of the DNA origami technique, another design demonstrated that it is possible to control the spacing and geometry of antigens on the surface of DNA origami in order to modulate immune responses and antibody production [222]. Recently, DNA origami vaccines have also shown promising potential in inducing strong cytotoxic T-cell-mediated immunity for cancer treatment [223].

5.6 Therapeutic Oligonucleotide Delivery Including siRNA, Antisense, CRISPR, and Gene Delivery

The ability of the structural DNA nanotechnology platform to attach nucleic acids moieties in a number and spatial arranged manner has enabled drug delivery platform development to deliver therapeutic oligonucleotides to diseased cells. A landmark demonstration of the potential of the technology was shown by Lee et al. where the authors developed molecularly self-assembled DNA nanoparticles to deliver siRNA specifically to intended target cells in vivo [175]. More recently, Zhang et al. extended this concept to demonstrate effective gene silencing in plants, which depended on the size and shape of the nanostructure as well as the location for attachment of siRNA [224] (Fig. 4d) Specific to the DNA origami platform, Rahman et al. revealed the potential of DNA origami nanostructures functionalized with antisense oligonucleotides to silence the antiapoptotic protein, Bcl-2, which correlated with cell growth inhibition both in vitro and in vivo [225], suggesting a promising systemic delivery system for antisense cancer therapy. In addition, combining conventional small molecule chemotherapeutic delivery with antisense oligonucleotides is straightforward using the DNA origami platform. Wang et al. recently showed that tubular DNA origami nanostructures may be effectively loaded with both doxorubicin and siRNA molecules within the inner cavity to achieve antitumor efficacy both in vitro and in vivo [211], demonstrating strong potential to combine small molecule and antisense chemotherapies on a single DNA origami drug delivery device. Furthermore, Liu et al. combined small molecule delivery (doxorubicin) with gene delivery (linear tumor therapeutic gene p53) on a triangle DNA origami nanostructure and showed efficient gene delivery and effective inhibition of tumor growth in vitro and in vivo without apparent systemic toxicity [24]. And although non-DNA origami, Liu et al. showed that DNA nanostructures can effectively deliver a gene editing system to demonstrate efficient inhibition of tumor growth without apparent systemic toxicity [25], suggesting that DNA origami can effectively deliver gene editing molecules in addition to antisense oligonucleotides to diseased cells. Importantly, the versatility of DNA nanotechnology including DNA origami is not limited to drug delivery systems in mammalian systems. Zhang et al. used a panel of DNA nanostructures including the DNA origami method to attach siRNA molecules for delivery in plant cells, where the authors observed that nanostructure internalization into plant cells and corresponding gene silencing efficiency depends on the DNA nanostructure size, shape, compactness, stiffness, and location of the siRNA attachment locus on the nanostructure [224]. Taken together, evidence from these studies suggests that DNA origami represents a promising and effective drug delivery device for therapeutic antisense oligonucleotides and gene editing molecules in preclinical cancer model systems.

6 Conclusions

DNA origami nanostructures have many desirable features for therapeutic delivery applications including unprecedented control over nanoscale geometry, precise functionalization with specific number and arrangement of molecules, and large loading capacity (e.g., ~10–100 nucleic acids or thousands of small molecule intercalating drugs). These characteristics have been applied to develop nanoplasmonic or nanophotonic devices [246, 247], molecular machines [146], and measurement devices [145], and a significant amount of research over the last decades has sought to leverage these features for therapeutic applications. Many studies have demonstrated exciting results in cell culture or other model systems, and recently, a number of studies have demonstrated exciting promise in animal model systems for applications like anthracycline, siRNA, and vaccine delivery [23, 209, 211, 223].

DNA NPs are still in relatively early stages of development, and a number of challenges remain. Major challenges include biological stability, appropriate pharmacokinetics profile, manufacturing scalability, barriers to precise intracellular delivery (e.g., targeting payload to cytoplasm or nucleus), and potential immunogenicity, which remains not well understood. More broadly, the nonspecific interactions in complex biological environments are not fully characterized. For example, what plasma proteins interact with DNA origami and how those interactions depend on surface coating or functionalization are open questions. As discussed earlier, significant efforts have begun to address some of these challenges, especially stability, but additional work will be required to study how these modified more stable DNA origami structures behave in vivo. In addition, while manufacturing and cost have been perceived as a barrier, new methods for scalable manufacturing and cost-effective scaling have been and continue to be developed [248]. In fact, recent work suggests that DNA origami could be cost-effective relative to other delivery systems [74, 75]. However, current efforts on scaling have focused on the DNA origami structures themselves, and it is likely that additional advances or new methodologies will be required for scaling a complete delivery system (e.g., functionalized with targeting moieties, modified for stability, and loaded with drug molecules).

Despite these challenges, the previously demonstrated and emerging advantages of DNA origami nanostructures are likely to lead to novel technologies such as diagnostics and a range of delivery platforms that could be applied to a variety of diseases. In particular, the precise control over DNA origami structure and functionalization, combined with the relative ease of modification and rapid manufacturing, can lead to a new paradigm for personalized precision medicine. Current research remains in the proof-of-concept and preclinical development stages, but it seems likely that DNA origami can begin to realize its biomedical potential in the coming decade.

References

1. Seeman, N. C. (1982). Nucleic acid junctions and lattices. *Journal of Theoretical Biology, 99*, 237–247.
2. Rothemund, P. W., & Folding, K. (2006). DNA to create nanoscale shapes and patterns. *Nature, 440*, 297–302.
3. Kearney, C. J., Lucas, C. R., O'Brien, F. J., & Castro, C. E. (2016). DNA origami: Folded DNA-nanodevices that can direct and interpret cell behavior. *Advanced Materials, 28*, 5509–5524.
4. Dey, S., et al. (2021). DNA origami. *Nature Reviews Methods Primers, 1*, 1–24.

5. Hong, F., Zhang, F., Liu, Y., & Yan, H. (2017). DNA origami: Scaffolds for creating higher order structures. *Chemical Reviews, 117*, 12584–12640.
6. Sameiyan, E., et al. (2019). DNA origami-based aptasensors. *Biosensors and Bioelectronics, 143*, 111662.
7. Wang, S., et al. (2020). DNA origami-enabled biosensors. *Sensors, 20*, 6899.
8. Loretan, M., Domljanovic, I., Lakatos, M., Rüegg, C., & Acuna, G. P. (2020). DNA origami as emerging technology for the engineering of fluorescent and plasmonic-based biosensors. *Materials, 13*, 2185.
9. Julin, S., Nummelin, S., Kostiainen, M. A., & Linko, V. (2018). DNA nanostructure-directed assembly of metal nanoparticle superlattices. *Journal of Nanoparticle Research, 20*, 119.
10. Fan, D., Wang, J., Wang, E., Dong, S., & Propelling, D. N. A. (2020). Computing with materials' power: Recent advancements in innovative DNA logic computing systems and smart bio-applications. *Advanced Science, 7*, 2001766.
11. Lv, H., Li, Q., Shi, J., Fan, C., & Wang, F. (2021). Biocomputing based on DNA strand displacement reactions. *Chemphyschem, 22*, 1151–1166.
12. Marras, A. E., Zhou, L., Su, H.-J., & Castro, C. E. (2015). Programmable motion of DNA origami mechanisms. *Proceedings of the National Academy of Sciences of the United States of America, 112*, 713–718.
13. Castro, C. E., Su, H.-J., Marras, A. E., Zhou, L., & Johnson, J. (2015). Mechanical design of DNA nanostructures. *Nanoscale, 7*, 5913–5921.
14. Chao, J., Zhu, D., Zhang, Y., Wang, L., & Fan, C. (2016). DNA nanotechnology-enabled biosensors. *Biosensors and Bioelectronics, 76*, 68–79.
15. Chen, T., et al. (2018). DNA nanotechnology for cancer diagnosis and therapy. *International Journal of Molecular Sciences, 19*, 1671.
16. Lu, X., Liu, J., Wu, X., & Ding, B. (2019). Multifunctional DNA origami nanoplatforms for drug delivery. *Chemistry – An Asian Journal, 14*, 2193–2202.
17. Shaw, A., Benson, E., & Högberg, B. (2015). Purification of functionalized DNA origami nanostructures. *ACS Nano, 9*, 4968–4975.
18. Jiang, Q., et al. (2012). DNA origami as a carrier for circumvention of drug resistance. *Journal of the American Chemical Society, 134*, 13396–13403.
19. Halley, P. D., et al. (2016). Daunorubicin-loaded DNA origami nanostructures circumvent drug-resistance mechanisms in a leukemia model. *Small, 12*, 308–320.
20. Zhang, Q., et al. (2014). DNA origami as an in vivo drug delivery vehicle for cancer therapy. *ACS Nano, 8*, 6633–6643.
21. Zhao, Y.-X., et al. (2012). DNA origami delivery system for cancer therapy with tunable release properties. *ACS Nano, 6*, 8684–8691.
22. Lee, H., et al. (2012). Molecularly self-assembled nucleic acid nanoparticles for targeted in vivo siRNA delivery. *Nature Nanotechnology, 7*, 389–393.
23. Pan, Q., et al. (2020). Aptamer-functionalized DNA origami for targeted codelivery of antisense oligonucleotides and doxorubicin to enhance therapy in drug-resistant cancer cells. *ACS Applied Materials & Interfaces, 12*, 400–409.
24. Liu, J., et al. (2018). A DNA-based nanocarrier for efficient gene delivery and combined cancer therapy. *Nano Letters, 18*, 3328–3334.
25. Liu, J., et al. (2019). A self-assembled platform based on branched DNA for sgRNA/Cas9/antisense delivery. *Journal of the American Chemical Society, 141*, 19032–19037.
26. Li, S., et al. (2018). A DNA nanorobot functions as a cancer therapeutic in response to a molecular trigger in vivo. *Nature Biotechnology, 36*, 258–264.
27. Douglas, S. M., Bachelet, I., & Church, G. M. (2012). A logic-gated nanorobot for targeted transport of molecular payloads. *Science, 335*, 831–834.
28. Wu, T., et al. (2019). A nanobody-conjugated DNA nanoplatform for targeted platinum-drug delivery. *Angewandte Chemie International Edition, 58*, 14224–14228.
29. Seeman, N. C., & Structural, D. N. A. (2005). Nanotechnology: An overview. *Methods in Molecular Biology, 303*, 143–166.

30. Winfree, E., Liu, F., Wenzler, L. A., & Seeman, N. C. (1998). Design and self-assembly of two-dimensional DNA crystals. *Nature, 394*, 539–544.
31. Seeman, N. C. (2016). *Structural DNA nanotechnology*. Cambridge: Cambridge University Press. https://doi.org/10.1017/CBO9781139015516.
32. Westhof, E., Masquida, B., & Jaeger, L. (1996). RNA tectonics: towards RNA design. *Folding and Design, 1*, R78–R88.
33. Yan, H., LaBean, T. H., Feng, L., & Reif, J. H. (2003). Directed nucleation assembly of DNA tile complexes for barcode-patterned lattices. *PNAS, 100*, 8103–8108.
34. Yin, P., et al. (2008). Programming DNA tube circumferences. *Science, 321*, 824–826.
35. Singh, H. R., Kopperger, E., & Simmel, F. C. (2018). A DNA nanorobot uprises against cancer. *Trends in Molecular Medicine, 24*, 591–593.
36. Kuzuya, A., & Komiyama, M. (2010). DNA origami: Fold, stick, and beyond. *Nanoscale, 2*, 309–321.
37. Tørring, T., Voigt, N. V., Nangreave, J., Yan, H., & Gothelf, K. V. (2011). DNA origami: A quantum leap for self-assembly of complex structures. *Chemical Society Reviews, 40*, 5636–5646.
38. Voigt, N. V., et al. (2010). Single-molecule chemical reactions on DNA origami. *Nature Nanotechnology, 5*, 200–203.
39. Liedl, T., Högberg, B., Tytell, J., Ingber, D. E., & Shih, W. M. (2010). Self-assembly of three-dimensional prestressed tensegrity structures from DNA. *Nature Nanotechnology, 5*, 520–524.
40. Hung, A. M., et al. (2010). Large-area spatially ordered arrays of gold nanoparticles directed by lithographically confined DNA origami. *Nature Nanotechnology, 5*, 121–126.
41. Rinker, S., Ke, Y., Liu, Y., Chhabra, R., & Yan, H. (2008). Self-assembled DNA nanostructures for distance-dependent multivalent ligand–protein binding. *Nature Nanotechnology, 3*, 418–422.
42. Douglas, S. M., Chou, J. J., & Shih, W. M. (2007). DNA-nanotube-induced alignment of membrane proteins for NMR structure determination. *PNAS, 104*, 6644–6648.
43. Andersen, E. S., et al. (2008). DNA origami design of dolphin-shaped structures with flexible tails. *ACS Nano, 2*, 1213–1218.
44. Andersen, E. S., et al. (2009). Self-assembly of a nanoscale DNA box with a controllable lid. *Nature, 459*, 73–76.
45. Ke, Y., et al. (2009). Scaffolded DNA origami of a DNA tetrahedron molecular container. *Nano Letters, 9*, 2445–2447.
46. Williams, S., et al. (2009). Tiamat: A three-dimensional editing tool for complex DNA structures. In A. Goel, F. C. Simmel, & P. Sosík (Eds.), *DNA computing* (pp. 90–101). Springer. https://doi.org/10.1007/978-3-642-03076-5_8
47. Douglas, S. M., et al. (2009). Rapid prototyping of 3D DNA-origami shapes with caDNAno. *Nucleic Acids Research, 37*, 5001–5006.
48. Ke, Y., Ong, L. L., Shih, W. M., & Yin, P. (2012). Three-dimensional structures self-assembled from DNA bricks. *Science, 338*, 1177–1183.
49. Perrault, S. D., & Shih, W. M. (2014). Virus-inspired membrane encapsulation of DNA nanostructures to achieve in vivo stability. *ACS Nano, 8*, 5132–5140.
50. Schreiber, R., et al. (2013). Chiral plasmonic DNA nanostructures with switchable circular dichroism. *Nature Communications, 4*, 2948.
51. Veneziano, R., et al. (2016). Designer nanoscale DNA assemblies programmed from the top down. *Science, 352*, 1534–1534.
52. Benson, E., et al. (2015). DNA rendering of polyhedral meshes at the nanoscale. *Nature, 523*, 441–444.
53. Matthies, M., Agarwal, N. P., & Schmidt, T. L. (2016). Design and synthesis of triangulated DNA origami trusses. *Nano Letters, 16*, 2108–2113.
54. Jun, H., Wang, X., Bricker, W. P., & Bathe, M. (2019). Automated sequence design of 2D wireframe DNA origami with honeycomb edges. *Nature Communications, 10*, 5419.

55. Huang, C.-M., Kucinic, A., Johnson, J. A., Su, H.-J., & Castro, C. E. (2021). Integrated computer-aided engineering and design for DNA assemblies. *Nature Materials*, 1–8. https://doi.org/10.1038/s41563-021-00978-5
56. Snodin, B. E. K., Schreck, J. S., Romano, F., Louis, A. A., & Doye, J. P. K. (2019). Coarse-grained modelling of the structural properties of DNA origami. *Nucleic Acids Research, 47*, 1585–1597.
57. Shi, Z., Castro, C. E., & Arya, G. (2017). Conformational dynamics of mechanically compliant DNA nanostructures from coarse-grained molecular dynamics simulations. *ACS Nano, 11*, 4617–4630.
58. Snodin, B. E. K., et al. (2016). Direct simulation of the self-assembly of a small DNA origami. *ACS Nano, 10*, 1724–1737.
59. Dunn, K. E., et al. (2015). Guiding the folding pathway of DNA origami. *Nature, 525*, 82–86.
60. Zhou, L., Marras, A. E., Su, H.-J., & Castro, C. E. (2014). DNA origami compliant nanostructures with tunable mechanical properties. *ACS Nano, 8*, 27–34.
61. Dietz, H., Douglas, S. M., & Shih, W. M. (2009). Folding DNA into twisted and curved nanoscale shapes. *Science, 325*, 725–730.
62. Castro, C. E., et al. (2011). A primer to scaffolded DNA origami. *Nature Methods, 8*, 221–229.
63. Kim, D.-N., Kilchherr, F., Dietz, H., & Bathe, M. (2012). Quantitative prediction of 3D solution shape and flexibility of nucleic acid nanostructures. *Nucleic Acids Research, 40*, 2862–2868.
64. Yoo, J., & Aksimentiev, A. (2013). In situ structure and dynamics of DNA origami determined through molecular dynamics simulations. *PNAS, 110*, 20099–20104.
65. Doye, J. P. K., et al. (2013). Coarse-graining DNA for simulations of DNA nanotechnology. *Physical Chemistry Chemical Physics, 15*, 20395–20414.
66. Snodin, B. E. K., et al. (2015). Introducing improved structural properties and salt dependence into a coarse-grained model of DNA. *The Journal of Chemical Physics, 142*, 234901.
67. Sharma, R., Schreck, J. S., Romano, F., Louis, A. A., & Doye, J. P. K. (2017). Characterizing the motion of jointed DNA nanostructures using a coarse-grained model. *ACS Nano, 11*, 12426–12435.
68. Maffeo, C., & Aksimentiev, A. (2020). MrDNA: a multi-resolution model for predicting the structure and dynamics of DNA systems. *Nucleic Acids Research, 48*, 5135–5146.
69. Douglas, S. M., et al. (2009). Self-assembly of DNA into nanoscale three-dimensional shapes. *Nature, 459*, 414–418.
70. Wagenbauer, K., et al. (2017). How we make DNA origami. *Chembiochem, 18*, 1873–1885.
71. Jungmann, R., Liedl, T., Sobey, T. L., Shih, W., & Simmel, F. C. (2008). Isothermal assembly of DNA origami structures using denaturing agents. *Journal of the American Chemical Society, 130*, 10062–10063.
72. Kopielski, A., Schneider, A., Csáki, A., & Fritzsche, W. (2015). Isothermal DNA origami folding: Avoiding denaturing conditions for one-pot, hybrid-component annealing. *Nanoscale, 7*, 2102–2106.
73. Sobczak, J.-P. J., Martin, T. G., Gerling, T., & Dietz, H. (2012). Rapid folding of DNA into nanoscale shapes at constant temperature. *Science, 338*, 1458–1461.
74. Praetorius, F., et al. (2017). Biotechnological mass production of DNA origami. *Nature, 552*, 84–87.
75. Halley, P. D., Patton, R. A., Chowdhury, A., Byrd, J. C., & Castro, C. E. (2019). Low-cost, simple, and scalable self-assembly of DNA origami nanostructures. *Nano Research, 12*, 1207–1215.
76. Warner, C. M., Barker, N., Lee, S.-W., & Perkins, E. J. (2014). M13 bacteriophage production for large-scale applications. *Bioprocess and Biosystems Engineering, 37*, 2067–2072.
77. Kick, B., Praetorius, F., Dietz, H., & Weuster-Botz, D. (2015). Efficient production of single-stranded phage DNA as scaffolds for DNA origami. *Nano Letters, 15*, 4672–4676.
78. Kick, B., Hensler, S., Praetorius, F., Dietz, H., & Weuster-Botz, D. (2017). Specific growth rate and multiplicity of infection affect high-cell-density fermentation with bacteriophage M13 for ssDNA production. *Biotechnology and Bioengineering, 114*, 777–784.

79. Urbaschek, B. & Urbaschek, R. (1979). The inflammatory response to endotoxins. *Bibliotheca Anatomica, 17,* 74–104.
80. Bastings, M. M. C., et al. (2018). Modulation of the cellular uptake of DNA origami through control over mass and shape. *Nano Letters, 18,* 3557–3564.
81. Said, H., et al. (2012). M1.3 – a small scaffold for DNA origami. *Nanoscale, 5,* 284–290.
82. Marchi, A. N., Saaem, I., Vogen, B. N., Brown, S., & LaBean, T. H. (2014). Toward larger DNA origami. *Nano Letters, 14,* 5740–5747.
83. Engelhardt, F. A. S., et al. (2019). Custom-size, functional, and durable DNA origami with design-specific scaffolds. *ACS Nano, 13,* 5015–5027.
84. Bellot, G., McClintock, M. A., Lin, C., & Shih, W. M. (2011). Recovery of intact DNA nanostructures after agarose gel–based separation. *Nature Methods, 8,* 192–194.
85. Schüller, V. J., et al. (2011). Cellular immunostimulation by CpG-sequence-coated DNA origami structures. *ACS Nano, 5,* 9696–9702.
86. Lin, C., Perrault, S. D., Kwak, M., Graf, F., & Shih, W. M. (2013). Purification of DNA-origami nanostructures by rate-zonal centrifugation. *Nucleic Acids Research, 41,* e40.
87. Stahl, E., Martin, T. G., Praetorius, F., & Dietz, H. (2014). Facile and scalable preparation of pure and dense DNA origami solutions. *Angewandte Chemie International Edition, 53,* 12735–12740.
88. How to Store Primers: 7 Things to Know | IDT. *Integrated DNA Technologies* https://www.idtdna.com/pages/education/decoded/article/storing-oligos-7-things-you-should-know.
89. Kielar, C., et al. (2019). Effect of staple age on DNA origami nanostructure assembly and stability. *Molecules, 24,* 2577.
90. Xin, Y., et al. (2020). Cryopreservation of DNA origami nanostructures. *Small, 16,* 1905959.
91. Kube, M., et al. (2020). Revealing the structures of megadalton-scale DNA complexes with nucleotide resolution. *Nature Communications, 11,* 6229.
92. Bai, X., Martin, T. G., Scheres, S. H. W., & Dietz, H. (2012). Cryo-EM structure of a 3D DNA-origami object. *PNAS, 109,* 20012–20017.
93. Lei, D., et al. (2018). Three-dimensional structural dynamics of DNA origami Bennett linkages using individual-particle electron tomography. *Nature Communications, 9,* 592.
94. Willner, E. M., et al. (2017). Single-molecule observation of the photoregulated conformational dynamics of DNA origami nanoscissors. *Angewandte Chemie International Edition, 56,* 15324–15328.
95. Endo, M., & Sugiyama, H. (2014). Single-molecule imaging of dynamic motions of biomolecules in DNA origami nanostructures using high-speed atomic force microscopy. *Accounts of Chemical Research, 47,* 1645–1653.
96. Suzuki, Y., Endo, M., & Sugiyama, H. (2015). Lipid-bilayer-assisted two-dimensional self-assembly of DNA origami nanostructures. *Nature Communications, 6,* 8052.
97. Raab, M., et al. (2018). Using DNA origami nanorulers as traceable distance measurement standards and nanoscopic benchmark structures. *Scientific Reports, 8,* 1780.
98. Stein, I. H., Schüller, V., Böhm, P., Tinnefeld, P., & Liedl, T. (2011). Single-molecule FRET ruler based on rigid DNA origami blocks. *Chemphyschem, 12,* 689–695.
99. Scheckenbach, M., Bauer, J., Zähringer, J., Selbach, F., & Tinnefeld, P. (2020). DNA origami nanorulers and emerging reference structures. *APL Materials, 8,* 110902.
100. Jungmann, R., et al. (2010). Single-molecule kinetics and super-resolution microscopy by fluorescence imaging of transient binding on DNA origami. *Nano Letters, 10,* 4756–4761.
101. Jungmann, R., et al. (2014). Multiplexed 3D cellular super-resolution imaging with DNA-PAINT and Exchange-PAINT. *Nature Methods, 11,* 313–318.
102. Strauss, S., & Jungmann, R. (2020). Up to 100-fold speed-up and multiplexing in optimized DNA-PAINT. *Nature Methods, 17,* 789–791.
103. Strauss, S., et al. (2018). Modified aptamers enable quantitative sub-10-nm cellular DNA-PAINT imaging. *Nature Methods, 15,* 685–688.
104. Hudoba, M. W., Luo, Y., Zacharias, A., Poirier, M. G., & Castro, C. E. (2017). Dynamic DNA origami device for measuring compressive depletion forces. *ACS Nano, 11,* 6566–6573.

105. Ketterer, P., Willner, E. M., & Dietz, H. (2016). Nanoscale rotary apparatus formed from tight-fitting 3D DNA components. *Science Advances, 2*, e1501209.
106. Marras, A. E., et al. (2018). Cation-activated avidity for rapid reconfiguration of DNA nanodevices. *ACS Nano, 12*, 9484–9494.
107. Baker, M. A. B., et al. (2018). Dimensions and global twist of single-layer DNA origami measured by small-angle X-ray scattering. *ACS Nano, 12*, 5791–5799.
108. Bruetzel, L. K., Walker, P. U., Gerling, T., Dietz, H., & Lipfert, J. (2018). Time-resolved small-angle X-ray scattering reveals millisecond transitions of a DNA origami switch. *Nano Letters, 18*, 2672–2676.
109. Stelson, A. C., et al. (2019). Label-free detection of conformational changes in switchable DNA nanostructures with microwave microfluidics. *Nature Communications, 10*, 1174.
110. Steinhauer, C., Jungmann, R., Sobey, T. L., Simmel, F. C., & Tinnefeld, P. (2009). DNA origami as a nanoscopic ruler for super-resolution microscopy. *Angewandte Chemie International Edition, 48*, 8870–8873.
111. Schmied, J. J., et al. (2012). Fluorescence and super-resolution standards based on DNA origami. *Nature Methods, 9*, 1133–1134.
112. Dong, R. *et al.* (2021). DNA origami patterning of synthetic T cell receptors reveals spatial control of the sensitivity and kinetics of signal activation. *bioRxiv* 2021.03.12.434905. https://doi.org/10.1101/2021.03.12.434905.
113. Huang, J., et al. (2020). Arranging small molecules with subnanometer precision on DNA origami substrates for the single-molecule investigation of protein–ligand interactions. *Small Structures, 1*, 2000038.
114. Verheyen, T., et al. (2020). Spatial organization-dependent EphA2 transcriptional responses revealed by ligand nanocalipers. *Nucleic Acids Research, 48*, 5777–5787.
115. Shaw, A., et al. (2014). Spatial control of membrane receptor function using ligand nanocalipers. *Nature Methods, 11*, 841–846.
116. Le, J. V., et al. (2016). Probing nucleosome stability with a DNA origami nanocaliper. *ACS Nano, 10*, 7073–7084.
117. Funke, J. J., et al. (2016). Uncovering the forces between nucleosomes using DNA origami. *Science Advances, 2*, e1600974.
118. Lauback, S., et al. (2018). Real-time magnetic actuation of DNA nanodevices via modular integration with stiff micro-levers. *Nature Communications, 9*, 1446.
119. Selnihhin, D., Sparvath, S. M., Preus, S., Birkedal, V., & Andersen, E. S. (2018). Multifluorophore DNA origami beacon as a biosensing platform. *ACS Nano, 12*, 5699–5708.
120. Fu, J., Liu, M., Liu, Y., Woodbury, N. W., & Yan, H. (2012). Interenzyme substrate diffusion for an enzyme cascade organized on spatially addressable DNA nanostructures. *Journal of the American Chemical Society, 134*, 5516–5519.
121. Aslan, H., Krissanaprasit, A., Besenbacher, F., Gothelf, K. V., & Dong, M. (2016). Protein patterning by a DNA origami framework. *Nanoscale, 8*, 15233–15240.
122. Busuttil, K., Rotaru, A., Dong, M., Besenbacher, F., & Gothelf, K. V. (2013). Transfer of a protein pattern from self-assembled DNA origami to a functionalized substrate. *Chemical Communications, 49*, 1927–1929.
123. Kuzyk, A., Laitinen, K. T., & Törmä, P. (2009). DNA origami as a nanoscale template for protein assembly. *Nanotechnology, 20*, 235305.
124. Liu, X., et al. (2018). Complex silica composite nanomaterials templated with DNA origami. *Nature, 559*, 593–598.
125. Nguyen, L., Döblinger, M., Liedl, T., & Heuer-Jungemann, A. (2019). DNA-origami-templated silica growth by Sol–Gel chemistry. *Angewandte Chemie International Edition, 58*, 912–916.
126. Thomas, G., et al. (2020). DNA origami for silicon patterning. *ACS Applied Materials & Interfaces, 12*, 36799–36809.
127. Shen, X., et al. (2012). Rolling up gold nanoparticle-dressed DNA origami into three-dimensional plasmonic chiral nanostructures. *Journal of the American Chemical Society, 134*, 146–149.

128. Zhou, C., Duan, X., & Liu, N. (2015). A plasmonic nanorod that walks on DNA origami. *Nature Communications, 6*, 8102.
129. Thubagere, A. J., et al. (2017). A cargo-sorting DNA robot. *Science, 357*, eaan6558.
130. Li, J., et al. (2018). Exploring the speed limit of toehold exchange with a cartwheeling DNA acrobat. *Nature Nanotechnology, 13*, 723–729.
131. Hao, Y., Li, Q., Fan, C., & Wang, F. (2021). Data storage based on DNA. *Small Structures, 2*, 2000046.
132. Dong, Y., Sun, F., Ping, Z., Ouyang, Q., & Qian, L. (2020). DNA storage: research landscape and future prospects. *National Science Review, 7*, 1092–1107.
133. Davis, J. (1996). Microvenus. *Art Journal, 55*, 70–74.
134. Church, G. M., Gao, Y., & Kosuri, S. (2012). Next-generation digital information storage in DNA. *Science, 337*, 1628–1628.
135. Leier, A., Richter, C., Banzhaf, W., & Rauhe, H. (2000). Cryptography with DNA binary strands. *Biosystems, 57*, 13–22.
136. Zhang, Y., et al. (2019). DNA origami cryptography for secure communication. *Nature Communications, 10*, 5469.
137. Banal, J. L., et al. (2021). Random access DNA memory using Boolean search in an archival file storage system. *Nature Materials*, 1–9. https://doi.org/10.1038/s41563-021-01021-3
138. Chatterjee, G., Dalchau, N., Muscat, R. A., Phillips, A., & Seelig, G. (2017). A spatially localized architecture for fast and modular DNA computing. *Nature Nanotechnology, 12*, 920–927.
139. Yurke, B., Turberfield, A. J., Mills, A. P., Simmel, F. C., & Neumann, J. L. (2000). A DNA-fuelled molecular machine made of DNA. *Nature, 406*, 605–608.
140. Adleman, L. M. (1994). Molecular computation of solutions to combinatorial problems. *Science, 266*, 1021–1024.
141. Zhang, D. Y., & Seelig, G. (2011). Dynamic DNA nanotechnology using strand-displacement reactions. *Nature Chemistry, 3*, 103–113.
142. Qian, L., Winfree, E., & Bruck, J. (2011). Neural network computation with DNA strand displacement cascades. *Nature, 475*, 368–372.
143. Wang, D., et al. (2014). Molecular logic gates on DNA origami nanostructures for MicroRNA diagnostics. *Analytical Chemistry, 86*, 1932–1936.
144. Stömmer, P., et al. (2021). A synthetic tubular molecular transport system. *Nature Communications, 12*, 4393.
145. Castro, C. E., Dietz, H., & Högberg, B. (2017). DNA origami devices for molecular-scale precision measurements. *MRS Bulletin, 42*, 925–929.
146. Nummelin, S., et al. (2020). Robotic DNA nanostructures. *ACS Synthetic Biology, 9*, 1923–1940.
147. DeLuca, M., Shi, Z., Castro, C. E., & Arya, G. (2020). Dynamic DNA nanotechnology: toward functional nanoscale devices. *Nanoscale Horizons, 5*, 182–201.
148. Mao, C., Sun, W., Shen, Z., & Seeman, N. C. (1999). A nanomechanical device based on the B–Z transition of DNA. *Nature, 397*, 144–146.
149. Shin, J.-S., & Pierce, N. A. (2004). A synthetic DNA walker for molecular transport. *Journal of the American Chemical Society, 126*, 10834–10835.
150. Pan, J., Li, F., Cha, T.-G., Chen, H., & Choi, J. H. (2015). Recent progress on DNA based walkers. *Current Opinion in Biotechnology, 34*, 56–64.
151. Xu, M., & Tang, D. (2021). Recent advances in DNA walker machines and their applications coupled with signal amplification strategies: A critical review. *Analytica Chimica Acta, 1171*, 338523.
152. Pan, J., Cha, T.-G., Chen, H., Li, F., & Choi, J. H. (2017). DNA walkers as transport vehicles of nanoparticles along a carbon nanotube track. In Y. Ke & P. Wang (Eds.), *3D DNA nanostructure: Methods and protocols* (pp. 269–280). Springer. https://doi.org/10.1007/978-1-4939-6454-3_19
153. Gu, H., Chao, J., Xiao, S.-J., & Seeman, N. C. (2010). A proximity-based programmable DNA nanoscale assembly line. *Nature, 465*, 202–205.

154. Zhang, F., Nangreave, J., Liu, Y., & Yan, H. (2012). Reconfigurable DNA origami to generate quasifractal patterns. *Nano Letters, 12*, 3290–3295.
155. Han, D., Pal, S., Liu, Y., & Yan, H. (2010). Folding and cutting DNA into reconfigurable topological nanostructures. *Nature Nanotechnology, 5*, 712–717.
156. Gerling, T., Wagenbauer, K. F., Neuner, A. M., & Dietz, H. (2015). Dynamic DNA devices and assemblies formed by shape-complementary, non–base pairing 3D components. *Science, 347*, 1446–1452.
157. Johnson, J. A., Dehankar, A., Winter, J. O., & Castro, C. E. (2019). Reciprocal control of hierarchical DNA origami-nanoparticle assemblies. *Nano Letters, 19*, 8469–8475.
158. Liu, M., et al. (2018). Rapid photoactuation of a DNA nanostructure using an internal photocaged trigger strand. *Angewandte Chemie International Edition, 57*, 9341–9345.
159. Kuzyk, A., et al. (2016). A light-driven three-dimensional plasmonic nanosystem that translates molecular motion into reversible chiroptical function. *Nature Communications, 7*, 10591.
160. Kopperger, E., et al. (2018). A self-assembled nanoscale robotic arm controlled by electric fields. *Science, 359*, 296–301.
161. Lauback, S., et al. (2018). Real-time magnetic actuation of DNA nanodevices via modular integration with stiff micro-levers. *Nature Communications, 9*, 1446.
162. Funke, J. J., Ketterer, P., Lieleg, C., Korber, P., & Dietz, H. (2016). Exploring nucleosome unwrapping using DNA origami. *Nano Letters, 16*, 7891–7898.
163. Wang, Y., et al. (2021). A nanoscale DNA force spectrometer capable of applying tension and compression on biomolecules. *Nucleic Acids Research*. https://doi.org/10.1093/nar/gkab656
164. Grossi, G., Dalgaard Ebbesen Jepsen, M., Kjems, J., & Andersen, E. S. (2017). Control of enzyme reactions by a reconfigurable DNA nanovault. *Nature Communication, 8*, 992.
165. Kuzyk, A., et al. (2014). Reconfigurable 3D plasmonic metamolecules. *Nature Materials, 13*, 862–866.
166. Hu, Y., et al. (2020). Dynamic DNA assemblies in biomedical applications. *Advanced Science, 7*, 2000557.
167. Wamhoff, E.-C., et al. (2019). Programming structured DNA assemblies to probe biophysical processes. *Annual Review of Biophysics, 48*, 395–419.
168. Chakraborty, K., Veetil, A. T., Jaffrey, S. R., & Krishnan, Y. (2016). Nucleic acid–based nanodevices in biological imaging. *Annual Review of Biochemistry, 85*, 349–373.
169. Weiden, J., & Bastings, M. M. C. (2021). DNA origami nanostructures for controlled therapeutic drug delivery. *Current Opinion in Colloid & Interface Science, 52*, 101411.
170. Griffith, D. M., Jayaram, D. T., Spencer, D. M., Pisetsky, D. S., & Payne, C. K. (2020). DNA-nanoparticle interactions: Formation of a DNA corona and its effects on a protein corona. *Biointerphases, 15*, 051006.
171. Chandrasekaran, A. R. (2021). Nuclease resistance of DNA nanostructures. *Nature Reviews Chemistry, 5*, 225–239.
172. Hahn, J., Wickham, S. F. J., Shih, W. M., & Perrault, S. D. (2014). Addressing the instability of DNA nanostructures in tissue culture. *ACS Nano, 8*, 8765–8775.
173. Wang, S.-T., et al. (2020). DNA origami protection and molecular interfacing through engineered sequence-defined peptoids. *PNAS, 117*, 6339–6348.
174. Stephanopoulos, N. (2019). Strategies for stabilizing DNA nanostructures to biological conditions. *Chembiochem, 20*, 2191–2197.
175. Lee, H., et al. (2012). Molecularly self-assembled nucleic acid nanoparticles for targeted in vivo siRNA delivery. *Nature Nanotechnology, 7*, 389–393.
176. Gerling, T., Kube, M., Kick, B., & Dietz, H. (2018). Sequence-programmable covalent bonding of designed DNA assemblies. *Science Advances*, **4**, eaau1157.
177. Rajendran, A., Endo, M., Katsuda, Y., Hidaka, K., & Sugiyama, H. (2011). Photo-cross-linking-assisted thermal stability of DNA origami structures and its application for higher-temperature self-assembly. *Journal of the American Chemical Society, 133*, 14488–14491.
178. Anastassacos, F. M., Zhao, Z., Zeng, Y., & Shih, W. M. (2020). Glutaraldehyde cross-linking of oligolysines coating DNA origami greatly reduces susceptibility to nuclease degradation. *Journal of the American Chemical Society, 142*, 3311–3315.

179. Ponnuswamy, N., et al. (2017). Oligolysine-based coating protects DNA nanostructures from low-salt denaturation and nuclease degradation. *Nature Communications, 8*, 1–9.
180. Bila, H., Kurisinkal, E. E., & Bastings, M. M. C. (2019). Engineering a stable future for DNA-origami as a biomaterial. *Biomaterials Science, 7*, 532–541.
181. The potential of DNA origami to build multifunctional materials - IOPscience. https://iopscience.iop.org/article/10.1088/2399-7532/ab80d5.
182. Zeng, Y., Nixon, R. L., Liu, W., & Wang, R. (2021). The applications of functionalized DNA nanostructures in bioimaging and cancer therapy. *Biomaterials, 268*, 120560.
183. Heck, C., et al. (2020). Label as you fold: methyltransferase-assisted functionalization of DNA nanostructures. *Nanoscale, 12*, 20287–20291.
184. Kong, G., et al. (2021). DNA origami-based protein networks: from basic construction to emerging applications. *Chemical Society Reviews, 50*, 1846–1873.
185. Xu, A., et al. (2017). DNA origami: The bridge from bottom to top. *MRS Bulletin, 42*, 943–950.
186. Madsen, M., & Gothelf, K. V. (2019). Chemistries for DNA Nanotechnology. *Chemical Reviews, 119*, 6384–6458.
187. Saccà, B., & Niemeyer, C. M. (2011). Functionalization of DNA nanostructures with proteins. *Chemical Society Reviews, 40*, 5910–5921.
188. Ge, Z., et al. (2020). DNA origami-enabled engineering of ligand–drug conjugates for targeted drug delivery. *Small, 16*, 1904857.
189. Gong, H., et al. (2016). Simple method to prepare oligonucleotide-conjugated antibodies and its application in multiplex protein detection in single cells. *Bioconjugate Chemistry, 27*, 217–225.
190. Dugal-Tessier, J., Thirumalairajan, S., & Jain, N. (2021). Antibody-oligonucleotide conjugates: A twist to antibody-drug conjugates. *Journal of Clinical Medicine, 10*, 838.
191. Jung, Y. J., et al. (2007). Dendron arrays for the force-based detection of DNA hybridization events. *Journal of the American Chemical Society, 129*, 9349–9355.
192. Ho, D., et al. (2009). Force-driven separation of short double-stranded DNA. *Biophysical Journal, 97*, 3158–3167.
193. Woodside, M. T., et al. (2006). Nanomechanical measurements of the sequence-dependent folding landscapes of single nucleic acid hairpins. *PNAS, 103*, 6190–6195.
194. Mosayebi, M., Louis, A. A., Doye, J. P. K., & Ouldridge, T. E. (2015). Force-induced rupture of a DNA duplex: From fundamentals to force sensors. *ACS Nano, 9*, 11993–12003.
195. Ma, V. P.-Y., & Salaita, K. (2019). DNA nanotechnology as an emerging tool to study mechanotransduction in living systems. *Small, 15*, e1900961.
196. Halley, P. D., et al. (2016). Daunorubicin-loaded DNA origami nanostructures circumvent drug-resistance mechanisms in a leukemia model. *Small, 12*, 308–320.
197. Yang, F., Teves, S. S., Kemp, C. J., & Henikoff, S. (2014). Doxorubicin, DNA torsion, and chromatin dynamics. *Biochimica et Biophysica Acta (BBA) - Reviews on Cancer, 1845*, 84–89.
198. Riahi, S., et al. (2008). A theoretical study on interactions between mitoxantrone as an anticancer drug and DNA: application in drug design. *Chemical Biology & Drug Design, 71*, 474–482.
199. Pérez-Arnaiz, C., Busto, N., Leal, J. M., & García, B. (2014). New insights into the mechanism of the DNA/doxorubicin interaction. *The Journal of Physical Chemistry. B, 118*, 1288–1295.
200. Kim, K.-R., et al. (2016). Self-assembled mirror DNA nanostructures for tumor-specific delivery of anticancer drugs. *Journal of Controlled Release, 243*, 121–131.
201. Bagalkot, V., Farokhzad, O. C., Langer, R., & Jon, S. (2006). An aptamer–doxorubicin physical conjugate as a novel targeted drug-delivery platform. *Angewandte Chemie International Edition, 45*, 8149–8152.
202. Sun, H., et al. (2014). Oligonucleotide aptamers: New tools for targeted cancer therapy. *Molecular Therapy - Nucleic Acids, 3*, e182.

203. Zhao, Y.-X., et al. (2012). DNA origami delivery system for cancer therapy with tunable release properties. *ACS Nano, 6*, 8684–8691.
204. Taghdisi, S. M., et al. (2016). Double targeting and aptamer-assisted controlled release delivery of epirubicin to cancer cells by aptamers-based dendrimer in vitro and in vivo. *European Journal of Pharmaceutics and Biopharmaceutics, 102*.
205. Ijäs, H., Hakaste, I., Shen, B., Kostiainen, M. A., & Linko, V. (2019). Reconfigurable DNA origami nanocapsule for pH-controlled encapsulation and display of cargo. *ACS Nano, 13*, 5959–5967.
206. Banerjee, A., et al. (2013). Controlled release of encapsulated cargo from a DNA icosahedron using a chemical trigger. *Angewandte Chemie International Edition, 52*, 6854–6857.
207. Takenaka, T., et al. (2014). Photoresponsive DNA nanocapsule having an open/close system for capture and release of nanomaterials. *Chemistry – A European Journal, 20*, 14951–14954.
208. Lane, L. A. (2020). Physics in nanomedicine: Phenomena governing the in vivo performance of nanoparticles. *Applied Physics Reviews, 7*, 011316.
209. Zhang, Q., et al. (2014). DNA origami as an in vivo drug delivery vehicle for cancer therapy. *ACS Nano, 8*, 6633–6643.
210. Palazzolo, S., et al. (2019). An effective multi-stage liposomal DNA origami nanosystem for in vivo cancer therapy. *Cancers (Basel), 11*, 1997.
211. Wang, Z., et al. (2021). A tubular DNA nanodevice as a siRNA/chemo-drug co-delivery vehicle for combined cancer therapy. *Angewandte Chemie International Edition, 60*, 2594–2598.
212. Kalyane, D., et al. (2019). Employment of enhanced permeability and retention effect (EPR): Nanoparticle-based precision tools for targeting of therapeutic and diagnostic agent in cancer. *Materials Science and Engineering: C, 98*, 1252–1276.
213. Chang, M., Yang, C.-S., & Huang, D.-M. (2011). Aptamer-conjugated DNA icosahedral nanoparticles as a carrier of doxorubicin for cancer therapy. *ACS Nano, 5*, 6156–6163.
214. Tasciotti, E. (2018). Smart cancer therapy with DNA origami. *Nature Biotechnology, 36*, 234–235.
215. Huang, Y., Huang, W., Chan, L., Zhou, B., & Chen, T. (2016). A multifunctional DNA origami as carrier of metal complexes to achieve enhanced tumoral delivery and nullified systemic toxicity. *Biomaterials, 103*, 183–196.
216. Zhang, H., et al. (2015). A controllable aptamer-based self-assembled DNA dendrimer for high affinity targeting, bioimaging and drug delivery. *Scientific Reports, 5*, 10099.
217. Bhatia, D., Surana, S., Chakraborty, S., Koushika, S. P., & Krishnan, Y. (2011). A synthetic icosahedral DNA-based host–cargo complex for functional in vivo imaging. *Nature Communications, 2*, 339.
218. Roki, N., et al. (2019). Unprecedently high targeting specificity toward lung ICAM-1 using 3DNA nanocarriers. *Journal of Controlled Release, 305*, 41–49.
219. Kumari, P., Ghosh, B., & Biswas, S. (2016). Nanocarriers for cancer-targeted drug delivery. *Journal of Drug Targeting, 24*, 179–191.
220. Cifuentes-Rius, A., Desai, A., Yuen, D., Johnston, A. P. R., & Voelcker, N. H. (2021). Inducing immune tolerance with dendritic cell-targeting nanomedicines. *Nature Nanotechnology, 16*, 37–46.
221. Shaw, A., et al. (2019). Binding to nanopatterned antigens is dominated by the spatial tolerance of antibodies. *Nature Nanotechnology, 14*, 184–190.
222. Veneziano, R., et al. (2020). Role of nanoscale antigen organization on B-cell activation probed using DNA origami. *Nature Nanotechnology, 15*, 716–723.
223. Liu, S., et al. (2021). A DNA nanodevice-based vaccine for cancer immunotherapy. *Nature Materials, 20*, 421–430.
224. Zhang, H., et al. (2019). DNA nanostructures coordinate gene silencing in mature plants. *PNAS, 116*, 7543–7548.
225. Rahman, M. A., et al. (2017). Systemic delivery of Bc12-targeting siRNA by DNA nanoparticles suppresses cancer cell growth. *Angewandte Chemie International Edition, 56*, 16023–16027.

226. Pedersen, R. O., Loboa, E. G., & LaBean, T. H. (2013). Sensitization of transforming growth factor-β Signaling by multiple peptides patterned on DNA nanostructures. *Biomacromolecules, 14*, 4157–4160.
227. Dai, L., et al. (2021). DNA origami: An outstanding platform for functions in nanophotonics and cancer therapy. *Analyst, 146*, 1807–1819.
228. Keefe, A. D., Pai, S., & Ellington, A. (2010). Aptamers as therapeutics. *Nature Reviews. Drug Discovery, 9*, 537–550.
229. Kaur, H., Bruno, J. G., Kumar, A., & Sharma, T. K. (2018). Aptamers in the therapeutics and diagnostics pipelines. *Theranostics, 8*, 4016–4032.
230. Jiang, D., et al. (2018). DNA origami nanostructures can exhibit preferential renal uptake and alleviate acute kidney injury. *Nature Biomedical Engineering, 2*, 865–877.
231. Angelin, A., et al. (2015). Multiscale origami structures as interface for cells. *Angewandte Chemie International Edition, 54*, 15813–15817.
232. Wang, Y., Baars, I., Fördös, F., & Högberg, B. (2021). Clustering of death receptor for apoptosis using nanoscale patterns of peptides. *ACS Nano, 15*, 9614–9626.
233. Fang, T., et al. (2021). Spatial regulation of T-cell signaling by programmed death-ligand 1 on wireframe DNA origami flat sheets. *ACS Nano, 15*, 3441–3452.
234. Shen, X. Y., Orson, F. M., & Kosten, T. R. (2012). Vaccines against drug abuse. *Clinical Pharmacology & Therapeutics, 91*, 60–70.
235. Butkovich, N., Li, E., Ramirez, A., Burkhardt, A. M., & Wang, S.-W. (2021). Advancements in protein nanoparticle vaccine platforms to combat infectious disease. *WIREs Nanomedicine and Nanobiotechnology, 13*, e1681.
236. Nguyen, B., & Tolia, N. H. (2021). Protein-based antigen presentation platforms for nanoparticle vaccines. *NPJ Vaccines, 6*, 1–11.
237. Pati, R., Shevtsov, M., & Sonawane, A. (2018). Nanoparticle vaccines against infectious diseases. *Frontiers in Immunology, 9*, 2224.
238. Kelly, H. G., Kent, S. J., & Wheatley, A. K. (2019). Immunological basis for enhanced immunity of nanoparticle vaccines. *Expert Review of Vaccines, 18*, 269–280.
239. Liu, X., et al. (2012). A DNA nanostructure platform for directed assembly of synthetic vaccines. *Nano Letters, 12*, 4254–4259.
240. Rincon-Restrepo, M., et al. (2017). Vaccine nanocarriers: Coupling intracellular pathways and cellular biodistribution to control CD4 vs CD8 T cell responses. *Biomaterials, 132*, 48–58.
241. He, W., et al. (2020). Nanocarrier-mediated cytosolic delivery of biopharmaceuticals. *Advanced Functional Materials, 30*, 1910566.
242. Mohri, K., et al. (2012). Design and development of nanosized DNA assemblies in polypod-like structures as efficient vehicles for immunostimulatory CpG motifs to immune cells. *ACS Nano, 6*, 5931–5940.
243. Li, J., et al. (2011). Self-assembled multivalent DNA nanostructures for noninvasive intracellular delivery of immunostimulatory CpG oligonucleotides. *ACS Nano, 5*, 8783–8789.
244. Rattanakiat, S., Nishikawa, M., Funabashi, H., Luo, D., & Takakura, Y. (2009). The assembly of a short linear natural cytosine-phosphate-guanine DNA into dendritic structures and its effect on immunostimulatory activity. *Biomaterials, 30*, 5701–5706.
245. Nishikawa, M., et al. (2011). Biodegradable CpG DNA hydrogels for sustained delivery of doxorubicin and immunostimulatory signals in tumor-bearing mice. *Biomaterials, 32*, 488–494.
246. Liu, N., & Liedl, T. (2018). DNA-assembled advanced plasmonic architectures. *Chemical Reviews, 118*, 3032–3053.
247. Kuzyk, A., Jungmann, R., Acuna, G. P., & Liu, N. (2018). DNA origami route for nanophotonics. *ACS Photonics, 5*, 1151–1163.
248. Coleridge, E. L., & Dunn, K. E. (2020). Assessing the cost-effectiveness of DNA origami nanostructures for targeted delivery of anti-cancer drugs to tumours. *Biomedical Physics & Engineering Express, 6*, 065030.

Index

Y. Yun, K. Yoder (eds.), *Biotechnologies for Gene Therapy*,
https://doi.org/10.1007/978-3-030-93333-3

N

MIX
Papier aus verantwortungsvollen Quellen
Paper from responsible sources
FSC® C105338

If you have any concerns about our products,
you can contact us on
ProductSafety@springernature.com

In case Publisher is established outside the EU,
the EU authorized representative is:
Springer Nature Customer Service Center GmbH
Europaplatz 3, 69115 Heidelberg, Germany

Printed by Libri Plureos GmbH
in Hamburg, Germany